ADVANCED MATHEMATICAL AND COMPUTATIONAL TOOLS IN METROLOGY AND TESTING XIII

Series on Advances in Mathematics for Applied Sciences – Vol. 94

ADVANCED MATHEMATICAL AND COMPUTATIONAL TOOLS IN METROLOGY AND TESTING XIII

Editors

F Pavese
IMEKO TC21, Italy

A Bošnjaković
Institute of Metrology of Bosnia and Herzegovina, Bosnia and Herzegovina

S Eichstädt
Physikalisch-Technische Bundesanstalt, Germany

A B Forbes
National Physical Laboratory, UK

J A Sousa
Instituto Português da Qualidade, Portugal

World Scientific

NEW JERSEY • LONDON • SINGAPORE • BEIJING • SHANGHAI • HONG KONG • TAIPEI • CHENNAI

Published by

World Scientific Publishing Co. Pte. Ltd.
5 Toh Tuck Link, Singapore 596224
USA office: 27 Warren Street, Suite 401-402, Hackensack, NJ 07601
UK office: 57 Shelton Street, Covent Garden, London WC2H 9HE

Library of Congress Cataloging-in-Publication Data
Names: Pavese, Franco, editor. | Bošnjaković, A., editor. | Eichstädt, Sascha, editor. | Forbes, Alistair B., editor. | Sousa, J A e, editor.
Title: Advanced mathematical and computational tools in metrology and testing XIII / editors, F Pavese, A Bošnjaković, S Eichstädt, A B Forbes, J A e Sousa.
Description: 1st edition. | New Jersey : World Scientific, [2025] | Series: Series on advances in mathematics for applied sciences, 1793-0901 ; vol. 94 | Includes bibliographical references and index.
Identifiers: LCCN 2024038850 | ISBN 9789819800667 (hardcover) | ISBN 9789819800674 (ebook)
Subjects: LCSH: Metrology. | Statistics.
Classification: LCC QC88 .A382 2025 | DDC 389/.10285--dc23/eng/20241023
LC record available at https://lccn.loc.gov/2024038850

British Library Cataloguing-in-Publication Data
A catalogue record for this book is available from the British Library.

For any available supplementary material, please visit
https://www.worldscientific.com/worldscibooks/10.1142/14037#t=suppl

F Pavese, A Bošnjaković, S Eichstädt, A B Forbes, J Alves e Sousa (eds.):
Advanced Mathematical and Computational Tools in Metrology and Testing XIII
Series on Advances in Mathematics for Applied Sciences, Vol. 94

https://doi.org/10.1142/9789819800674_fmatter

Foreword

This volume, the thirteenth in the series, contains original, refereed contributions from researchers representing institutions and laboratories across the world involved in metrology and testing. The contributions are associated with presentations made at the 2023 edition of the Advanced Mathematical and Computational Tools in Metrology conference hosted online by the Institute for Metrology Institute of Metrology of Bosnia and Herzegovina (IMBiH), Sarajevo, in September 2023. The conference was co-organised by the International Measurement Confederation (IMEKO, www.imeko.org) technical committees TC21, Mathematical Tools for Measurements, TC6, Digitization, the European Metrology Network for Mathematics and Statistics (MATHMET, www.euramet.org/european-metrology-networks/mathmet) and IMBiH.

The goals of the conference were to:

- Provide a better understanding of mathematical and computational tools for the needs of metrology,
- Develop new tools for metrology and testing applications,
- Promote regional and international cooperation in metrology and testing,
- Support young researchers in metrology and related fields,
- Transfer knowledge to industry and stakeholders.

The conference also celebrated the 30th anniversary of the first Advanced Mathematical Tools for Metrology conference, which was held in Turin in 1993, and the contributions by Franco Pavese, now honorary Chairman of TC21, for his leading role in all editions of the AMCTM conference series and the associated published volumes.

The themes in this volume address the goals of the conference, support the aims of TC6, TC21 and MATHMET, and reflect the importance of mathematical, statistical and numerical tools and techniques in metrology and testing, their role in addressing the challenges associated with the Metre Convention relating to the demand for measurement standards of ever-increasing accuracy, range and diversity, and the need to demonstrate equivalence between national measurement standards.

Teddington, June 2024 *The Editors*

Book Editors

F Pavese (Honorary Chair, IMEKO TC21, Torino, Italy
frpavese@gmail.com
A Bošnjaković (Institute of Metrology of Bosnia and Herzegovina, Bosnia and Herzegovina)
alen.bosnjakovic@met.gov.ba
S Eichstädt (Chair IMEKO TC6, Physikalisch-Technische Bundesanstalt, Germany)
sascha.eichstaedt@ptb.de
A B Forbes (Chair, IMEKO TC21, National Physical Laboratory, UK)
alistair.forbes@npl.co.uk
J A Sousa (Instituto Português Da Qualidade, Portugal)
jasousa@ipq.pt)

For the first twelve Volumes, see this Series vol. 16 (1994), vol. 40, (1996), vol. 45 (1997), vol. 53 (2000), vol. 57 (2001), vol. 66 (2004), vol. 72 (2006), vol. 78 (2009), vol. 84 (2012), vol. 86, (2015), vol. 89 (2018) and vol. 90 (2022).

This volume was produced as a collaboration of:

IMEKO TECHNICAL COMMITTEE 21 "MATHEMATICAL TOOLS FOR MEASUREMENTS".
https://www.imeko.org/index.php/tc21-homepage

IMEKO TECHNICAL COMMITTEE 6 "DIGITALISATION"
https://www.imeko.org/index.php/tc6-homepage

MATHMET: EUROPEAN METROLOGY NETWORK FOR MATHEMATICS AND STATISTICS
https://www.euramet.org/european-metrology-networks/mathmet

following the *Advanced Mathematical and Computation Tools for Metrology 2023* conference, co-organised and hosted by the Institute of Metrology of Bosnia and Herzegovina.

The editors are grateful to the officers, members and friends of IMEKO TC21, TC6 and MATHMET for their contribution to the peer review process associated with this volume.

F Pavese, A Bošnjaković, S Eichstädt, A B Forbes, J Alves e Sousa (eds.):
Advanced Mathematical and Computational Tools in Metrology and Testing XIII
Series on Advances in Mathematics for Applied Sciences, Vol. 94

https://doi.org/10.1142/9789819800674_fmatter

Contents

Papers

F Pavese, A Bošnjaković, S Eichstädt, A B Forbes, J Alves e Sousa (eds.):
Advanced Mathematical and Computational Tools in Metrology and Testing XIII
Series on Advances in Mathematics for Applied Sciences, Vol. 94
 (pp. 1–24)
https://doi.org/10.1142/9789819800674_0001

From uncertain time-series data can one infer an extrapolated deterministic trend? Taking uncertainty correctly into account in (measurement) science

F. Pavese[†]

Independent Scientist
Torino, Italy
E-mail: frpavese@gmail.com

This paper addresses the issue of forward extrapolation in time of the trend of existing experimental datasets for forecast purposes. It aims to introduce difficulties and limitations intrinsic to that task also in science, quite studied but often ignored, with the consequent risk of propagating false information. Making a forecast is one of the most popular expectations from science in economics, politics, Society, and scientific and non-scientific Communities. Science is assumed to be the safest way to allow reliable predictions. However, science too, being based on experimental data or inference, can only provide such predictions within a limited level of confidence — according to its statistical meaning — so never provide certainty. Scientists, in fact, never ignore that *uncertainty* is always associated with all their current findings or thinking, including the possibility of (further) evolution in time. As a consequence, from uncertain time-series data, one should not be able to get a deterministic extrapolated trend supplying the full correct information. A few examples are reported taken from climate analysis results.

Keywords: extrapolation; uncertainty; truth; inference; uncertainty budget; process-based models.

1. Introduction

One of the most popular expectations of non-scientific communities is that science allows reliable *prediction*, but quite often without understanding that prediction reliability is always limited within a level of confidence — in the statistical meaning, never providing certainty. Scientists, on the other hand, are expected to not ignore the latter fact, as they are supposed to never ignore the uncertainty associated with any of their findings or thinking, and the possibility for mistakes in them, or of evolution affecting them.

However, as amply discussed especially in philosophy of science, the

[†]Former Research Director at the National Research Council of Italy (CNR-IMGC; then INRiM since 2006).

problem is about how to take the doubt or the (conditional) certainty into account. This issue is particularly critical when concerning the top level of the knowledge pyramid: the recognized "laws of nature", even when expressed in their less ambiguous form, the mathematical one.[1]

Science is always a complex frame, being the crossroads of different disciplines. According to the Britannica definition: "*Scientific method: mathematical and experimental technique employed in the sciences. More specifically, it is the technique used in the construction and testing of a scientific hypothesis. The process of observing, asking questions, and seeking answers through tests and experiments is not unique to any one field of science. In fact, the scientific method is applied broadly in science, across many different fields. Many empirical sciences ... use mathematical tools borrowed from probability theory and statistics, together with outgrowths of these ...*" [1].

One crossroad is between experimental observations and theoretical inferences, two frames that are interlaced, one providing experimental data, the other being responsible for the inference of the underlying "laws": it is difficult to state in general who comes first in the scientific process. Another crossroad is with philosophy, where "*... Philosophers of science have addressed general methodological problems, such as the nature of scientific explanation and the justification of induction*" (here called "inference") [1].

In both cases, one needs inter-communication in a lexical form (oral or written) within the Community of the findings and hypotheses [2]. Only the "language" is different: in the case of theory, it is the mathematical one, considered a universal symbolic tool (carrying no ambiguity); in the case of data, it is a universally defined and accepted symbolic language (typically algebra); in the case of philosophical foundations, it is a logical symbolic language or the use of its idiom — i.e., of a specific local-language. In all cases, it is assumed that the next generations will be able to correctly understand past communications — directly or through historical education — and compare them with their novel findings.

These assumptions mean that:

(a) Scientists must be all "on the same page", i.e. the transmitted information is understood and shared by the Community, as spelled out in [3]: "*The concept of measurand should be shared by the relevant Community, because the same measurand is supposed to be the object of replicated measures that must be*

---

[1]Concerning the laws in themselves, according to Popper's principle of falsification, an *a posteriori* evaluation tool, a forecast on a law *must* assume *ipso facto* its future validity, differently from the case of a *model*.

comparable, i.e., it should be recognised as a quantity having a current recognisable meaning for the Community. In the dialect of the science philosophers, this means that it should be projected into a social framework".
In the scientific frame, this means that the measurand model must be of the "prescriptive" type, meaning "*giving directions or injunctions*" [4], which does not always mean being a "physical model".
"[*One consequence is that] the design of an experiment (DoE) must start from this initial conceptual model of the measurand, "socially shared", not from building up the descriptive model of the measuring system, which is specific to each measurement arrangement*" [3].
(b) Past knowledge is invariant in time (not meaning "true"), *but neither* meaning that the *whole* previous knowledge remains invariant in the future because one of the main science goals is to progress in knowledge as much as possible, also correcting, or even contrasting (e.g., according to Kuhn's revolutions), the previous one.

2. The truth in science

It is a fact that science, namely measurement science, basically is not looking for "truth",[2] but simply for a consistent explanation of facts — considered satisfactory by the Community — consisting of a sufficiently long roadmap of observations sufficient to show an accepted degree of repeatability — i.e., an uncertainty sufficiently small — and of theoretical inferences considered *valid.*

In addition, along the knowledge-acquisition roadmap, a *diversity of positions* almost invariably, and intrinsically in science, confronts each other, requiring time and often adjustment for advancing knowledge, until the issue can converge univocally and be considered acceptable for the whole scientific Community.[3]

In treating the specific case of experimental data, the studies at a high level in metrology allow the scientist to acquire a special competence, and understanding of data features and the level of confidence attached to them, especially when related to a key parameter, *uncertainty*, necessary to infer a confidence level and to compare different positions. Scientists in other disciplines often do not have the necessary feeling of the need to use specific

---

[2]How can we understand even if we are "approaching" truth, considering that we do not know *where* it stands, as also observed by Kuhn?
[3]Or, according to a popular joke, "or is the position of the last survivor ...".

tools and refined procedures, when their data analysis goes beyond simple examination, to take uncertainty into due account.

This under-evaluation is not infrequent and often affects very critically a procedure not founded on sufficient bases, like, in particular, the case of *data extrapolation outside the available data interval.* Excessive length of extrapolation — though often appealing in non-scientific frames — or insufficient precision of the portion constraining the tracking function, is often confidently used. Instead, it should not, especially when the good practice to take in due account the measurement uncertainty of the available data is not correctly followed.

Forecast is a popular tool, often also used by the fans of each scientific position in order to support their own, to show how (possibly) good a *prediction* can be obtained from the asserted position — i.e. how well the yet "unknown" looks to be consistent with the assumed "known". The scientific tool is called the "forecast model" and is a mathematical tool supported by observations. Basically, it consists of the evolution, most often in time, of some parameters of the *to-date* available law(s) considered to underlie the specific case.

The relevant available data are obviously limited to a *past-known* period — admittedly in most cases a limited one — where the "law" underlying the trend is considered valid and verified with sufficient precision. That is a *risky assumption*, e.g., because in most cases confidence in the *precision of the function* fitted to these *data* may be (also) affected by *un-homogeneity* due to *data precision* increasing with time — being that a goal of experimental science. However, rarely data are "weighted" for their precision, so confidence in the precision of the function fitted to these data could be, e.g., affected also by precision un-homogeneity. Then the mathematical (set of) function(s) is extrapolated ahead to show to the present and next generations what is supposed to happen in the future (*so forecast is an inference*): long since this is commonly done for the weather forecast, but safely only for a subsequent period of *a few days* [5].

This paper addresses the issue of the *extrapolation of experimental data* ahead for forecast purposes (i.e., where *no information* — direct knowledge — is still available), recalling the main difficulties and limitations intrinsic in that task and the consequent risk of propagating false information, especially in the field of thermodynamics. As examples, this paper will explicitly address the specific discipline of climate forecast, for its present importance.[4]

---

[4]Only climate-related examples will be used in the following. See, e.g., [6–8]. See Appendix.

The basic assumption in this paper is that from uncertain data one cannot limit the forecast to a deterministic function of the extrapolated trend. This paper in fact assumes that the available knowledge is affected by partial ignorance due to the uncertainty affecting the available data on which the model(s) need(s) be based. This paper also notes the different situations arising from different amounts of knowledge gained during given periods of different lengths covered by the existing data with respect to the proportion between its length and the extrapolated period length.

Some important conclusions are drawn on the possible confidence level attached to the extrapolations, by also showing some specific examples — concerning a current subject matter of extreme importance, climate.

3. From partial knowledge to inference and back

There is an extremely vast literature on this subject matter with very different solutions for this difficult problem, based on mathematical/statistical models, neural networks, hybrid models, etc.

It is a basic rule in Science to make a distinction between theoretical studies, mind constructions based on *inference*, and experimental studies, based generally on the acquisition of (numerical) values by *observations* obtained by means of instrumental experimental setups — mostly human-made or -designed.

The *latter* can be the preliminary step, or it can follow the former, but in both cases, it acts as a validation of the *former*, within a given *uncertainty*.

Uncertainty is therefore an unavoidable component that makes *partial* the gained knowledge (exceptions exist, e.g., today about the basic roundness of the Earth! But it took centuries of studies), and any current conclusion has a confidence/credibility level attached.

The results of the theoretical studies are generally expressed in the symbolic language of mathematics. They are called "laws" when assumed to describe a general feature of Nature, or they are called "models" when their function is merely to track experimental findings.

Both laws and models are assumed to be consistent with the experimental findings (considered as "evidence") within the uncertainty of the latter. However, much too often *no uncertainty is found associated with models or laws*, possibly except when they are obtained through "data-driven" inference, i.e. they are based only on datasets.

Most often today a model is *not* (fully) based on data only but arises also from an *inference* about the underlying "laws" governing the observed phenomenon.

Here comes the already cited issue of "diversity" in science, [9] arising from the fact that knowledge is generally *partial*. However, there is a basic difference between the *diversity of experimental results* in different datasets and the *diversity of inferences* in theoretical (mathematical) models.

In the case of *data*, though when they are individually considered as "objective facts"[5] contributing to extend or confirm our current knowledge, diversity may also arise from different levels of variability in a dataset or from inconsistencies between datasets, both due to "errors", or to experimental setup imperfections, or to inference limited capability. *These factors can bring different conclusions. Often, they are taken as the only causes of data uncertainty.*

In the case of *theoretical inference*, diversity arises from different "positions" of scientists in evaluating the possible underlying laws to be included in the model, and their respective relationships and cross-influence. Thus, the construction of a model is to be considered fully a mind construction (inference), related to human perception[6] of the external world, valid only via its consistency with (part of) the available experimental (uncertain) data.

Here comes another feature attributed to Science, especially by frames outside science like society, politics, ...: the ability to provide a *forecast* based on the *validity of the assumptions* underlying the observed trends of the current models, assuming the lack of subsequent change in the assumed influence factors — or via a supplementary forecast on changes — and their validity.

However: How from uncertain time-series data can one get a deterministic extrapolated function? Can an inference be back-used to contribute to increasing partial knowledge?

4. Uncertainty of inference

Also, this issue is vastly treated in the literature, especially in applications to *economic forecasts*, where decisions have to be taken based on the inference of future trends.

Limiting here the issue to *experimental science* and leaving now apart the case of the *data-driven* method (see below), the inference is made by means of a

---

[5]But most of the influence factors might have origin from human mind, nevertheless uncertain data are considered as *objective* facts.

[6]... for a "standard human" (i.e. equipped with "standard" sensorial and brain capabilities).

model consisting of mathematical function(s), i.e. by using a *deterministic tool.* It is chosen:

(a) By using a (set of) functions (called "model") inferred to represent (be the cause of) the observed (underlying) physical trend, by "best" reproducing the current data trend, or

(b) By making the "best fitting" of the data representing the current knowledge (see above).

In both cases, an uncertainty *must* be associated with inference, arising from; (i) the data uncertainty, and (ii) the degree of confidence in the correctness/suitability of the inference. The two cases require very different tools for their evaluation, the first basically of a random nature, while the second derives mostly from the assumptions and systematic effects. Thus:

(a) *Uncertainty of data best fit and confidence in its forecast*

The suitability and credibility of the function used for the fit are most generally evaluated by means of the *standard deviation* of the fit, care is also taken to avoid overfitting. However, the uniqueness of the fit is *not ensured,* thus differences between fitting functions can also contribute to fit uncertainty through their "trend". This is more critical when the function is to be extrapolated ahead, because *no constraints* to the functions exist anymore due to the lack of data, except those placed by the mathematical functions. The current *partial* knowledge is instead extrapolated ahead by usually assuming that the fitting-function trend is still represented, as already said, by the current *process-based* trend, or via suppressing-old/introducing-new model functions from inference of modified future influence factors.

See in the following the issue of *single-data uncertainty.*

(b) *Uncertainty from model parameters, and forecast credibility*

Data uncertainty is always involved, a parameter that instead is often absent from the information supplied with the inferences. Data uncertainty must always be taken into account in fitting and the relevant information always be supplied, since the quality of the fitting (on the available data) is vital for the quality of the subsequent extrapolation. By lacking it, basic information is lost and there are instances where the data uncertainty is so large that the quality of their fit is too bad to be already sufficient to consider any extrapolation meaningless.

Most frequently, supplying the results from more than one *deterministic* model is preferred, as their multiplicity may be able to allow an *indirect* evaluation of the possible variability of the forecast. This comparison of different *deterministic* models can hopefully mitigate the risk of false extrapolations, at least when made with different (set of) equations — and of different complexity — on fitting the same data, to alleviate the ignorance intrinsic in a single model. The final forecast will rather consist of an *area* (typically increasing its width in the forecasted period) where the future determinations are assumed to be comprised, within an assigned probability.

As it should now be quite clear from the above, the extrapolation of past data to a future period, i.e. the information that should be passed to the next generations, is affected by a high risk. Risk level is a tool often used in predictions to make a measure of the reliability that one can assign to them. However, in many fields of prediction, this important criterium is not available, though it can result to be extremely different from case to case.

From another point of view, a mathematical modelling function in itself can only be represented as a *deterministic "line"*. That is one of the usual drawbacks of a *process-driven model*, where one can only evaluate the *statistical dispersion trend* with respect to the *available set* of data, through fitting quality. A second drawback, more important in this case, is that one cannot quantify the adequacy of the chosen function *since subsequent data are not (yet) available*: i.e., the extrapolation adequacy of a forecast cannot be evaluated. One might only evaluate, with a sensitivity analysis or with a method like, e.g., the Monte Carlo one, or else, its capability to be robust enough with respect to the values assigned to the function parameters.[7]

Less frequently, a prediction can be based on a *variety of dataset simulations* based on existing data (*data-driven models*, see later).

Much less frequently, the information consists of an estimate of the *uncertainty* necessarily associated with *each model or dataset (or single data)*. The issue of uncertainty in theoretical studies is often omitted at all, but will not be completely in this discussion.[8]

Except in the case of *data-driven* inference, *continuous* functions are mostly used with the forecast, since data are obviously not (yet) available there.

The meaning of the uncertainty of such a function is not univocally defined:
(a) Function parameters are assumed having a "flexibility" (sensitivity analysis);
(b) Data ahead may be simulated several times and the change in the function behaviour evaluated accordingly;

---

[7]Very complicated methods are reported in the literature allowing a climate 50-years ahead forecast estimate for process-driven models.

[8]It is known that Popper initially was quite embarrassed about including uncertainty in his philosophical treatments — falsification is univocal only in mathematics — until he resorted to a related concept that he called "propensity".

(c) Scatter of functions trends arising from different inferences can be analysed.

Thus:

(a) The "sensitivity method" of the function trend evaluates the effect of the variation of the value of each parameter across a given range of values. It allows showing the different critical effects of the parameters (or of a single function-component of the inference).

(b) New sets of data are inferred ahead and the suitability of the function is then evaluated, in both cases with probabilistic tools e.g., non-parametric ones, like Bootstrap, Monte-Carlo, ...[9]

(c) Several functions are inferred, in two ways: by using the same set of inferred influence factors; or, by inferring (partially) different influence factors. This method is popular and used, for example by IPCC in climate analyses ("*In cases where observations are lacking, we resort to inter-comparison of model results to provide at least some quantification of model uncertainty via inter-model spread*") [7]. Strictly speaking, this method does *not* concern the uncertainty of *any* specific function.

In the above respect, a *deterministic* function, in a *process-driven model*, normally only consists of approximating the data according to its own mathematical representational capability — except the completely different case, not considered here, when the function is aimed instead at accurately tracking a data trend, e.g., when the aim is of exactly tracking a (geometrical) "profile". So, normally, a deterministic model can ensure only an "*exact*" description of an experimental dataset, not necessarily the *best* one and irrespective of data uncertainty.

5. Uncertainty of data-driven inference

This kind of inference makes use of non-parametric statistical methods and is not expressed by functional relationships like in the previous case called "process-based" models. It is the focal point, e.g., of the paper [10] with its application of the *bootstrap method* applied "*by creating random subsets of the data*". Bootstrap is a method that the author has applied in cooperation with mathematicians for several years [11, 15] (less than the Monte Carlo method). The only difference in [10] is that the randomization of the subsets is performed

[9]Simulation can also be applied by changing existing data. This is not arbitrary since any datum can possibly have been "omitted" as not all the planned ones might effectively have been measured — there are many reasons, during any experimental procedure for being forced to alter the initially planned measurement schedule — or because any value might have been possibly different within its measurement uncertainty.

by means of an algorithm that, by using machine learning, *subsequently* "learns" from this set how to become able to predict a trend for an increased-size set of data, also (possibly) extending to a *subsequent* period.

Here comes a clear advantage of the data-driven method, in particular when, like in the paper [10], a *non-parametric* way is used to check the dispersion of the resulting trend, by computing it, e.g., with the *bootstrap method.* This method can take into account not only the effect of different data subsets (either overlapping or not overlapping), but also the fact that *each* data in itself does not have a fixed value but is determined only within its *measurement uncertainty* — there are cases where this uncertainty is even larger than the one arising from the dispersion of the dataset. Thus, the final information becomes a *band* "enveloping" *possible trends* that can be treated very effectively with statistical means for supplying a reliable estimate of the *imprecision* affecting the knowledge of the overall trend.

On the other hand, in process-driven procedures, a *function can* be computed also over a subsequent period, where data are not (yet) available — so making an extrapolation of it — though implicitly embedding the *very strong* assumption that the underlying law has not changed — or instead inferring specific (*deterministic*) changes of it. Instead, the same possibility for the data-driven method looks impossible!

6. The uncertainty budget

The term *uncertainty budget* (UB) is not only meant here for a tool that reports in its full composition the reasons for the uncertainty of an inference: (i) due to inconsistent or imperfect "fitting" of the observed data behaviour, or (ii) related to model uncertainty, or (iii) as the *only two* components of uncertainty.

On the contrary, it takes into account the *fact* that *no experimental datum* is intrinsically exempt from having a *large(r) number of components* of uncertainty. This is not a mere metrological scepticism or claim, but is awareness!

The (so far here) two previous uncertainty components may be *one of the reasons* why the observed data have a dispersion of their values — not necessarily *all* the reasons why a set of data may show a dispersion in its observed trend — but *each* datum has *also its own "original" measurement uncertainty*, which cannot be mitigated by any subsequent means, statistical or else.

In this respect, a *deterministic function* — as a *process-mode model* normally only consists of — only *approximates* the data according to its own mathematical *representational capability*. So normally *no deterministic model can ensure a complete description of the quality of an experimental data set.*

The two components of uncertainty treated so far are, in fact, only the ones that reflect on the observed *scatter* of data values — not necessarily the main causes why a *set* of data of different origins may show dispersion in their observed trends.

However, confounding *data scatter* with *data precision* is a gross mistake (data could show excellent consistency but be all wrong). This mistake is more frequent in process-driven than in data-driven practice.

In fact, in addition, any datum is affected almost invariably by a *larger number* of uncertainty components, which cannot be mitigated by any subsequent means, statistical or else. Because of that, current practice is to decorate each datum, when represented graphically by its own uncertainty bar; or, when reported numerically (as ± or in parentheses) by its own uncertainty, as commonly done in strictly metrological papers.

The ensemble of the uncertainty components of a datum is called the "uncertainty budget" (UB), a *strictly internationally required feature in measurement science*, but not so frequently used in other measurement fields. According to the worldwide metrological definition [16], the UB is the solely tool explicitly and actually providing the total (or interval of) uncertainty, for a given confidence level, to be assigned to the data value — the only allowing a user of these data to make their own evaluation on scientific bases.

A UB consists of a Table, reporting on single lines each measurement component having an uncertainty affecting the results — a numerical estimate arising from the measurement procedure. The overall uncertainty is then computed by combination in quadrature of the uncertainty values in all lines (and also according to the classification of the type of uncertainty).

One striking example of a *lack of use* of a UB has been found in the procedure used by IPCC and other Climate Organisations for the purpose of treating the surface-air mean temperature (SAMT) — also called global mean surface temperature (GMST) — on the whole Earth.

In that case, the first line of a UB should report the uncertainty attributed to the temperature value assigned by each Meteorological Station. The overall uncertainty should then be computed by a combination of root squares of the values in all the subsequent lines. Then, each manipulation of these data made by the relevant Central Bodies, introducing new uncertainty components, makes the final UB, which also contains one line for each overall component of uncertainty arising from each type of manipulation. The Report due from each measuring Station is used by the Central Bodies as the *initial input* information from each

Station. The final uncertainty of the SAMT value should then be computed by a combination of root-squares of all these component values. Instead, the one reported in [17–19], is the standard deviation for a 95% confidence level of the final one resulting from the *final statistical treatment* only (e.g., in IPCC's last Report is (0.95–1.20) °C, but in the past even ± 0.05 °C was declared). However, the communication provided to third parties of the collected SAMT values does *not include uncertainty*. Notice that surface-air temperature values from all the meteorological stations include the first decimal digit, except in the METAR system [5] where temperature is supplied as an integer number.

Instead, the WMO has published a Guide about the evaluation of uncertainty [20–21], though without including a formal UB scheme, and is strongly engaged in taking uncertainty into account for all the meteorological parameters it is engaged worldwide with, in its recent revision of the goals, *for the next decade and beyond.*

The Guide conforms to the general methods used in measurement science, specifically in those concerning metrological good practice [22–28]. However, they are not yet implemented uniformly and *with the same conformity around the world*, as it would be strictly necessary in this case.

A consistent scientific literature exists about the precision of the meteorological stations, a largely standardised instrumental setup, where the precision of the thermometer — which can be as good as ± 0.02 °C — only brings a small contribution to the *total uncertainty* in the delivered measurement of air. In fact, their UBs consistently reported total value (rounded here) by WMO is ± 0.5 °C for each delivered temperature measurement (e.g., [29, 30]), with a perspective for reducing it *over several decades* in the future to 0.2 °C.

That uncertainty is to be considered the "original" contribution to the computation of the SAMT by IPCC, which cannot be mitigated by any elaboration procedure then applied to the dataset, to obtain, according to the IPCC, what is intended to be the *worldwide* SAMT, whose published value and uncertainty of the last determination has been as low as ± 0.05 °C for a reported value of +0.12 °C.[10] [31] No uncertainty budget has ever been made available.

It must be noted that in science, measurements are based on two basic procedures, originating from the metrological frame of measurement science: (a) Calibration of the thermometers; (b) Worldwide *traceability* of the *calibration.*

[10]In all instances, a fit of the annual changes of this parameter in the last 40 years brings to an expanded uncertainty of ± 0.21 °C.

(a) *Calibration* ensures that the measured values pertain, within a specified uncertainty, to a specific and unique scale of measurement, in this case, the Celsius scale based on the Celsius unit °C (considering also the kelvin scale and ITS-90 is irrelevant here). The thermometer stability in time is not indefinite, meaning that recalibration is necessary at regular time intervals.
(b) *Traceability* means that the measured values can be compared with each other because it is known that the standards used to calibrate the thermometers are, within a certain uncertainty, realizations of the same temperature unit (so that they are consistent with each other).

For the numerical values to be collected in the above case of the Earth overall database, *conformity* to the *same* standard of the measurement technique and the apparatus used should be ensured (often the thermometers used today are of the electrical type, e.g. measuring electrical resistance, but not all, as mercury-in-glass thermometers, are still widely used). This fact introduces additional uncertainty, as is the way to report the *local* data — e.g., by making means from different thermometers, or measurements at different times when the provided value concerns a full (or part of a) day. Consequently, the originally supplied data are already affected by a (large) number of uncertainty components. Then, their elaboration follows, performed, as said, by the Central Bodies responsible for obtaining the global mean value. By reading the Climate Final Reports summarizing the elaboration, one becomes informed about the procedures adopted (e.g., normalization, interpolation of data when not existing, smoothing, homogenization, introduction of new assumptions, ...), normally based on sound statistical procedures, but very rarely numerical information is also provided. Since a decision must eventually be taken to provide a single Earth mean value, the above procedures as assumed to take into account also the risk of false components of the decision, so they are elaborated being based on "known probabilities" [32].

Instead, "uncertainties" are affected by "unknown probabilities", [32] originally affecting the originally supplied data, not mitigated by the risk evaluation. Actually, these uncertainties are apparently not included in the computation of the final standard deviation eventually quoted as the uncertainty of the supplied Earth mean value — the latter being the only numerical datum included in most final Reports. If so, the provided result looks *to confound data consistency with data uncertainty*. Consistency can be improved by the above manipulations, but, as said, the original uncertainty (plus all the above indicated supplementary uncertainty components) cannot be mitigated.

Similarly, most of the other climate parameters were found reported without their respective uncertainties.[11] For example, IPCC stated in the reported reference: "*The rate of meltwater release from the Greenland and Antarctic ice sheets in response to climate change remains a major source of uncertainty in projections of sea level rise*", [7] but without quantification.

7. Effects of current knowledge uncertainty on the inference forecast for "data-based" models

This type of model does not resort to theoretical bases but is totally empirical for using only available experimental (or simulated) data and their distribution and trend, and the uncertainty of the latter, depending both on their quality, distribution and dispersion.

Non-parametric well-established methods exist to represent and qualify their characteristics, like the bootstrap or Monte-Carlo ones. As already pointed out (see References), they allow us to obtain an overall *band of* uncertainty of the dependent variable for a given confidence level over the full range of the independent variable where the *current* measurement exists.

8. Some relevant facts about the Nature of the Earth

A particularly risky field of prediction is that for *thermodynamic* phenomena, e.g., of those dominant in Nature. The required extension of studies to "many bodies" often randomly sparse, with respect to mechanics, makes physics totally a different affair, namely when considering the colligative behaviour of discrete ensembles of bodies, i.e. depending only on the body numberosity and not from their chemical-physical nature. This is the case of thermodynamics, where the difficulties are usually somewhat mitigated by considering its dynamics as a sequence of *equilibrium* states, in which case granularity is usually ignored and continuous mathematical functions are used to describe its behaviour (*stationary systems*). However, that is a simplification that cannot hold in very large systems, in which cases the validity of models, extrapolated ahead, becomes more and more questionable, especially in non-homogeneous systems and in case of complex (physical-chemical) interactions among bodies, or the case of discrete systems. The development of science for the case of discreteness in

[11]The fact that they might be considered as inter-laboratory "consensus values" does not justify the absence of the uncertainty indication. IPCC is not a normative Organisation. See also [33].

physics and chemistry is accelerating, but at present it looks still insufficient. This theme has already been partially treated in a previous paper [34].

Thus, in the current situation, the present development of sound mathematical and statistical tools cannot be considered sufficient to match the increasing importance of reliable forecast for a much longer span than presently available for climate forecast, and it still needs considerable progress to cover the expectations of the next generation.

Forecast often includes also *actions* that are proposed to be taken in the future to *modify* the present law governing data extrapolation, thus introducing "changing points" in the modelling equations. This feature is *not* discussed in this paper.

9. Conclusions

The author observes in the general literature insufficient attention and care to two basic issues in the treatment of the experimental data that constitute the foundation of knowledge:

(A) The distinction between the *uncertainty* that can be evaluated from the dispersion/consistency of the dataset, from the existing *intrinsic original uncertainty* of *each* experimental datum arising from the measurement process. The latter may be comparable in size or may not, e.g. due to the influence of systematic effects (i.e., data may look very consistent with each other, but all be wrong). In addition, the systematic components cannot be mitigated by measurement repetition, or by statistical tools like the mean. Only a few methods, not so popular, can directly treat an "interval" as such.

(B) Evaluation of the *uncertainty* of the forecast. Though an extremely vast and varied literature exists on this issue, there is a lack in most cases of a real assessment, especially when based on the uncertainty associated with a single *extrapolated deterministic* function in process-based evaluation — and especially when the aimed extrapolated portion is (much) longer than the period where data exist, the only portion allowing evaluation evidence.

In order to inform about the possible estimation variety of the future, any Report necessarily would need to add to the *estimated value* of the relevant parameter and also its evaluated uncertainty — only the full uncertainty evaluation process can be fully described in a separate report, including the information on the means used to guarantee the calibration of the sensors used, to assess their stability in time, and to declare the estimated level of traceability, all issues that must correspond to a component of uncertainty in the UB Report.

Only *independent authorized* international Standard Organizations can omit the uncertainty indication, by eventually providing "stipulated values" of those parameters, when considered useful. IPCC is not one of them. The above rules were provided, and agreed internationally, in order to obtain a top-level *validation* of the experimental results, especially when the scientific findings have to be used also in other frames for other purposes.

References

1. www.britannica.org.
2. F. Pavese, Measurement in science: between Evaluation and Prediction, AMCTM XII (Pavese F., Forbes A.B., Zhang N.F., Chunovkina A.G., Eds.), Series on Advances in Mathematics for Applied Sciences, 2022, Singapore, World Scientific Publishing Co., pp. 346–363.
3. F. Pavese, "On the classification in random and systematic effects", AMCTM XI, 2018, in A.B. Forbes, N.F. Zhang, A.G. Chunovkina, S. Eichstädt, F. Pavese (Eds.): "Advanced Mathematical and Computational Tools in Metrology and Testing XI", Vol. 11, Series on Advances in Mathematics for Applied Sciences Vol. 89, World Scientific, Singapore, October 2018, pp. 58–69.
4. N. de Courtenay, F. Grégis, The evaluation of measurement uncertainty and its epistemological ramifications, Studies in History and Philosophy of Science **65-66** 2017. https://doi.org/10.1016/j.shpsa.2017.05.003.
5. METAR (METeorological Aerodrome Report) on https://www.orbifly.com/.
6. C. Baumberger, R. Knutti, G. Hirsch Hadorn, Building confidence in climate model projections: an analysis of inferences from fit, WIREs Clim. Change 2017, 8re454.
7. Flato, G., J. Marotzke, B. Abiodun, P. Braconnot, S.C. Chou, W. Collins, P. Cox, F. Driouech, S. Emori, V. Eyring, C. Forest, P. Gleckler, E. Guilyardi, C. Jakob, V. Kattsov, C. Reason and M. Rummukainen, 2013: AR5, Evaluation of Climate Models. In: Climate Change 2013: The Physical Science Basis. Contribution of Working Group I to the Fifth Assessment Report of the Intergovernmental Panel on Climate Change [Stocker, T.F., D. Qin, G.-K. Plattner, M. Tignor, S.K. Allen, J. Boschung, A. Nauels, Y. Xia, V. Bex and P.M. Midgley (Eds.)]. Cambridge University Press, Cambridge, United Kingdom and New York, NY, USA.
8. Mastrandrea, M.D., C.B. Field, T.F. Stocker, O.Edenhofer, K.L. Ebi, D.J. Frame, H. Held, E. Kriegler, K.J. Mach, P.R. Matschoss, G.W. Yohe

and F.W. Zwiers: Guidance Note for Lead Authors of the IPCC Fifth Assessment Report on Consistent Treatment of Uncertainties. 2010 IPCC.

9. F. Pavese, P. De Bièvre, Fostering diversity of thought in measurement, in Advanced Mathematical and Computational Tools in Metrology and Testing X, Vol. 10 (F. Pavese, W. Bremser, A.G. Chunovkina, N. Fischer, A.B. Forbes, Eds.), Series on Advances in Mathematics for Applied Sciences Vol. 86, World Scientific, Singapore, 2015, pp. 1–8. ISBN: 978-981-4678-61-2, ISBN: 978-981-4678-62-9(ebook).
10. B. Knüsel, C. Baumberger, Understanding climate phenomena with data-driven models. Stud. in History & Philosophy of Science 84 (2020) 46–56.
11. P. Ciarlini, F. Pavese, Computational and statistical analysis of the thermodynamic data which form the basis of the ITS-90 between 13.8 K and 273.16 K – I Computational basis, Temperature, Its Measurement and Control in Science and Industry (J.F. Schooley ed.) **6**, Amer. Inst. of Phys., New York (1992) 75–78.
12. P. Ciarlini, F. Pavese, Application of special reduction procedures to metrological data, in: Algorithms for Approximation III, Numerical Algorithms **5** (1993), 479–490.
13. P. Ciarlini, G. Regoliosi, F. Pavese, Non-parametric bootstrap with applications to metrological data, in Advanced Mathematical Tools in Metrology, Vol. 1 (P. Ciarlini, M.G. Cox, R. Monaco F. Pavese, Eds.), Series on Advances in Mathematics for Applied Sciences Vol. 16, World Scientific, Singapore, 1994, 219–230.
14. P. Ciarlini, G. Regoliosi, F. Pavese, A bootstrap algorithm for mixture models and interval data in inter-comparisons, Algorithms for Approximation IV (edited by I. J. Anderson, J. Levesley, J. C. Mason), University of Huddersfield, UK, 2002, 138–145.
15. D. Ichim, P. Ciarlini, F. Pavese, A. Premoli, M.L. Rastello, Comparison of LS techniques for the linear approximation of data affected by heteroschedastic errors in both variables with uncertainty estimation, in AMCTM VI (P. Ciarlini, M.G. Cox, F. Pavese, G.B. Rossi, Eds.) Series on Advances in Mathematics for Applied Sciences vol. 66, World Scientific, Singapore, 2004, 294–299.
16. VIM, Vocabulaire international de métrologie – Concepts fondamentaux et généraux et termes associés, 3rd edition, 2012, BIPM, Sèvres, France.
17. IPCC, Climate Change 2021, Physical Science Basis, Fin Report AR6 WG1.
18. C. P. Morice, J. J. Kennedy, N. A. Rayner, J. P. Winn, E. Hogan, R. E. Killick, R. J. H. Dunn, T. J. Osborn, P. D. Jones, I. R. Simpson, An

Updated Assessment of Near-Surface Temperature Change From 1850: The HadCRUT5 Data Set (2021) Journal of Geophysical Research: Atmospheres, 126. https://doi.org/10.1029/2019JD032361.

19. NOAA, National Centers for Environmental Information, Global Climate Report – Annual Report 2020, https://www.ncdc.noaa.gov/sotc/global/202013.
20. J. Duvernoy, Guidance on the computation of calibration uncertainties, Instruments and Observing Methods, Report No. 119 WMO (2015) p. 34 http://www.wmo.int/pages/prog/www/IMOP/publ6307-6326ications-IOM-series.html.
21. https://public.wmo.int/en/bulletin/communicating-forecast-uncertainty.
22. Koichi Nakashima, Metrological traceability and Uncertainty of measurement, 2013 Meteorological Instrument Center, Japan Meteorological Agency, presentation p. 81.
23. GUM, Guide to the expression of uncertainty in measurement, JCGM-WG1, JCGM_100_2008, BIPM, Sèvres, France.
24. B.N. Taylor, C.E. Kuyatt, Technical Note 1297, Guidelines for Evaluating and Expressing the Uncertainty of NIST Measurement Results, 1994 Edition p. 25.
25. EMUE, Good practice in evaluating measurement uncertainty, Compendium of examples (A.M.H. van der Veen and M.G. Cox, Eds.) 27 July 2021, UE EMPIR Project, p. 640.
26. EURACHEM/CITAC Guide, Measurement uncertainty arising from sampling, Produced jointly with Eurolab, Nordtest, and RSC Analytical Methods Committee p. 119.
27. F. Pavese, Mathematical and statistical tools in metrological measurement, 2013, Chapter in Physical Methods, Instruments and Measurements, [Ed. UNESCO-EOLSS Joint Committee], in Encyclopedia of Life Support Systems(EOLSS), Developed under the Auspices of the UNESCO, Eolss Publishers, Oxford, UK, p. 41. http://www.eolss.net.
28. Hogan R., Sources of Uncertainty in Measurement Uncertainty for every Uncertainty Budget, 2015 ISOBudgets LLC. P.O. Box 2455 Yorktown, VA 23692, USA. http://www.isobudgets.com.
29. M. Dobre, D. Sestan, A. Merlone, Air temperature measurement uncertainty associated to a mounting configuration temperature sensor-radiation shield, 2018, WMO Technical Conference on Meteorological and Environmental Instruments and Methods of Observation, https://www.wmocimo.net/wpcontent/uploads/O1_4_Dobre_Air_Temp_MU_Extended-Abstract-TECO2018-Dobre-Sestan-Merlone.pdf.

30. P. Frank, Uncertainty in the global average surface temperature index: a representative lower limit, Energy & Environment, Vol. 21 No. 8, 2010–21.
31. Nathan J. L. Lenssen, Gavin A. Schmidt, James E. Hansen, Matthew J. Menne, Avraham Persin, Reto Ruedy, Daniel Zyss, Improvements in the GISTEMP Uncertainty Model, Journal of Geophysical Research: Atmospheres Vol. 124 (12) pp. 6307–6326, https://doi.org/10.1029/2018JD029522.
32. G. Heal, A. Millnery, Reflections: Uncertainty and Decision Making in Climate Change Economics, Review of Environmental Economics and Policy, volume 8, issue 1, 2014, pp. 120–137. doi:10.1093/reep/ret023.
33. W. Rehg, K. Staley, "Agreement" in the IPCC Confidence measure, Studies in History and Philosophy of Modern Physics 57 (2017) 126–134.
34. F. Pavese, About continuity and discreteness of quantities: examples from physics and metrology, Ukrainian Metrological Journal (2021) n.2, UDC 53.02:006.91 DOI:10.24027/2306-7039.2.2021.236054.
35. IPCC, 2019: Summary for Policymakers. In: IPCC Special Report on the Ocean and Cryosphere in a Changing Climate [H.-O. Pörtner, D.C. Roberts, V. Masson-Delmotte, P. Zhai, M. Tignor, E. Poloczanska, K. Mintenbeck, M. Nicolai, A. Okem, J. Petzold, B. Rama, N. Weyer (Eds.)].
36. Global Climate Report - Annual 2020. National Centers for Environment, https://www.ncdc.noaa.gov/sotc/global 202013.
37. C.P. Morice, J.J. Kennedy, N.A. Rayner, J.P. Winn, E. Hogan, R.E. Killick, R.J.H. Dunn, T.J. Osborn, P.D. Jones, I.R. Simpson (2019) An updated assessment of near surface temperature change from 1850: the HadCRUT5 dataset. J. Geophys. Res. Atmos. https://doi.org/10.1029/2019/2019JD032361/.
38. GISS Surface Temperature Analysis (GISTEMP v4), https://data.giss.nasa.gov/gistemp/.
39. B. Huang, M.J. Menne, T. Boyer, E. Freeman, B. E. Gleason, J. H. Lawrimore, C.Y. Liu, J.J. Rennie, C.J. Schreck III, F.Y. Sun, R. Vose, C.N. Williams, X.G. Yin, H.-M. Zhang, Uncertainty Estimates for Sea Surface Temperature and Land Surface Air Temperature in NOAAGlobalTemp Version 5, J. of Climate 2020 pp. 1351–1479.
40. F. Pavese, Analysis of current scientific data on climate and on their extrapolation beyond 2100, Seminar at University of Florence, DINFO, Florence, February 2020, p. 31.

41. M. Mengel, A. Levermann, K. Frieler, A. Robinson, B. Marzeion, R. Winkelmann, Future sea level rise constrained by observations and long-term commitment, www.pnas.org/cgi/doi/10.1073/pnas.1500515113.
42. IPCC, 2019: Summary for Policymakers. In: IPCC Special Report on the Ocean and Cryosphere in a Changing Climate [H.-O. Pörtner, D.C. Roberts, V. Masson-Delmotte, P. Zhai, M. Tignor, E. Poloczanska, K. Mintenbeck, M. Nicolai, A. Okem, J. Petzold, B. Rama, N. Weyer (Eds.)].

Appendix A: Some examples of Climate analyses

As indicated above in Footnote 4, the following Figs. 1–4 were selected to illuminate some of the above problems. The few reported Figures mainly aim at bringing evidence of how critical the forecast can be, especially over long periods of future natural behaviour, and especially when basically ignoring data uncertainty. They concern some popular parameters in climate forecast, such as Earth's global mean temperature and global Ocean level increases, here elaborated from data by some of the Climate Organisations, namely Ref. 45, with *process-based algorithms* — comments are reported as Figure captions.

1) *Reliability of the currently available data on SAMT (GMST)*

Figure 1. SAMT estimate from NOAA. [36]
Note 1: The industrial era actually started in the middle 1800s, but a non-significant temperature increase in 100 years was observed until the 1970s.
Note 2: Fitting from 1980 to 2021 (s.d. 0.21 °C) has a slope of 0.17 °C/decade. However, it is possible to fit from 1910 to 2021 with a better s.d. (0.14 °C) and indicating a *lower* slope (0.10 °C/decade). *Thus, the most recent reported slope value is overestimated by 75%.*
Note 3: The indicated uncertainty of the data is purely estimated from the s.d. is that of the fits, it is *not* the data precision (not less than 0.5 °C).

2) *Reliability of extrapolation of SAMT (GMST) models in subsequent years*

Figure 2. Models extrapolation for SAMT. [36]
Note 1: The forecast is very much dependent on the model used.
Note 2: No uncertainty is reported deriving from the data (those of Fig. 1). For the models the estimated uncertainty is shown on the right.

3) *Reliability of SAMT (GMST) estimated when using Word maps*

Figure 3. Maps from different sources: (a) HadCRUT [37]: the darker the higher *t* is; (b) L-OTI (NASA) [38]; (c) NOAA [39].

Note 1: Earth temperature maps allow a simpler and better evaluation of reliability, as they show also the kind of distribution the SAMT is based on.

Note 2: Maps of different origins show significantly different temperature distributions.

Note 3: The temperature distribution is extremely varied, a fact in itself that prevents us from reporting the SAMT (GMST) with low uncertainty.

4) *Reliability of extrapolation of Ocean mean level increase in the long term*

Figure 4(a). Changes in IPCC forecast in the last 17 years until 2013. Modified from G. Wrighstone, 2017. [40]

Figure 4(b). PNAS extrapolation upward to 2100. [41]

Figure 4(c). SROCC-SPM extrapolation upward to 2300. [42]

Note 1: The trend is constrained to (the forecast of) the trends of Figure 1.

Note 2: The forecast in 1990 of +35 cm was modified by lowering it to the +5 cm observed in 2020.

Note 3: Figures 4(b) and 4(c) are reported to show the difference in the two forecasts until 2100.

Note 4: With a decrease like that in Note 1, still of a few centimetres until 2050, the extrapolation to 2300 is *totally* a mathematical effect of the *trend* of the used model functions. The estimate of the indicated uncertainty is not justified.

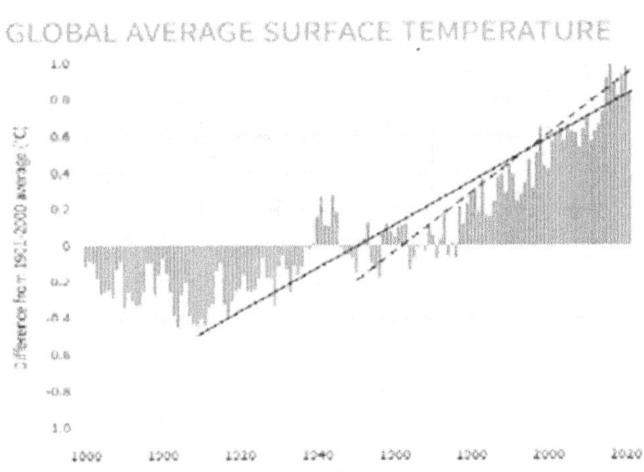

Fig. 1.

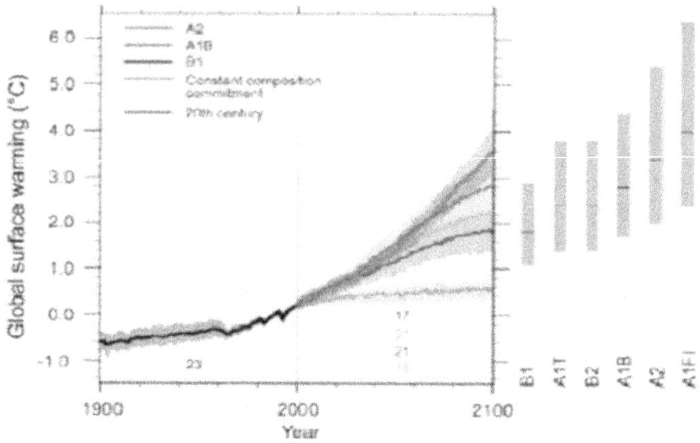

Fig. 2.

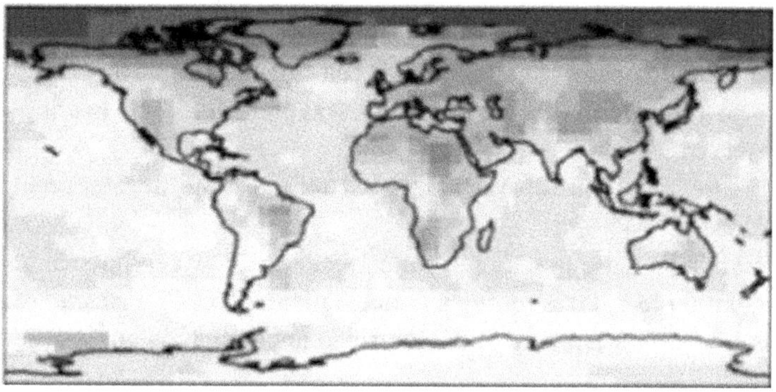

Fig. 3(a).

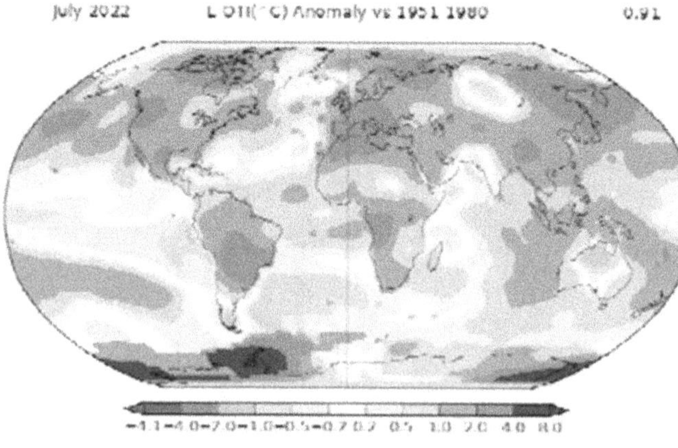

Fig. 3(b).

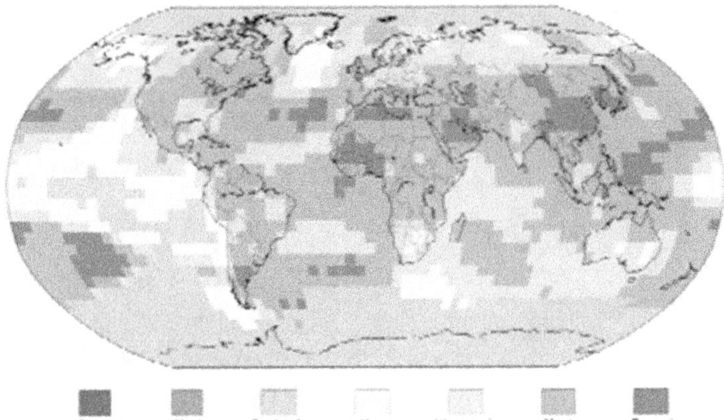

Fig. 3(c).

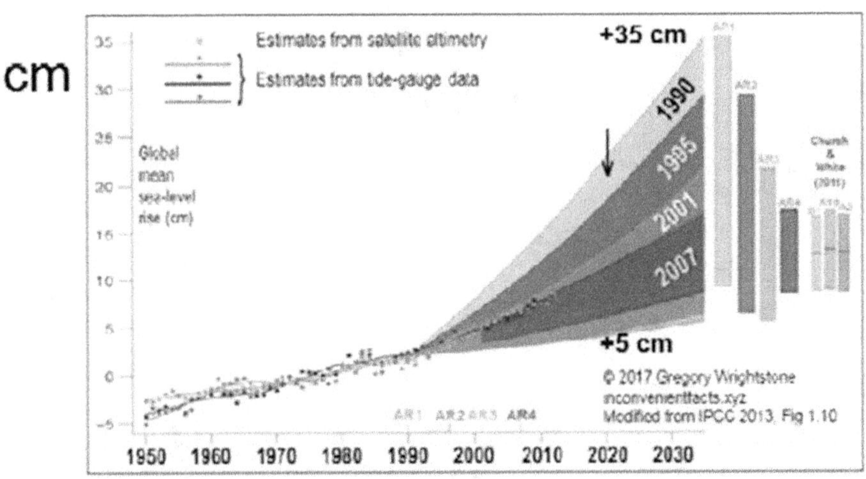
cm
35
30
25
20
15
10
5
0
-5
Estimates from satellite altimetry
Estimates from tide-gauge data
Global mean sea-level rise (cm)
+35 cm
1990
1995
2001
2007
+5 cm
AR1
AR2
AR3
AR4
Church & White (2011)
© 2017 Gregory Wrightstone
inconvenientfacts.xyz
Modified from IPCC 2013, Fig 1.10
1950
1960
1970
1980
1990
2000
2010
2020
2030

Fig. 4(a).

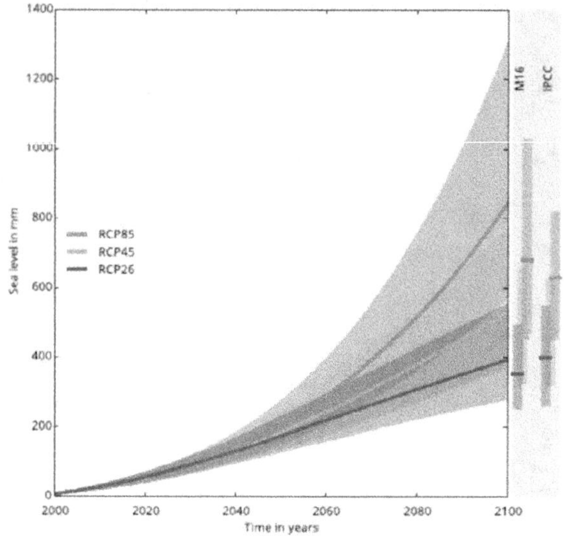
RCP85
RCP45
RCP26
Sea level in mm
Time in years
M16
IPCC

Fig. 4(b).

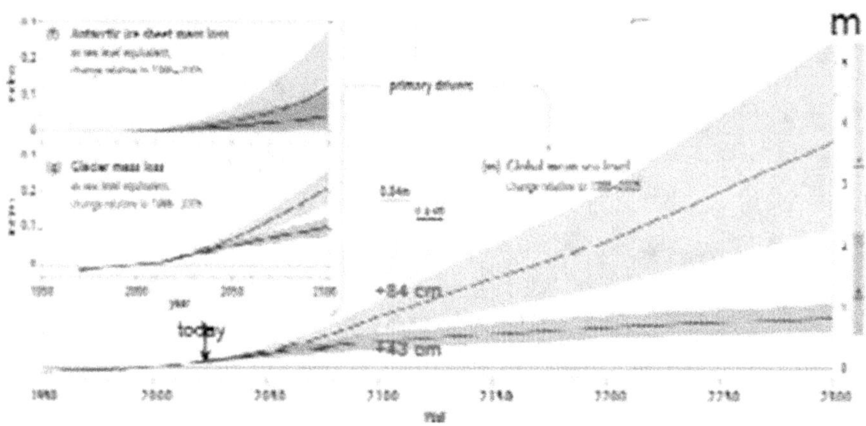
today
+84 cm
+43 cm

Fig. 4(c).

F Pavese, A Bošnjaković, S Eichstädt, A B Forbes, J Alves e Sousa (eds.):
Advanced Mathematical and Computational Tools in Metrology and Testing XIII
Series on Advances in Mathematics for Applied Sciences, Vol. 94

https://doi.org/10.1142/9789819800674_0002

Evaluation of repeatability measurement procedures in gravimetric volume calibrations of glassware

E. Batista and J. Alves e Sousa

Instituto Portugues da Qualidade,
Caparica, 2829-513, Portugal
E-mail: ebatista@ipq.pt
www.ipq.pt

Andrea Malendo and Davide Torchio

INRIM
Torino, 10135, Italy

Olaf Schnelle-Werner

ZMK & Analytik GmbH
Bitteerfeld-Wolfen, D 06766, Germany

Urška Turnšek

MIRS
Celje, 3000, Slovenia

Georgia Harris

NIST-OWM
Gaithersburg, 2600, USA

The EURAMET collaboration project 1525 on evaluation of repeatability measurement procedures in gravimetric volume calibrations of glassware was undertaken by six National Metrology Laboratories and Accredited laboratories, with the purpose of investigating the different approaches regarding repeatability measurements of glassware calibration using the procedures described in two volume standards (ISO 4787 and ASTM E542), and its influence on the determined volume and uncertainty. Six flasks of different volume capacity were calibrated. The consistency of the data was determined using several statistical methods and all results are presented in this paper, along with reproducibility analysis and uncertainty calculation.

Keywords: volume measurements; repeatability; uncertainty; glassware.

1. Introduction

Following the cooperation between EURAMET Technical Committee for Flow and ASTM Committee E41 it was decided to investigate the different approaches regarding repeatability measurements of glassware calibration performed by different national metrology laboratories, accredited laboratories and manufactures around the world and its influence on the determined volume and uncertainty.

Suitable repeatability estimates are needed for evaluating measurement results and for determining the measurement uncertainties. Lack of repeatability agreement within a set of measurement results can lead to significant problems with the operating characteristics of the volumetric instrument. The number of repeats affects the impact of the repeatability to the overall uncertainty and can be decided with respect to the scope of the specific volume determination (batch, verification, calibration). This paper describes the different approaches in repeatability measurements from each participant and its influence on the final uncertainty calculation and volume determination of three sets of flasks of 100 mL, 500 mL and 1000 mL. Two common approaches for obtaining repeatability statistics were investigated: The first approach uses standard deviation control charts to monitor the measurement process at the time of calibration by ensuring that observed and accepted standard deviations agree. This method requires maintaining standard deviation control charts to monitor the measurement process and to determine a pooled standard deviation. It enables a smaller number of repeated measurements for each calibration. The second approach uses a larger number of repeated measurements for each calibration (5 or 10 replicates).

2. Calibration method

Calibration of the flasks consists of the determination of the amount of water contained in the flask at reference temperature of 20 °C, using the gravimetric method. The following Eq. (1) described in ISO standard 4787:2021 [1] and in ASTM E 542–01: 2022 [2] was used to determine the volume (V_{20}) of each flask.

$$V_{20} = (I_L - I_E) \times \frac{1}{\rho_W - \rho_A} \times \left(1 - \frac{\rho_A}{\rho_B}\right) \times [1 - \gamma(t - 20)] \quad (1)$$

The value of the density of mass pieces (ρ_B) and expansion of the material of the flask (γ) must be known in advanced.

The participating laboratories used their own test procedure of calibration. The simplified calibration procedure is:

- Measure ambient conditions in order to determine the density of air (ρ_A).
- Weigh the empty dry flask and record the mass, I_E.
- Fill the flask up to the reference line and measure the liquid temperature (t) in order to determine the density of the water (ρ_W).

- Adjust the meniscus and wipe out (drying) any water drops above the reference line. The meniscus of the flasks shall be set so that the plane of the upper edge of the graduation line is horizontally tangential to the lowest point of the meniscus. The shape of the meniscus is set such that the surface of the liquid forms a curve that meets the glass tangentially.
- Weigh the filled flask and record the mass, I_L.

3. Measurement results

3.1. *Evaluation of the results*

To determine the reference value the formula of the weighted mean is used, by means of the inverses of the squares of the associated standard uncertainty are the weighing factors [3].

$$y = \frac{x_1/u^2(x_1)+\ldots+x_n/u^2(x_n)}{1/u^2(x_1)+\ldots.+1/u^2(x_n)} \tag{2}$$

To determine the standard uncertainty $u(y)$ associated with y is used the following expression:

$$u(y) = \sqrt{\frac{1}{1/u^2(x_1)+\ldots+1/u^2(x_n)}} \tag{3}$$

To identify an overall consistency of the results the E_n value is also calculated [4].

$$E_n = \frac{V_{Lab} - V_{\mathrm{Re}f}}{\sqrt{U_{Lab}^2 - U_{\mathrm{Re}f}^2}} \tag{4}$$

3.2. *Stability of the artifacts*

In order to identify the stability of the artifacts IPQ performed measurements in the beginning of the comparison (IPQ-1) and in the end of the comparison (IPQ-2). The results are presented in Table 1. For two of the flasks this stability could not be assessed because they were broken during transportation.

From the obtained results, it can be verified that all the flasks were stable, the volume variation is smaller than the uncertainty for all cases. For the determination of the reference value only the first results from IPQ were used.

3.3. *Volume results*

The obtained volume results and uncertainty for the two 100 mL flasks are presented in Table 2.

Table 1. Volume variation measurement results.

		Nominal volume	Volume (µl)	U_{exp} (µl)	ΔV (µl)
Flask 9572	IPQ – 1	100	99,955	0,013	0,002
	IPQ – 2	100	99,953	0,012	
Flask 9573	IPQ – 1	100	99,962	0,015	0,002
	IPQ – 2	100	99,960	0,012	
Flask 9577	IPQ – 1	500	499,863	0,053	0,025
	IPQ – 2	500	499,838	0,050	
Flask 9578	IPQ – 1	1000	999,71	0,10	0,05
	IPQ – 2	1000	999,66	0,10	

Table 2. Volume results of 100 mL flasks.

	Flask 9572		Flask 9573	
Participant	Volume (mL)	U (mL)	Volume (mL)	U (mL)
IPQ	99,955	0,013	99,962	0,015
INRIM	99,942	0,013	99,963	0,013
MIRS	99,955	0,031	99,987	0,023
ZMK	99,928	0,050	99,964	0,050
Artel	99,937	0,031	99,969	0,031
NIST OWM	99,911	0,046	99,920	0,046
Reference value	99,946	0,008	99,965	0,008

The normalized error was determined for both flasks. The results are described in Fig. 1.

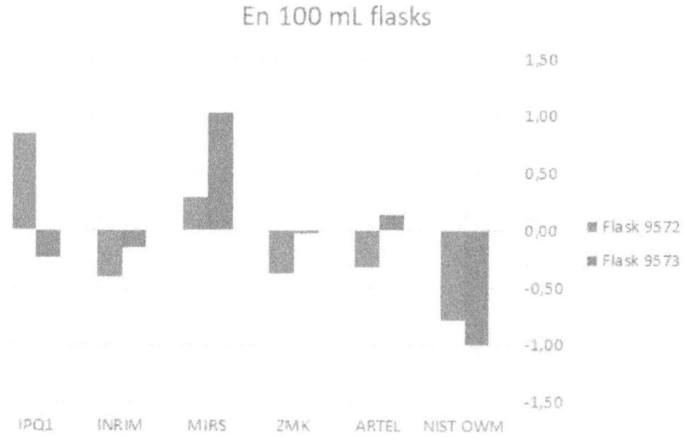

Fig. 1. En values for the 100 mL flask.

From Fig. 1, it can be observed that all the volume results are consistent but two of them very close to 1. The obtained volume results and uncertainty for the three 500 mL flasks are presented in Table 3.

Table 3. Volume results of 500 mL flasks.

	Flask 9577		Flask 9576		Flask 1926	
Participant	Volume (mL)	*U* (mL)	Volume (mL)	*U* (mL)	Volume (mL)	*U* (mL)
IPQ	499,863	0,053	499,818	0,057	499,951	0,050
INRIM	499,879	0,024	499,847	0,024	-	-
MIRS	499,918	0,052	499,897	0,051	500,032	0,042
ZMK	499,85	0,23			499,95	0,23
Artel	499,990	0,077			499,95	0,08
NIST OWM	499,814	0,070			499,77	0,07
Reference value	499,851	0,020	499,877	0,019	49,990	0,029

The normalized error was determined for the three flasks. The results are described in Fig. 2.

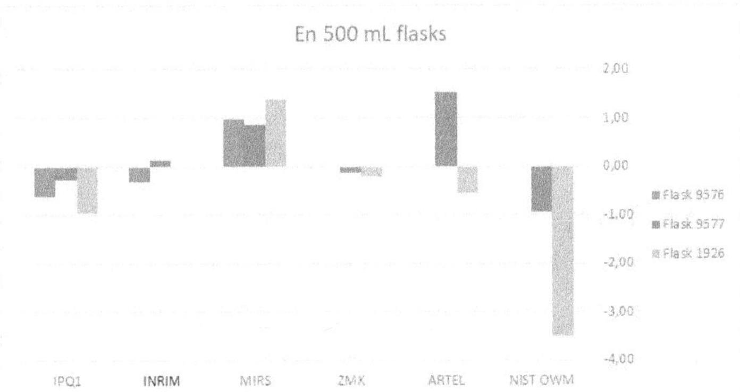

Fig. 2. En values for the 500 mL flask.

Table 4. Volume results of 1000 mL flasks.

	Flask 9578		Flask 9579		Flask 1	
Participant	Volume (mL)	*U* (mL)	Volume (mL)	*U* (mL)	Volume (mL)	*U* (mL)
IPQ	999,71	0,10	1000,12	0,11	1000,16	0,10
INRIM	999,77	0,04	1000,165	0,036	-	-
MIRS	999,71	0,12	1000,14	0,14	-	-
ZMK	999,78	0,45	999,91	0,45	-	-
NIST OWM	999,52	0,17	-	-	999,92	0,17
Reference value	999,763	0,033	1000,158	0,033	1000,93	0,085

It can be observed from Fig. 2 that there were three inconsistent values, one from MIRS, another from Artel and other from NIST OWM. The obtained volume results and uncertainty for the three 1000 mL flasks are presented in Table 4.

The normalized error was determined for the three flasks. The results are described in Fig. 3.

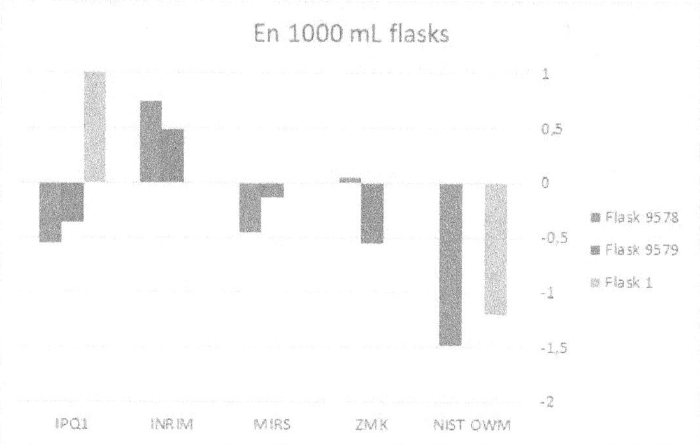

Fig. 3. En values for the 1000 mL flasks.

For the 3 flasks measured it can be observed that there were two inconsistent values from NIST OWM.

4. Consistency results determination based on the NIST Consensus Builder tool

The NIST Consensus Builder web tool [5] was used to combine measurement results obtained by different laboratories, into a consensus estimate of the value of a scalar measurand and for the determination of consistency. Three different methods for data analysis were used:

- DerSimonian-Laird;
- Hierarchical Bayes;
- Linear Pool.

The results obtained for the DerSimonian-Laird method [6] can be found in Fig. 4 for the 500 mL flask number 9577.

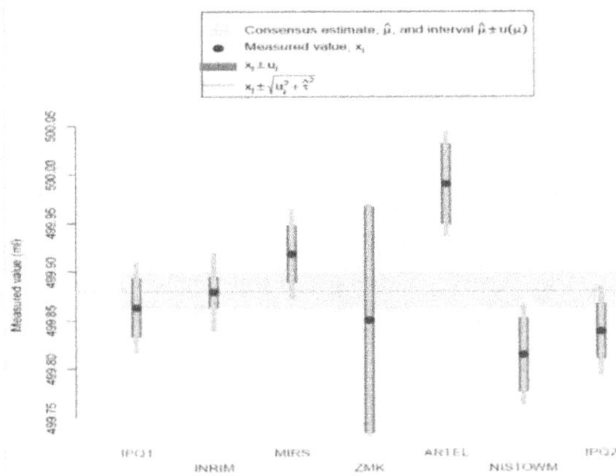

Fig. 4. Consistency check using DerSimonian-Laird method.

The mean value determined using the DerSimonian-Laird method was 500,00 mL and uncertainty of 0,018 mL. The results obtained by the Weighed mean method [3] are found in Fig. 5 for the same flask.

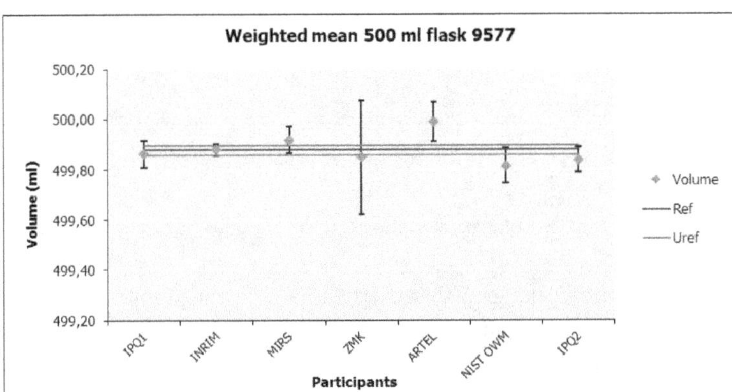

Fig. 5. Consistency check using the weighted mean.

The mean value determined using the weighted mean was 499,88 mL and uncertainty of 0,019 mL.

The results were obtained by the NIST Consensus Builder tool [5] are consistent with the ones obtained from the weighing mean analysis [3] but a dark uncertainty, or laboratory random effect (τ) [7] of 0,01% value was found for the 500 mL flask 9577 and this value is in same order of magnitude as the participants expanded uncertainty. For the 100 mL flasks and 1000 mL flask no significant difference was found regarding the weighing mean analysis and also, no dark uncertainty was found.

5. Reproducibility of results

Two sets of each flask that were circulated by the participants, this allowed to verify if there was any variation within the laboratories.

In relative terms, the errors of the measurement's Di/V obtained by the laboratories are shown in Fig. 6, where the error bars correspond to the relative expanded uncertainties.

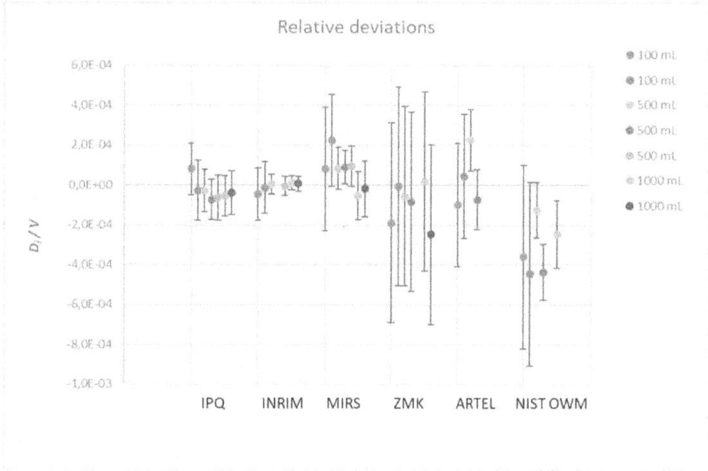

Fig. 6. Relative deviations of the volume flasks.

From Fig. 6, it can be seen that the results of each laboratory have similar relative uncertainty values and same relative error value when compared with the reference, with a strong correlation. It can also be verified that the majority of NIST OWM results are always lower values than the other laboratories due to the aforementioned meniscus reading differences but also consistent within each other. Assessing the standard deviation of the relative errors, the reproducibility of the measurements for each laboratory was evaluated, which, as expected, is proportional to the declared uncertainties, always remaining below about 0.8 times the expanded uncertainty. The analysis is shown in Fig. 7, where the relative standard deviation is plotted against the mean value of the expanded relative uncertainties. This shows that the stated uncertainties are adequately evaluated, it is noted that the ZMK laboratory overestimated their uncertainty, and this was due to the CMC claim values.

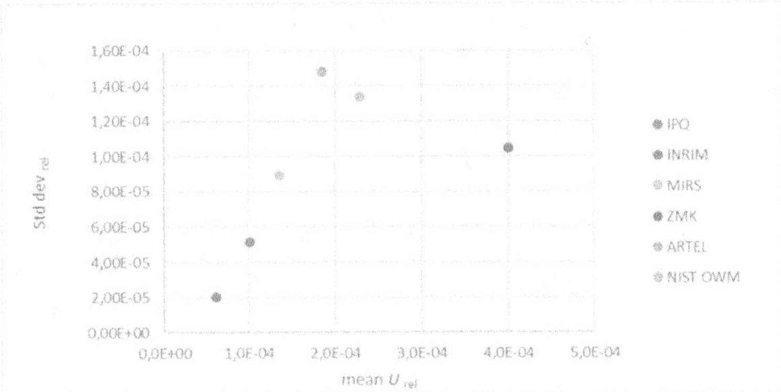

Fig. 7. Uncertainty analysis of the volume variation.

6. Uncertainty calculations

Each laboratory described the uncertainty components. Both values, i.e. standard uncertainty and expanded uncertainty were stated, along with the relevant coverage factor *k*. The repeatability measurement procedure was described in detail; documented procedures were referenced (EURAMET cg-19 and NIST SOP 14 provide details).

For the evaluation of the measurement uncertainty, reference was made to the Guide to the Expression of Uncertainty in Measurement at approximately 95% confidence interval [8]. In general, the laboratories that used EURAMET cg 19 [9] had lower uncertainty than the one using the NIST SOP 14 (NIST OWM). The largest source of uncertainty was the repeatability and meniscus for the European countries and the meniscus reading and control limits for the Artel and NIST OWM laboratories., respectively. For the repeatability measurements all laboratories used 10 replicated except for NIST OWM that used 5 repetitions.

7. Discussion

This project had the purpose to investigate the different approaches regarding repeatability measurements of glassware calibration performed by different NMIs, accredited laboratories and manufactures around the world and its influence on the determined volume and uncertainty. Two sets of flasks of the following 3 nominal volumes were calibrated: 100 mL, 500 mL, 1000 mL.

NIST OWM has used the procedure described in NIST GMP 3 (Option A: setting bottom of the meniscus between upper and lower portion of graduation line ellipse), this led to lower volume values than the laboratories that use the ISO

4787 meniscus adjustment procedure resulting in inconsistent (biased) results. The consistency of results was also assessed using the NIST consensus builder tool and the results were different from the weighted mean analysis. A dark uncertainty of 0,01% was found for the 500 mL flask 9577 being this value in same order of magnitude as the participants expanded uncertainty. NIST OWM used the repeatability five runs and compared to historical process standard deviations with control charts to assess repeatability for each flask compared to an accepted value. In general, this approach led to a higher uncertainty value than the ones based on the EURAMET cg 19 and ISO 4787 Annex E, a) repeatability procedure determination, probably due to a larger dark uncertainty component.

8. Conclusions

This work performed the evaluation of repeatability measurement procedures of six laboratories (NMIs and accredited laboratories) around the world. Some inconsistency was verified within volume measurements results obtained along with different repeatability determination techniques that influenced the measurement uncertainty.

References

1. ASTM E 542:2022 - Standard practice for calibration of laboratory volumetric apparatus.
2. ISO 4787:2021; Laboratory glassware – Volumetric glassware – Methods for use and testing of capacity.
3. M.G. Cox, The evaluation of key comparison data, *Metrologia*, 39:589–595, 2002.
4. ISO 13528 - Statistical methods used in proficiency testing by interlaboratory comparisons.
5. https://consensus.nist.gov/app/nicob.
6. A. Koepke, T. Lafarge, A. Possolo and B. Toman, 2016 NIST Consensus Builder user's manual (Gaithersburg, MD: National Institute of Standards; Technology).
7. M. Thompson, S.L.R. Ellison, Dark uncertainty. *Accred Qual Assur* 16: 483–487, 2011.
8. BIPM, IEC, IFCC, ISO, IUPAC, IUPAP, OIML; Guide to the expression of uncertainty in measurement (GUM), Geneva, 1995.
9. EURAMET calibration guide-19 Version 3.0, Guidelines on the determination of uncertainty in gravimetric volume calibration.

F Pavese, A Bošnjaković, S Eichstädt, A B Forbes, J Alves e Sousa (eds.):
Advanced Mathematical and Computational Tools in Metrology and Testing XIII
Series on Advances in Mathematics for Applied Sciences, Vol. 94

https://doi.org/10.1142/9789819800674_0003

Signal interval to classification: Supervised machine learning for ECG*

A. Bošnjaković[1,*] and P.J. Aston[2,3]

[1]*Institute of Metrology of Bosnia and Herzegovina*
[2]*National Physical Laboratory, Teddington, UK*
[3]*University of Surrey, Guildford, UK*
**E-mail: alen.bosnjkovic@met.gov.ba*

Supervised machine learning methods were used to predict patient heart disease in this work. The dataset was extracted from an open, traceable dataset for research purposes. 1600 electrocardiogram (ECG) records were extracted, and only single label records were used. ECG analysis was performed by determining particular fiducial points for all the ECG signals from which seven intervals and five amplitudes were obtained. Furthermore, these features were individually calculated for each of the 12 ECG leads, except for the RR interval, which was calculated only once using all 12 leads. However, it was included in the feature vectors for all leads. The amplitude and interval features were used to form the feature vectors. The accuracy scores for each lead were calculated using two supervised machine learning methods and using five-fold cross-validation. Taking into account that the dataset was balanced, the results were studied using the accuracy score. The results were represented by confusion matrices and diagrams. All captured results were compared. Finally, the sensitivity of the machine learning methods was reviewed to consider variations in the classification performances.

Keywords: ECG signals; fiducial points; machine learning; accuracy score.

1. Introduction

The electrocardiogram (ECG) records the electrical activity of the heart and is widely used by cardiologists worldwide for diagnosis of many cardiac conditions. An ECG is usually analysed by finding various intervals and amplitudes which are derived from certain fiducial points on the signal, and many textbooks are devoted to the analysis and interpretation of the ECG (see for example [1, 2]).

An alternative to manual interpretation by a cardiologist is to use machine learning for diagnosis of various conditions using features derived from the signal using software such as ECGdeli [3]. In this work, we consider a classification of

*This project (18HLT07) has received funding from the EMPIR programme co-financed by the Participating States and from the European Union's Horizon 2020 research and innovation programme.

12-lead ECG signals as one of three cardiac conditions or healthy. Two supervised machine learning algorithms are used and the results are compared.

2. ECG signals

The publicly available PTB-XL database [4] was used as the source of the ECG signals used for this study, which is described in [5]. The database was developed through the scientific research project MedalCare [6] and, when it was published, represented the most extensive clinical database for 12-lead ECG signals. This database includes 21,837 10 second 12-lead ECG records sampled at 500 Hz and is labelled with many diagnostic classes including many healthy records. The ECG signals were annotated by up to two cardiologists as a multi-label dataset, where the diagnostic labels were further aggregated into super-classes and sub-classes. The combination, with additional metadata such as demographics, diagnostic statements, diagnosis likelihoods, manually annotated signal properties, and suggested folds for splitting training and testing sets, make this dataset a valuable resource for developing and evaluating automated ECG interpretation algorithms.

3. ECG features

Electrical activity in the heart is responsible for the waveform shape of an ECG signal. ECG analysis is done by finding fiducial points on an ECG signal from which various intervals and amplitudes can be derived and which can be used to diagnose different clinical conditions [1]. Any deviation from the standard signal leads to possible pathological conditions [7, 8].

The ECGdeli software [3], which is an open source ECG delineation toolbox, was used to determine the fiducial points. Nine fiducial points were extracted which were P wave onset (P_o), P wave peak (P), P wave offset (P_e), QRS onset (QRS_o), QRS peak (QRS), QRS offset (QRS_e), T wave onset (T_o), T wave peak (T), and T wave offset (T_e). The seven intervals and five amplitudes [9] that were derived from these points were P duration ($P_e - P_o$), PR interval ($QRS_o - P_o$), QRS interval ($QRS_e - QRS_o$), QT interval ($T_e - QRS_o$), T interval ($T_e - T_o$), ST segment ($T_o - QRS_o$), RR interval, P amplitude, Q amplitude, R amplitude, S amplitude, and T amplitude.

These 12 features were calculated individually for each of the 12 leads of the ECG, except for the RR interval, which was calculated only once using all 12 leads. However, it is included in the feature vector for all leads. The intervals and amplitudes are shown in Fig. 1. We construct a feature vector from these 12 features for each ECG record.

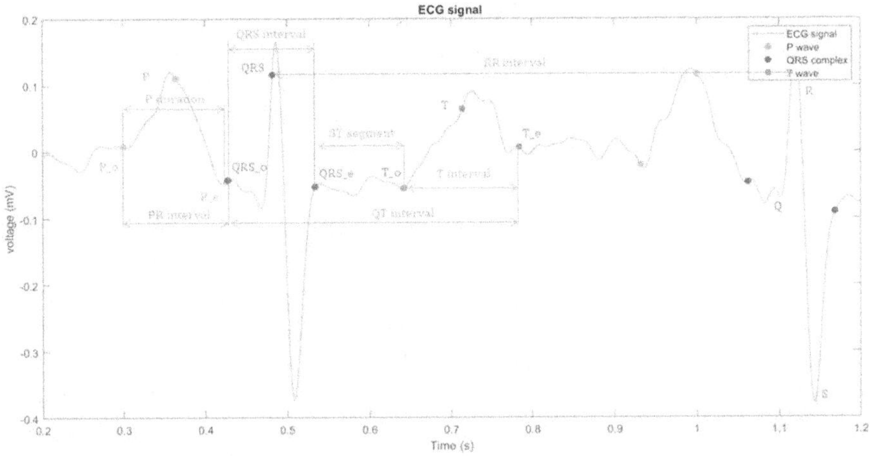

Fig. 1. Fiducial points, intervals and amplitudes.

ECG features include important and relevant information extracted from the raw ECG signal, which relate to cardiac abnormalities. Clinicians can diagnose over 150 distinct heart conditions using these and other ECG features [1]. Furthermore, clinicians meticulously study these key features to make decisions regarding the patient's conditions.

The intervals and amplitudes provide features that are typically analysed and understood by clinicians. They quantify key (large-scale) features and may be averaged over all beats in the signal or may be provided for each individual beat. They provide a quantitative waveform summary in a small number of features. However, some minor changes in the wave morphology may not be captured by these features. Moreover, while finding the fiducial points may be a straightforward task for a clean signal, it can be difficult for an algorithm to correctly identify these points on a noisy signal. The P duration is particularly important in the diagnosis of a left atrial abnormality. The PR interval is between 0.12 s and 0.2 s in healthy adults. The QRS interval is generally less than 0.10 s, and greater than this may show some pathology. Some changes in the QT interval can show a predisposition to dangerous ventricular arrhythmias. The RR interval represents the differences between the consecutive R peaks and is equivalent to heart rate. The P amplitude is the amplitude of the P wave. The QRS peak is the amplitude of the R peak. The T amplitude is the amplitude of the T wave. More information about these features can be found in the book [1].

4. Pathologies

Three pathologies were selected to be studied in addition to Normal, which were First degree atrioventricular block (AV block), Complete right bundle branch block (RBBB), and Complete left bundle branch block (LBBB). From the PTB-XL database [4], 1600 records were extracted, with 400 records for each class (Normal, AV Block, RBBB, and LBBB), giving a balanced dataset. Only single label records were used.

In patients with the healthy case, electrical activity is spontaneously initiated in the pacemaker cells at the senatorial node in the right atrium. After spreading throughout the atrial myocardial tissue and causing the P wave in the ECG, the excitation is delayed at the atrioventricular node [2, 6].

In patients with atrioventricular (AV) block, excitation conduction between the atria and ventricles is altered. In first-degree AV block, which is studied in this paper, conduction is markedly delayed and leads to PR intervals greater than 0.2 s in the ECG. However, all atrial impulses are still transferred to the ventricles and every P wave is followed by a QRS complex, unlike the second or third degree AV block [2]. This class represents pathological subjects.

Complete right bundle branch block is characterised by a blockage along the conduction pathways in the right fascicle running along the ventricular septum. In this case, the right ventricles are activated by impulses that are carried out through branches of the left bundle that reach the right ventricle through the ventricular myocardial tissue [2]. This class represents pathological subjects.

Analogously to right bundle branch block described above, complete left bundle branch block describes the condition of a blockage in the electrical conduction in the left bundle branch. As the left bundle branches into an anterior and a posterior fascicle, the term left bundle branch block refers to a conduction block prior to bifurcation. This class represents pathological subjects. For both RBBB and LBBB, the QRS complex is larger than 0.12 s.

5. Methods

In general terms, machine learning methods can be divided into two groups, namely supervised and unsupervised machine learning methods. For supervised machine learning, we start with a set of labelled feature vectors, and our goal is to find a model that predicts the label for previously unseen feature vectors. In addition, supervised machine learning can be divided into two classes, one is regression, where a model predicts a real number, and the second is classification, where a model predicts the class from a finite, discrete set. A simple example of binary classification is the prediction of healthy and or unhealthy subjects from their ECG.

Machine learning models can be white-box, such as decision trees, or black-box, such as neural networks. Furthermore, we can think about this in terms of generative and predictive modelling. The first goal is to understand the relationships between variables. The second goal is to make predictions based on a particular model. It could be stated that machine learning models are used for predictions, but algorithms could be used in both cases with different final objectives. For example, a linear regression model could be used to learn the relationships between inputs and outputs, but it could also be used to make predictions for new data.

Based on this short elaboration, it could be concluded that machine learning models can be white-box or black-box models. The white-box models represent relationships between variables, where these relationships are well understood. Furthermore, the black-box model relationships between variables where these relationships are not understood well.

When we are comparing traditional programming with machine learning, it could be stated that there are two approaches. For the first one, there is data, there is a program, and these are fed into a computer (traditional programming). For the second one, there is data, there is output associated with the data, and a model is generated (machine learning). This paradigm is based on Arthur Samuel's view of machine learning which he defined as the "Field of study that gives computers the ability to learn without being explicitly programmed" [10]. For example, there are independent data (input) and dependent data (output) that are fed into the computer to learn the model and the model is used for data prediction. In short, machine learning aims to design, understand, and apply computer programs that learn from experience (i.e., data) for modelling, prediction, or control.

Two supervised machine learning methods were used for ECG classification, namely Support Vector Machine (SVM) and Multinomial Logistic Regression (SoftMax) [11]. The first method (SVM) tries to (1) minimise the average Hinge loss on the training data, while (2) pushing margin boundaries apart. These two goals are opposite and need to be balanced. When losses are emphasised, the resulting classifier becomes heavily influenced by the few examples located near the decision boundary. In contrast, when the margin boundaries are extended, losses are incurred for examples that are farther away, and the solution is defined by the most positive and negative examples. The objective function for the SVM method is given by the following equation:

$$J(\theta, \theta_0) = \frac{1}{n}\sum_{i=1}^{n} \mathrm{Loss_h}\left(y^{(i)}\left(\theta \cdot x^{(i)} + \theta_0\right)\right) + \frac{\lambda}{2}||\theta||^2, \quad (1)$$

where the first part represents the loss function, and the second part represents the regularisation term where λ is the regularisation parameter. The optimal parameters (θ, θ_0) are obtained by minimising the objective function (1).

The second method applied is SoftMax. The objective function for the SoftMax method is given by the following equation:

$$J(\theta) = -\frac{1}{n}\left[\sum_{i=1}^{n}\sum_{j=0}^{k-1}[\![y^{(i)} = j]\!] \log\frac{e^{\theta_j \cdot x^i/\tau}}{\sum_{l=0}^{k-1} e^{\theta_l \cdot x^i/\tau}}\right] + \frac{\lambda}{2}\sum_{j=0}^{k-1}\sum_{i=0}^{d-1}\theta_{ji}^2, \quad (2)$$

where the first term again represents the loss function, and the second term represents the regularisation term where λ is the regularisation parameter. The optimal value of the parameter θ is obtained by minimising the objective function (2) using the gradient descent algorithm [11].

All approaches to machine learning have something in common — we need to have our features and (accurate) labels. These features are often extracted from raw data, as with our features that are extracted from the raw ECG signals. We represent them as vectors, known as feature vectors. Once we have a feature table, made up of many feature vectors, and the label for each feature vector, we need to apply an optimisation process that trains the model. Finally, we must confirm that everything works smoothly and test the model on unseen data.

6. Results

We used five-fold cross-validation in order to eliminate variability that can arise from the data. The dataset is split into five folds and four folds are used for training with the fifth fold used for testing.

This is repeated with each of the folds being the test data. Thus, 80% of the data is used for training and 20% for testing each time. At the end of the cross-validation process, each of the records has been used for testing precisely once and so the predictions are compared with the true labels for all records and a confusion matrix generated.

We present a number of metrics derived from the confusion matrix, including accuracy, precision, recall and F_1 score. For a balanced dataset, the accuracy is often considered when comparing performance whereas, for an imbalanced dataset, the F_1 score is considered a more appropriate metric to use.

For lead one, the confusion matrix is shown in Fig. 2 and the metrics are given in Table 1 for the SVM method. Similarly, for lead one, the confusion matrix and metrics for the SoftMax method are given in Fig. 3 and Table 2, respectively.

Based on these results, it can be observed that the results are very similar using the different machine learning methods with an accuracy score for the two methods of 0.81 or 0.82.

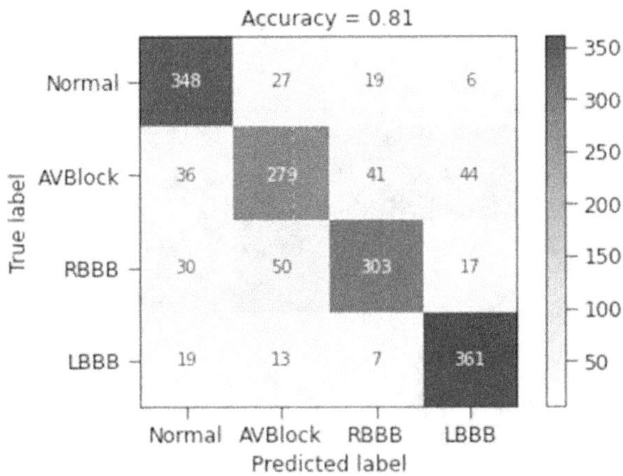

Fig. 2. Confusion matrix for lead one (all vs. all) - SVM.

Table 1. Classification metrics for lead one - SVM.

Classes	Precision	Recall	F_1 score
Normal	0.80	0.87	0.84
AVBlock	0.76	0.70	0.73
RBBB	0.82	0.76	0.79
LBBB	0.84	0.90	0.87

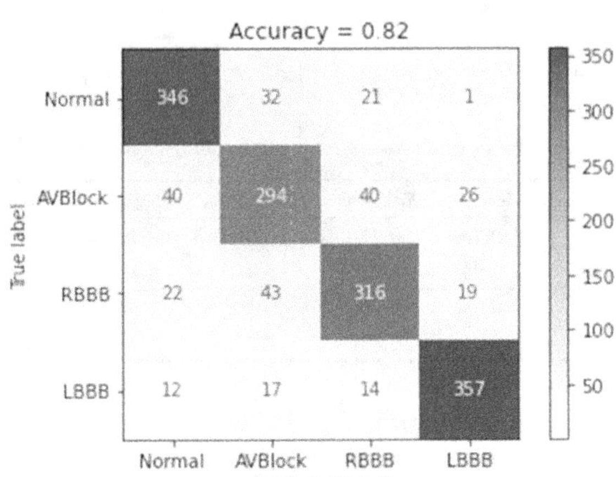

Fig. 3. Confusion matrix for lead one (all vs. all) - SoftMax.

Table 2. Classification metrics for lead one - SoftMax.

Classes	Precision	Recall	F_1 score
Normal	0.82	0.86	0.84
AVBlock	0.76	0.73	0.75
RBBB	0.81	0.79	0.80
LBBB	0.89	0.89	0.89

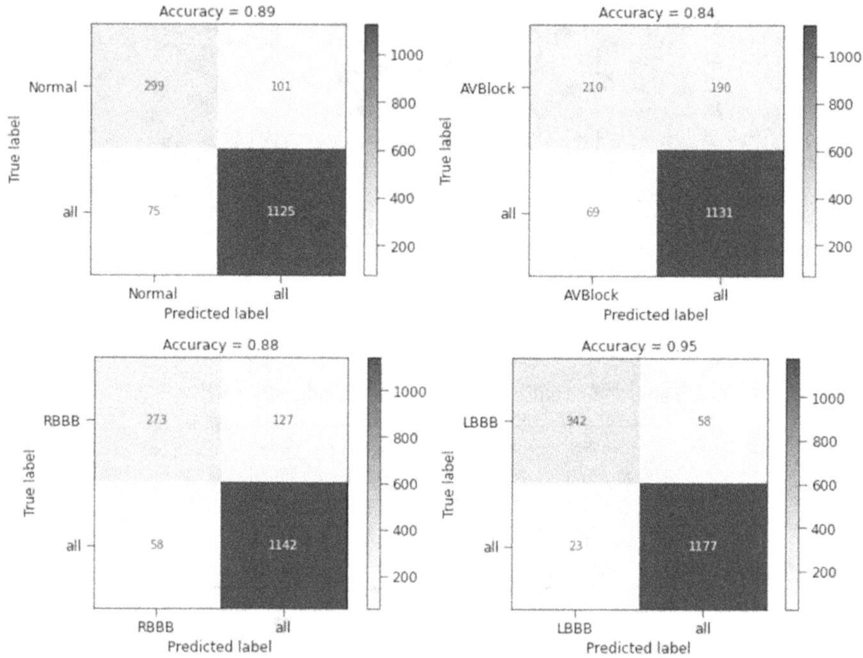

Fig. 4. Confusion matrices for lead one (one vs. all) - SVM.

Table 3. Lead one confusion metrics (one vs. all) - SVM.

Classes	Precision	Recall	F_1 score	Classes	Precision	Recall	F_1 score
Normal	0.80	0.75	0.77	AVBlock	0.75	0.53	0.62
All	0.92	0.94	0.93	All	0.86	0.94	0.90
RBBB	0.82	0.68	0.75	LBBB	0.94	0.85	0.89
All	0.90	0.95	0.93	All	0.95	0.98	0.97

We then considered the binary classification problems where each class in turn was considered separately versus all the other classes. The results are shown in Fig. 4 and Table 3 for the SVM method and in Fig. 5 and Table 4 for the

SoftMax method. Again, it can be observed that there is little difference in these results between the two methods. We note that the accuracy score is much better for some classes than for others, with the highest being 0.95 for LBBB using both methods and the lowest being 0.84 for AV block for both methods. We observe that for each class, the F_1 score for the 'All' category is significantly higher than for the one chosen class.

Results for all 12 leads of the ECG for the classification of all 4 classes (all vs. all) are shown in Figs. 6 and 7. We note that the best performance was obtained using lead V1, while lead aVF gave the worst performance. Furthermore, results for all 12 leads of the ECG for the classification of all 4 classes (one vs. all) are shown in Figs. 8 and 9. The performance for lead V1 was again very good, particularly in the cases of RBBB vs. All and LBBB vs. All.

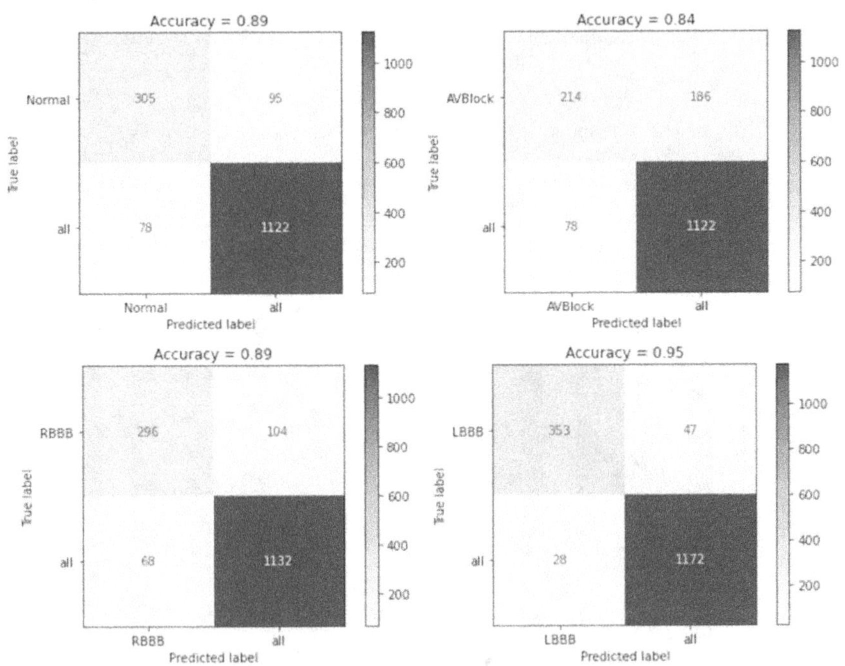

Fig. 5. Confusion matrices for lead one (one vs. all) - SoftMax.

Table 4. Lead one confusion metrics (one vs. all) - SoftMax.

Classes	Precision	Recall	F_1 score	Classes	Precision	Recall	F_1 score
Normal	0.80	0.76	0.78	AVBlock	0.73	0.54	0.62
All	0.92	0.94	0.93	All	0.86	0.94	0.89
RBBB	0.81	0.74	0.77	LBBB	0.93	0.88	0.90
All	0.92	0.94	0.93	All	0.96	0.98	0.97

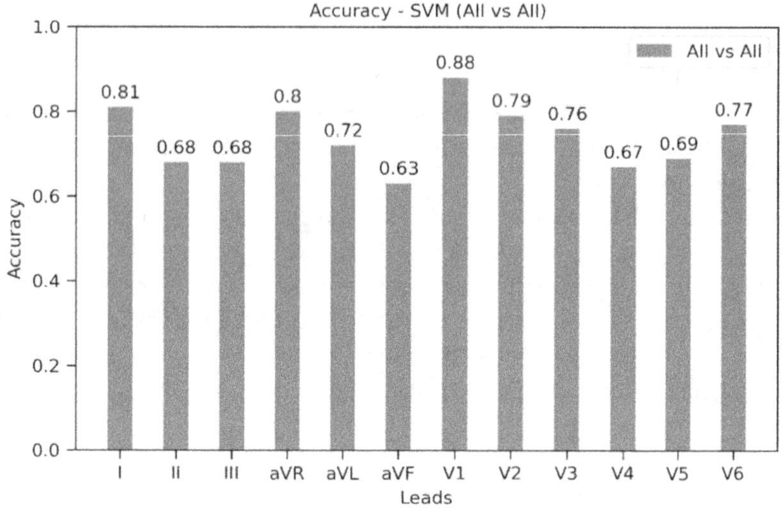

Fig. 6. The average accuracy scores for all leads (all vs. all) - SVM.

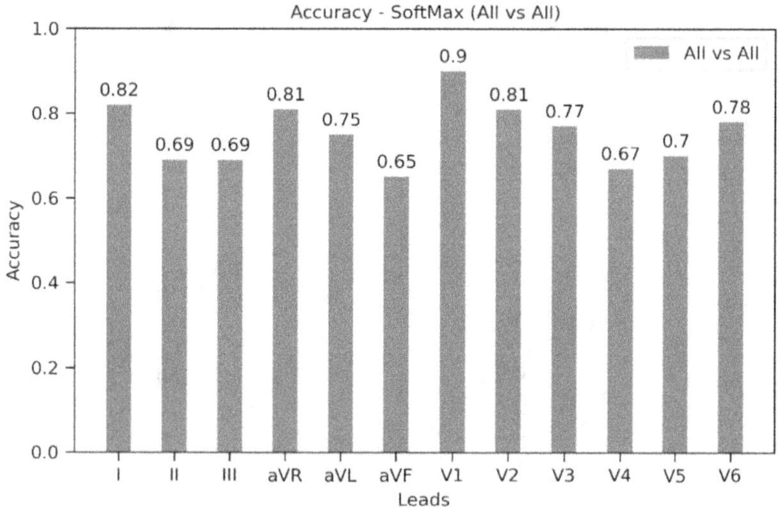

Fig. 7. The average accuracy scores for all leads (all vs. all) - SoftMax.

7. Conclusion

The ECG feature dataset has some advantages over the raw ECG dataset in terms of computation time, as it contains fewer data points than the raw one. For example, the feature vector for the raw signal has dimension $1600 \times 5000 \times 12$,

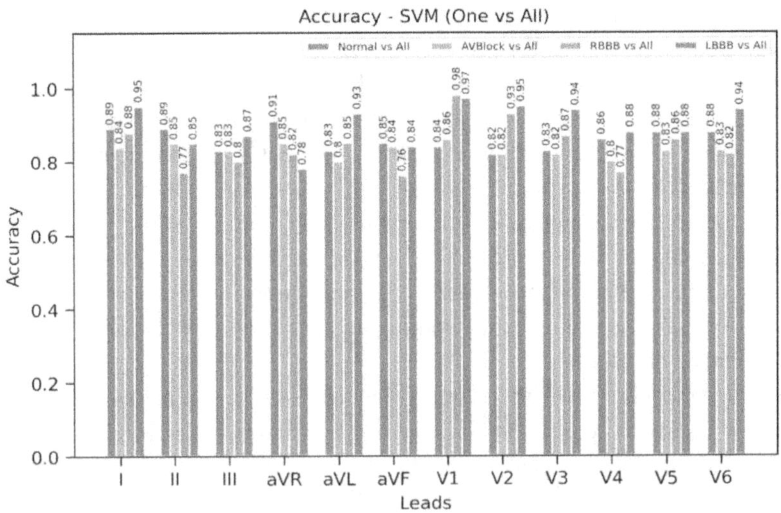

Fig. 8. The average accuracy scores for all leads (one vs. all) - SVM.

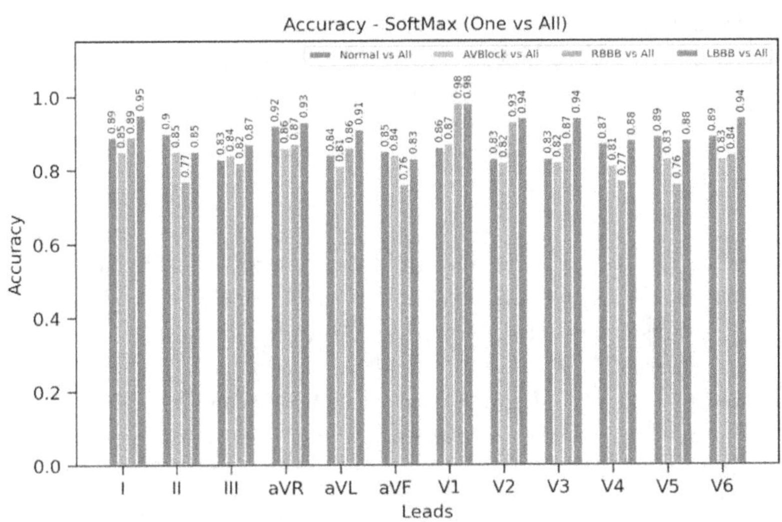

Fig. 9. The average accuracy scores for all leads (one vs. all) - SoftMax.

while the ECG feature vector has dimension 1600 × 12 × 12. The important machine learning characteristics are model building, validation and testing on unseen data from a metrology conformity perspective.

For the four-class dataset, five-fold cross-validation was used to eliminate any variability that may arise from the data used. Furthermore, it is used for the model sensitivity quantification. Taking into account that classes are balanced, the accuracy score is used for presenting results. The accuracy score is regarded as a more reliable performance metric than the F_1 score when classes are balanced. The results for all leads are given. Furthermore, the results show that performance of the machine learning methods varies for different leads, so the choice of lead is significant with the accuracy score for lead V1 being the highest.

For the one vs. all binary classifications, the F_1 score for the 'All' class was significantly higher than for the single pathology class in all cases, which is probably due to the imbalanced classes. In these cases, either more data could be selected for the single class or the single class could be oversampled to generate more balanced datasets.

As for measurement uncertainty quantification from a machine learning perspective, this is a relatively new research area in metrology, and t is underpinned through different research projects within the metrology community [6, 12]. Regarding future work, the focus will be placed on other machine learning methods (supervised and unsupervised) including raw records. Furthermore, the synthetic records will be compared with the real ones to estimate their usability for confirming assessment purposes in this field. Understanding uncertainty in machine learning is a big and important field. To have a greater impact, more effort is needed to quantify uncertainties more thoroughly, thus making them of more practical use.

References

1. A. L. Goldberger, Z. D. Goldberger and A. Shvilkin, *Clinical Electrocardiography: A Simplified Approach*, 9th edn. (Elsevier, 2018).
2. H.-P. Schuster and H.-J. Trappe, *EKG-Kurs f¨ur Isabel*, 7th edn. (eorg Thieme Verlag, 2017).
3. N. Pilia, C. Nagel, G. Lenis, S. Becker, O. D¨ossel and A. Loewe, *ECGdeli – An open source ecg delineation toolbox for matlab*, SoftwareX **13**, p. 100639(2021).
4. P. Wagner, N. Strodthoff, R.-D. Bousseljot, W. Samek and T. Schaeffter, PTB-XL, a large publicly available electrocardiography dataset (version 1.0.2), *PhysioNet* (2022).
5. P. Wagner, N. Strodthoff, R.-D. Bousseljot, D. Kreiseler, F. I. Lunze, W. Samek and T. Schaeffter, PTB-XL, a large publicly available electrocardiography dataset, *Scientific data* 7, 1 (2020).
6. Publishable summary: Metrology of automated data analysis for cardiac arrhythmia management (18HLT07) (2018).

7. P. J. Aston, T. Mehari, A. Bosnjakovic, P. M. Harris, A. Sundar, S. E. Williams, O. D¨ossel, A. Loewe, C. Nagel and N. Strodthoff, Multi-class ECG feature importance rankings: Cardiologists vs. algorithms, *Computing in Cardiology (CinC)* 2022, IEEE **498**, 1 (2023).
8. T. Mehari, A. Sundar, A. Bosnjakovic, P. Harris, S. E. Williams, A. Loewe, O. Doessel, C. Nagel, N. Strodthoff and P. J. Aston, ECG feature importance rankings: Cardiologists vs. algorithms, *IEEE Journal of Biomedical and Health Informatics* (2024).
9. C. Nagel, N. Pilia, A. Loewe and O. D¨ossel, Quantification of interpatient 12-lead ECG variabilities within a healthy cohort, *Current Directions in Biomedical Engineering* **6**, 493 (2020).
10. A. L. Samuel, Some studies in machine learning using the game of checkers, *IBM Journal of Research and Development* **44**, 210 (1959).
11. R. Barzilay, T. Jaakkola and K. Chu, Introduction. 6.86x: Machine Learning with Python: from Linear Models to Deep Learning: MIT (Last access date: 2023).
12. Publishable summary: Uncertainty quantification for machine learning models applied to photoplethysmography signals (22HLT01) (2023).

F Pavese, A Bošnjaković, S Eichstädt, A B Forbes, J Alves e Sousa (eds.):
Advanced Mathematical and Computational Tools in Metrology and Testing XIII
Series on Advances in Mathematics for Applied Sciences, Vol. 94

https://doi.org/10.1142/9789819800674_0004

On how to measure new non-physical quantities

R. Emardson
*Measurement Technology, RISE Research Institutes of Sweden,
Borås, 501 15, Sweden
E-mail: ragne.emardson@chalmers.se
www.ri.se*

This paper aims to address the challenges posed by the evolving landscape of measurements, especially in non-physical domains. With a growing need for informed decision-making, the landscape of measurements has expanded significantly. To address this, measurement procedures for non-physical measurements need to be developed. This paper proposes a framework for achieving traceability in non-physical measurements. A six-step process for the creation of novel measurements is suggested by the following steps. Define the underlying concept, dissect the concept into its constituent components, identify suitable measurement techniques, ensure the validity and reliability of chosen techniques, determine appropriate measurement scales and establish a reference for traceability.

Keywords: scales; quantity; Rasch; metrology; traceability.

1. Introduction

With a growing need for informed decision-making, the landscape of measurements has expanded significantly, reflecting a trend towards their increasing indispensability. While we have a well-established tradition of quantifying physical attributes and established procedures, exemplified by references like the Guide to the Expression of Uncertainty in Measurement (GUM) [1] and the International Vocabulary of Metrology (VIM) [2], situations frequently arise where the conventional methods fall short. In these instances, a discrepancy occurs between the demand for measurements and their available supply, often leading to a solution focusing on easily measurable quantities rather than the sought-after attributes.

To address this, the establishment of measurement standards for non-physical measurements is needed. This paper proposes a framework for achieving traceability in non-physical measurements. Following earlier work on measurability conditions by Rossi [3], he consider a quantity to be measurable when a systematic four-step procedure can be applied successfully, i.e., we can define a set of objects that have the specific property, identify its empirical properties, develop a reference measurement scale, and design at least one measurement system for this reference scale.

In the context of this study, below is introduced, in similar fashion, a six-step process for the creation of novel measurements. The steps you need to go through to create new measurements are:

- Define the underlying concept.
- Dissect the concept into its constituent components.
- Identify suitable measurement techniques.
- Ensure the validity of the chosen technique.
- Determine appropriate measurement scales.
- Establish a reference for traceability.

This paper aims to address the challenges posed by the evolving landscape of measurements, especially in non-physical domains, by providing a structured very practical approach for establishing and ensuring measurability that goes beyond the conventional physical attributes. Theoretical and more comprehensive descriptions on measurement can be found elsewhere [4]. This paper aims at responding to society's increasing demands for accurate and dependable measurements that span a range of products and services affecting our quality of life, including aspects such as satisfaction, comfort, taste, smell, and appearance often related to the increase in global trade. In the following sections, the six proposed steps are described together with a few examples of how they can be implemented.

2. Define the underlying concept

The initial phase in the process of quantifying a particular attribute involves developing a comprehension of that specific attribute. While a rigid definition might not be necessary, the critical aspect lies in constraining the parameters that define the quantity. Our main concern is to establish a shared understanding of the terminology employed — a consensus on what the quantity means to us holds greater significance than a rigid definition. The conceptualization of the attribute can manifest through various approaches, affording us the flexibility to establish its definition in diverse manners. In the process of developing techniques for measuring comfort, for example, comfort was defined by Farbrot *et al.* [5] as: "A pleasant and enabling state of a human being, as one crucial part of a system including also the product and important tasks and their mutual interactions, in a range of contexts, spanning from the psychological, physiological, psychosocial to the wider environmental." Here we can see comfort as a quantity that includes the three aspects, the person using the product, the product itself and the tasks the person using the products is performing. Hence all these three aspects and the combination of them need to be considered when forming the quantity comfort.

A definition of a quantity can also be specified through a mathematical expression. Emardson *et al.* [6] defined accessibility to rail transport as:

$$A_{ij} = \prod_b (1 - p_{jb} \varphi_i(d_b)) \tag{1}$$

where

A_{ij}: is the accessibility of person i with regard to destination j
d_b: barrier number b
φ: the probability that the barrier will prohibit the journey
p_{jb}: the probability to face barrier b going to destination j.

Here we can see accessibility defined as the absence of barriers for a traveler on a journey. This is formed as a combination of two aspects, the different degrees of difficulty if facing those barriers for a traveler and the probability of actually facing them on their journey.

3. Dissect the concept into its constituent components

New quantities frequently manifest as complex constructs, comprising a number of diverse elements. The previous section showcases just two such examples in comfort and accessibility. Based on the definition, we want to identify these constituent components, a process necessary for enabling their individual quantification. Churchill [7] suggests various methodologies for generating components. These methods include diverse approaches, such as literature searches, surveys, focus groups, and critical incident analyses.

Sundling [8] used the critical incident technique to identify the different barriers that collectively form the concept of accessibility, as described in the previous section. The Critical-Incident Method [9] (also known as Critical Incident Technique or CIT) serves as a qualitative tool aimed at portraying reality in its natural form. This methodology includes an array of techniques aimed at the direct observation of human behavior. Central to the Critical-Incident method is its emphasis on information that can only be described by the informants themselves. The method focuses on critical events, that is, those situations or observations in which the informants were personally engaged. For an event to be labeled as critical, it needs to be central with respect to the informants' experiences. In their investigation, Sundlings [8] examination of accessibility to rail transport found a compilation of 78 distinct incidents that led to alterations in travel behavior.

As described by Churchill, the identification of components can also be achieved through the collaborative efforts of expert groups. When undertaking measurements of different aspects of, for example, sustainable development, it is recommendable to base it on the groundwork laid during the formulation of the Sustainable Development Goals (SDGs), a set of 17 goals established by the United Nations in 2015, also referred to as the Global Goals. Another illustrative example is the concept of usability, often dissected into its constituent elements of efficiency, effectiveness, and satisfaction. Rice *et al.* [10] used the insights of expert reference groups to identify the main components of comfort for users of incontinence products.

4. Identify suitable measurement techniques

The concept of measurement can be described as the process of allocating numerical values, on a measurement scale, to a specific set of objects [2]. From this procedure, an observable outcome is generated, the measurand is assigned a value, or with other words the specific quantity is being measured.

There exists a wide array of measurement instruments, ranging from the widely recognized to the more obscure, tailored for physical measurements. In the context of measurements of new quantities, it often involves human engagement. Pendrill [11] categorizes measurement systems involving humans into two distinct groups:

- **Measuring Man**: This refers to situations where a human being serves as subject of measurement.
- **Man as a Measurement Instrument**: In this scenario, human senses replace conventional equipment for measurement of the object.

Here, for example, quantifying accessibility can involve either observing the choices and efforts made by individuals (Measuring Man), or assessing the perceived effort through individual ratings, using for example inquiries, (Man as a Measurement Instrument) for different routes.

Several quantities consist of a large variety of different components. Sustainable development, as described in the previous section, encompasses a large number of varying elements. This includes aspects like waste, which can be measured using normal scales or through perceived stress level at work which can be measured through responses to inquiries. In the assessment of incontinence care products, Rice *et al.* [10] adopted an approach, measuring usability through effectiveness (product fit), efficiency (time and physical workload), and satisfaction. The evaluation of these components was accomplished through a combination of inquiries, observations, and time interval measurements.

5. Ensure the validity of chosen techniques

No matter how precise our ability to measure a property, the measurement becomes futile if it does not relate to the specific attribute we are seeking to quantify. In this context, content validity emerges as an essential concept [12]. It reflects the degree to which the elements within a measurement procedure hold relevance and accurately reflect the underlying construct they are intended to measure [13]. Thus, the determination of validity is usually based on a judgement process supported by a variety of evidence. This evaluative phase often involves using a panel of experts to assess the adequacy with which relevant content has been incorporated. In their work, Rice *et al.* [10] pursued validation through engagement with a reference group of caretakers, a strategy aimed at ensuring the authenticity of the usability measurements of incontinence products.

In order to ensure validity when assessing accessibility, a main concern is to assemble a diverse group of individuals capable of evaluating the challenges associated with the identified barriers. Emardson *et al.* suggested a clustering methodology [14] based on principal component analysis to identify distinct demographic groups. Members of the response groups could then quantify the difficulties they would experience when facing the different barriers. An additional approach involves the utilization of simulation techniques. In instances where measurement models are established, we can use for example Monte Carlo simulations in order to investigate the model sensitivity. In the context of accessibility, for instance, concerns were raised that respondents could have a specific agenda and thus rate the barriers in a way to obtain a measurement result favorable for that specific agenda. Here a model can be subjected to simulations designed to trace the propagation of various types of errors and how different hidden agendas can affect the measurement result.

It is worth noting that, in practice, a combination of these methods is often considered necessary to ensure the validity of our assessments. Assessing the validity of a measurement is often difficult, particularly in cases involving non-physical attributes. A common mistake lies in measuring something tangential to the actual target of interest due to its apparent simplicity.

6. Determine appropriate measurement scales

Attention must be directed towards the selection of appropriate measurement scales. This consideration applies not only to the scales utilized for quantifying individual components but also extends to the determination of the final measurement scale to be employed for our devised metric. Stevens [15] described four basic scales:

- Nominal – Groups – Counting
- Ordinal – Grades – Median
- Interval – Temperature (C, F) – Mean
- Ratio – Mass, Length – Multiplication.

Here, the nominal scale employs numerals purely as designators devoid of inherent order. This scale can, for example, be compared to the numbering system for football players, aiding in convenient identification but lacking any inherent sequencing. An ordinal scale, on the other hand, relies on ranking to establish order among measurements; however, it does not provide insight into the distances between these measurements. The interval scale introduces intervals between measurements, yet it lacks a true zero point, restricting our possibility to use operations such as multiplication and division on measurements on the scale. In contrast, measurements on a ratio scale, a cornerstone of physics, are covered by four fundamental permissible properties: equality, rank order, interval comparison, and ratios. These properties enable operations that align with our intuitive understanding of measurement.

Very often we are faced with the situation of having the constituents forming the quantity on different measurement scales. We then need to convert the original measurement values, initially scattered across disparate scales, into a standardized format, using a common scale. Through this alignment, these values can be combined to yield a measurement for each area or a total measurement result. A good example as lifted above is the field of sustainable development with aspects like waste, measured normally in kg and perceived stress level at work, measured through responses to inquiries, that is, a large group of people rating different aspects on an ordinal scale.

Item response theory [16] is a useful technique when we are faced with measurements on different scales. Using the many facet Rasch formulation, the probability of respondent i answering x on question j about a service or product q can be written as a function of a number of latent variables as [17]:

$$\pi_{x,i,j,q} = P\{X_{x,i,j,q} = x\} = \frac{exp[\sum_{m=1}^{x} \theta_i - \delta_j - \tau_m - \gamma_q]}{\sum_{k=1}^{K} exp[\sum_{m=1}^{k} \theta_i - \delta_j - \tau_m - \gamma_q]} \tag{2}$$

where K is the number of response alternatives, θ represents the person location parameters, δ, the question location parameters, τ, the thresholds, and γ, the service specific parameters. The technique is based on an assumption that the answers given by a person on specific questions are based on these latent variables, which we then can estimate.

The application of the Rasch formulation can be understood as a transformative process, drawing parallels to the Fourier and Laplace transformations frequently employed within electronics and physics. Analogous to the Fourier transform, which often is used to translate data between the time domain and the frequency domain, the Rasch formulation serves as a powerful tool for performing a similar transformation. However, instead of dealing with the domains of time and frequency, the Rasch formulation can operate on data originally on an ordinal scale and transform them to an interval or ratio scale. With a large enough dataset comprising responses from multiple questions and respondents, we can estimate the latent parameters — δ, τ, and γ. The Rasch model is, however, highly non-linear. Hence, the parameter estimation is preferably performed using a maximum likelihood technique. In essence, this technique identifies parameter values that optimally align with the observed survey responses. Within the scope of this paper, primary focus rests on the parameter γ because it contains the information about the product, service or experience that we want to measure using a questionnaire.

Eventually, our aim is to strategically position the measurement scale, ensuring its alignment with intuitive comprehension and facilitating a meaningful connection to measurements conducted on the scale to the user. Additionally, our aim is to establish useful intervals between successive scale steps. To achieve this, we can use the logistic function, which proves to be useful in setting up the transformation of the scale to desired parameters. By employing this function, we can effectively map physical values onto a selected scale ranging between 0 and the maximum value, L. This transformation facilitates a scale that corresponds to an intuitive understanding.

$$f(\gamma) = \frac{L}{1+e^{-s(\gamma-\gamma_0)}} \tag{3}$$

where

- L: The maximum value of the scale
- γ_0: Reference value where the curve reaches its midpoint
- s: The slope or steepness of the curve.

7. Establish a reference for traceability

Ensuring traceability is of crucial importance, particularly when decisions are at stake. VIM [2] describes traceability as the scenario when a result of a measurement can be linked to a reference through an unbroken chain of comparisons. Each step within this chain typically involves the calibration using a standard with superior precision [18]. Ideally, the outcomes of measurements should trace back to relevant primary standards of SI units. However, should SI traceability not be attainable, alternative reference points can be employed to serve this purpose.

In scenarios involving measurements associated with individuals this is generally the situation. Regrettably, there exists no equivalent of globally accepted standards. This prompts the need for innovative strategies. The search for a consistent reference demands the identification of quantities that are constant over time, independent of the measured attribute. While the ideal scenario would involve the discovery of unchanging values, such eternal truths remain elusive [19]. Master Scaling [20], offers an alternative way forward. This approach facilitates the conversion of individual perceptual measurements into a comparable format, enabling inter-personal comparisons. As an example, Berglund *et al.* [21] employed pink noise as a reference when measuring odor sensitivity among smokers and non-smokers.

The Rasch formulation, as described in a previous section, can also be useful when creating references. The parameter θ, representing the individual location parameters, may serve as a reference point. It reflects an individual's tendency to rate responses. When measuring a specific parameter over time with consistent participants or a similar group of respondents, the parameter θ may be held constant, allowing for comparative analyses across temporal segments. Similarly, related to item response theory such as the Rasch formulation we can use item banks [22]. Item banks can be described as a collection of test items with calibrated data on their measurement characteristics. Such item banks can be stored and used as references for specific measurements.

Acknowledgements

This work was supported by Vinnova, project 2020-04330, Verksamhetsplan 2021-2023 för basverksamheten och investeringar vid RISE Riksmätplatser.

References

1. BIPM, IEC, IFCC, ILAC, ISO, IUPAC, IUPAP, OIML, *Evaluation of measurement data — Guide to the expression of uncertainty in measurement. Joint Committee for Guides in Metrology*, JCGM 100 (2008).
2. BIPM, IEC, IFCC, ILAC, ISO, IUPAC, IUPAP, OIML, *International Vocabulary of Metrology — Basic and General Concepts and Associated Terms. Joint Committee for Guides in Metrology*, JCGM 200 (2012).
3. G. B. Rossi, *Measurability*. Measurement, 40(6): pp. 545–562 (2007).
4. D. J. Hand, *Statistics and the theory of measurement*. Journal of the Royal Statistical Society: Series A (Statistics in Society) 159(3), pp. 445–473 (1996).
5. A. Farbrot, F. S. Abbas, A. Nihlstrand, J. Dagman, R. Emardson, S. Kanerva, L. R. Pendrill, *Defining comfort for heavily-incontinent patients assisted by health care products in several contexts*. The Simon Foundation for Continence's Innovating for Continence Conference Series (Chicago (US), April 2013).
6. R. Emardson, P. Jarlemark, L. Pendrill, C. Sundling, M. E. Nilsson, B. Berglund, *Measurement of accessibility to rail transport systems*. Advanced Mathematical and Computational Tools in Metrology and Testing, pp. 136–142, doi: 10.1142/9789814397957_0017 (2012).
7. G. A. Churchill Jr., *A paradigm for developing better measures of marketing constructs*. Journal of Marketing Research 16(1), pp. 64–73 (1979).
8. C. Sundling, *Travel behavior change in older travelers: Understanding critical reactions to incidents encountered in public transport*. International Journal of Environmental Research and Public Health 12(11), pp. 14741–14763 (2015).
9. J. C. Flanagan, *The critical incident technique*. Psychological Bulletin 51(4), pp. 327–358 (1954).
10. S. K. Rice, L. Pendrill, N. Petersson, J. Nordlinder, A. Farbrot, *Rationale and design of a novel method to assess the usability of body-worn absorbent incontinence care products by caregivers*. Journal of Wound, Ostomy, and Continence Nursing, 45(5), pp. 456–464 (2018).
11. L. Pendrill, *Man as a measurement instrument*. NCSLi Measure 9(4), pp. 24–35 (2014).
12. S. N. Haynes, D. Richard, E. S. Kubany, *Content validity in psychological assessment: A functional approach to concepts and methods*. Psychological assessment 7(3), pp. 238–247 (1995).

13. I.-C. Chiang, A. Rajiv, S. Jhangiani, P. C. Price, *Overview of nonexperimental research*. Research Methods in Psychology-2nd Canadian Edition (2015).
14. M. Halkidi, Y. Batistakis, M. Vazirgiannis. *Clustering algorithms and validity measures*. Proceedings Thirteenth International Conference on Scientific and Statistical Database Management. SSDBM 2001. IEEE (2001).
15. S. S. Stevens, *On the theory of scales of measurement*. Science 103(2684), pp. 677–680 (1946).
16. R. K. Hambleton, H. Swaminathan, H. J. Rogers, *Fundamentals of item response theory*. Vol. 2. Sage (1991).
17. D. Andrich, *A rating formulation for ordered response categories*. Psychometrika 43(4), pp. 561–573 (1978).
18. M. G. Cox, P. M. Harris, *Measurement uncertainty and traceability*. Meas. Sci. Technol, 17(533) (2006).
19. J. Fowles, *The problem of values in futures research*. Futures 9(4), pp. 303–314 (1977).
20. B. Berglund, E.-L. Harju, *Master scaling of perceived intensity of touch, cold and warmth*. European Journal of Pain, 7(4), pp. 323–334, Special Issue dedicated to Ulf Lindblom doi:10.1016/S1090-3801(03)00043-0 (2003).
21. B. Berglund, S. Nordin, *Detectability and perceived intensity for formaldehyde in smokers and non-smokers*. Chem. Senses, 17(3), pp. 291–306. doi: 10.1093/chemse/17.3.291 (1992).
22. B. Choppin, *Developments in item banking*. First European Contact Workshop (Windsor, UK, 1976).

F Pavese, A Bošnjaković, S Eichstädt, A B Forbes, J Alves e Sousa (eds.):
Advanced Mathematical and Computational Tools in Metrology and Testing XIII
Series on Advances in Mathematics for Applied Sciences, Vol. 94

https://doi.org/10.1142/9789819800674_0005

Analysis of data from experiments to estimate time to failure

Alistair B. Forbes

National Physical Laboratory, Teddington, Middlesex, UK
E-mail: alistair.forbes@npl.co.uk

This paper looks at the analysis of data from a set of experiments designed to estimate the time to failure of an engineering component. These experiments involve subjecting the component to a series of increasing loads/stresses, those greater than would be expected during the normal working life of the component, and recording the time to failure. On the basis of the data, an estimate of the time to failure of the component under typical loads is made. We consider the case in which the logarithm of the time to failure varies linearly with the applied load. Inferences are based on fitting a linear model to the data using generalised distance regression. This paper evaluates the information that can be derived if the component has yet to fail when the experiment is halted and how this information can be used in predicting time to failure.

Keywords: generalised regression; Bayesian inference; time to failure.

1. Introduction

A common experimental task in assessing the performance of a component used in civil engineering is to estimate the time to failure of the component. These experiments involve subjecting the component to a series of increasing loads/stresses, those greater than would be expected during the normal working life of the component, and recording the time to failure. On the basis of the data, an estimate of the time to failure of the component under typical loads is made. Often standards will require that the expected lifetime should exceed a pre-assigned threshold. In this paper, we consider models of failure for which the logarithm of the time to failure is expected to vary linearly with the applied load. The length of time over which the experiment is usually a small fraction of the expected life time of the component and experiments are designed so that a component fails just at the end of the experiment. This paper looks at what information can be derived if the component has yet to fail when the experiment is halted. This paper is motivated by standards relating to test methods for evaluating structures used in water management.[5] The rest of the paper is organised as follows. In Sec. 2, we describe the model and generalised distance regression (or error-in-variables) approaches to determine best estimates of the model parameters. In Sec. 3, we show how information from an experiment yet to fail can be incorporated into the posterior distribution for the model parameters in a Bayesian framework. Section 4 give a numerical example of the GDR and Bayesian approaches for three scenarios. Our concluding remarks are given in Sec. 5.

2. Generalised regression and estimating time to failure

2.1. *Model*

The modelled relationship between load $y^*$and time to failure t^* is given by

$$y^* = a + bx^*, \quad x^* = \log t^*, \tag{1}$$

where y^* is the load applied and t^* represents the mean time to failure for a particular batch of components. The observations associated with a set of measurements involving a sample of a components and a range of applied loads are modelled as

$$y_i^* = a + bx_i^*, \quad x_i = x_i^* + \delta_i, \quad y_i = y_i^* + \epsilon,$$

where y_i is the measured applied load, x_i is the logarithm of the recorded time t_i to failure and ϵ_i and δ_i are random effects modelled according to

$$\epsilon_i \in \mathcal{N}(0, \sigma_i^2), \quad \delta_i \in \mathcal{N}(0, \tau_i^2). \tag{2}$$

We assume that estimates of σ_i^2 and τ_i^2 are available.

2.2. *Objective*

Of interest is the ability of a material to sustain a load for a specified minimum time t_M. Given estimates of a and b, then the estimate of the load that will lead to failure at $x_M = \log t_M$ on average is $y_M = a + bx_M$. If the knowledge about $\boldsymbol{a} = (a, b)^\top$ is summarised as

$$\boldsymbol{a} \sim \mathcal{N}(\boldsymbol{a}_0, V_{\boldsymbol{a}}),$$

then the knowledge about the load y_M^* that will produce failure at time t_M is summarised by

$$y_M^* \sim \mathcal{N}(y_M, u^2(y_M)), \quad u^2(y_M) = [1\ x_M] V_{\boldsymbol{a}} [1\ x_M]^\top. \tag{3}$$

More generally, if $\boldsymbol{a}_q$ is a sample from the distribution $p(\boldsymbol{a})$ then $y_q = a_q + b_q x_M$ is a sample from the distribution $p(y_M^*)$.

2.3. *Generalised distance regression for straight line fits*

Given data (x_i, y_i), $i = 1, \dots, m$, of measured loads and $x_i = \log t_i$ derived from recorded times to failure t_i, estimates of the linear parameters $\boldsymbol{a} = (a, b)^\top$) can be found be solving the generalised distance regression (GDR) problem[1,6,11,12]

$$\min_{\boldsymbol{a}, \boldsymbol{x}^*} \sum_{i=1}^{m} \left\{ \left(\frac{x_i - x_i^*}{\tau_i} \right)^2 + \left(\frac{y_i - a - b x_i^*}{\sigma_i} \right)^2 \right\}, \tag{4}$$

involving the parameters $\boldsymbol{a}$ and $\boldsymbol{x}^* = (x_1^*, \dots, x_m^*)^\top$. The GDR estimates of the parameters are those that maximise the likelihood $p(\boldsymbol{x}, \boldsymbol{y}|\boldsymbol{a})$ with respect to $\boldsymbol{a}$. An estimate of the variance matrix $V_{\boldsymbol{a}}$ associated with $\boldsymbol{a}$ can be determined using the law of propagation of uncertainty applied to the nonlinear least squares problem[7,8] in (4), following the GUM methodology.[2,4]

3. Information from an experiment yet to fail

An experiment may be set up with an applied load y_i^* with an expected failure at or near x_E, with the aim of predicting performance at $x_M > x_E$. At log time at or near x_E, what information about a and b (and therefore y_M) is available given that no failure has been observed? How does this information compare with that gained from an observed failure? The GDR approach cannot be applied directly as it can only be used to fit a model to observed data points. However, it is still possible to evaluate a likelihood function and use Bayesian inference[9,13] to determine the posterior distributions for the parameters of interest, as discussed below.

3.1. *Incorporating information from a new experiment*

Suppose failures have been observed at $x_{1:m-1} = \{x_i, i = 1, \dots, m-1\}$, corresponding to measured loads $y_{1:m-1}$, yielding a posterior distribution $p(\boldsymbol{a}|\boldsymbol{x}_{1:m-1}, \boldsymbol{y}_{1:m-1})$ for $\boldsymbol{a} = (a, b)^\top$ approximated by

$$\boldsymbol{a}|x_{1:m-1}, y_{1:m-1} \sim \mathcal{N}(\boldsymbol{a}_{m-1}, V_{\boldsymbol{a},m-1}). \tag{5}$$

We denote this density by $p_{m-1}(\boldsymbol{a})$.

A new experiment has been set up with measured load y_m. If the test piece fails at log time x_m, then the posterior distribution for $\boldsymbol{a}$ is such that

$$p(\boldsymbol{a}, x_m^*|x_{1:m}, y_{1:m}) \propto p(x_m, y_m|\boldsymbol{a}, x_m^*) p(x_m^*) p(\boldsymbol{a}|x_{1:m-1}, y_{1:m-1}),$$

where

$$p(x_m, y_m|\boldsymbol{a}, x_m^*) \propto \exp\left\{-\frac{1}{2}\frac{(y_m - a - bx_m^*)^2}{\sigma_m^2}\right\}\exp\left\{-\frac{1}{2}\frac{(x_m - x_m^*)^2}{\tau_m^2}\right\}.$$

Below, we assume the non-informative prior distribution $p(x_m^*) \propto 1$. The marginalised distribution $p(\boldsymbol{a}|\boldsymbol{x}_{1:m}, \boldsymbol{y}_{1:m})$ for $\boldsymbol{a}$ can be approximated by

$$\boldsymbol{a}|x_{1:m}, y_{1:m} \sim \mathcal{N}(\boldsymbol{a}_m, V_{\boldsymbol{a},m}). \tag{6}$$

We will denote this distribution as $p_m(\boldsymbol{a})$.

We now consider the case where at time x_E, the load y_m has yet to produce a failure. On the basis of $p_{m-1}(\boldsymbol{a})$, the estimate $\hat{x}_m^*$ of x_m^* is given by

$$\hat{x}_m^* = \frac{y_m - a_{m-1}}{b_{m-1}}, \tag{7}$$

regarding y_m as the best estimate of y_m^* and $\boldsymbol{a}_{m-1}$ as the best estimate of $\boldsymbol{a}$. The uncertainty $u(x_m^*)$ associated with this estimate is given by

$$u^2(x_m^*) = \boldsymbol{g}^\top \begin{bmatrix} \sigma_m^2 & 0 \\ 0 & V_{\boldsymbol{a},m-1} \end{bmatrix} \boldsymbol{g}, \quad \boldsymbol{g}^\top = \frac{1}{b_{m-1}}[1,\ -1,\ -(y_m - a_{m-1})/b_{m-1}]. \tag{8}$$

Since $\hat{x}_m^*$ depends nonlinearly on b_{m-1}, the distribution for $\hat{x}_m^*$ is only approximately Gaussian. Given samples $y_q \in \mathcal{N}(y_m, \sigma_m^2)$ and $\boldsymbol{a}_q \in \mathcal{N}(\boldsymbol{a}_{m-1}, V_{\boldsymbol{a},m-1})$ then $x_q = (y_q - a_q)/b_q$ is a sample from the distribution for x_q^* and can be compared with the approximating Gaussian distribution.[3]

From (2), given x_m^*, the likelihood of observing a failure at $x_m < x_E$, that is, during the experiment, is given by

$$C(x_m^*|x_E) = \Pr(x_m < x_E|x_m^*) = \int_{-\infty}^{x_E} \exp\left\{-\frac{1}{2}\frac{(x_m - x_m^*)^2}{\tau_m^2}\right\} \mathrm{d}x_m, \tag{9}$$

and can be evaluated using the cumulative density function associated with a Gaussian distribution. Hence the likelihood $p(x_m > x_E, y_m|\boldsymbol{a}, x_m^*)$ is such that

$$\begin{aligned} p(x_m > x_E, y_m|\boldsymbol{a}, x_m^*) &= p(y_m|x_m > x_E, \boldsymbol{a}, x_m^*)p(x_m > x_E|\boldsymbol{a}, x_m^*), \\ &= p(y_m|\boldsymbol{a}, x_m^*)p(x_m > x_E|x_m^*), \\ &\propto \exp\left\{-\frac{1}{2}\frac{(y_m - a - bx_m^*)^2}{\sigma_m^2}\right\}(1 - C(x_m^*|x_E)). \end{aligned}$$

The first equality above follows from the product rule in probability theory while the second follows from the facts that the likelihood of observing y_m depends only on $\boldsymbol{a}$ and x_m^*, and the likelihood of $x_m > x_E$ depends only on

x_m^*. The last equality follows from (2) and (9). The posterior distribution $p(\boldsymbol{a}, x_m^* | x > x_E, y_m)$ is such that

$$p(\boldsymbol{a}, x_m^* | x_m > x_E, y_m) \propto p(x_m > x_E, y_m | \boldsymbol{a}, x_m^*) p(\boldsymbol{a} | x_{1:m-1}, y_{1:m-1}), \quad (10)$$

taking into account the fact that $p(x_m^*) \propto 1$. The expression on the right in (10) above can be evaluated directly using distribution functions associated with a Gaussian distribution. The distribution in (10) is the joint posterior distribution for the three parameters a, b and x^* and can be normalised and marginalised using numerical quadrature to determine $p(\boldsymbol{a} | x_m > x_E, y_m)$, the information about $\boldsymbol{a}$ given the data, including the fact that the (measured) load y_m has not resulted in failure before time t_E. Similarly $p(a, x_m^* | x_m > x_E, y_m)$, $p(b, x_m^* | x_m > x_E, y_m)$, $p(a | x_m > x_E, y_m)$, $p(b | x_m > x_E, y_m)$ and $p(x_m^* | x_m > x_E, y_m)$ can be determined using numerical quadrature. We will use the notation $p_E(\boldsymbol{a}, x_m^*)$ to denote the distribution $p(\boldsymbol{a}, x_m^* | x_m > x_E, y_m)$ and similarly $p_E(a)$ for $p(a | x_m > x_E, y_m)$, etc.

3.2. *Point estimates of the parameters*

Point estimates $\hat{\boldsymbol{a}}$ and $\hat{x}_m^*$ of $\boldsymbol{a}$ and x_m^* can be found by finding the point that maximises $p(\boldsymbol{a}, x_m^* | x_m > x_E, y_m)$ or, equivalently, minimises

$$E(\boldsymbol{a}, x_m^*) = -\log p(\boldsymbol{a}, x_m^* | x_m > x_E, y_m).$$

These point estimates can be determined using standard optimisation.[10] The Hessian matrix of second partial derivatives of E with respect to the parameters can be used to determine a Gaussian approximation to the distribution $p(\boldsymbol{a}, x_m^* | x_m > x_E, y_m)$. We will denote this distribution by $p_{E,0}$. A potentially significant advantage of working with $p_{E,0}$ is that it can be evaluated without the need to apply numerical quadrature techniques. An alternative point estimate of $\boldsymbol{a}$ can be found by the point that maximises the marginalised distribution $p(\boldsymbol{a} | x_m > x_E, y_m)$. In general, these two estimates of $\boldsymbol{a}$ will differ, with the second having somewhat better statistical properties as it takes into account all plausible values of x_m^*.

3.3. *Estimating the load required to cause failure at a future time*

The expression for y_M^* in (3) is valid if the knowledge of $\boldsymbol{a}$ is summarised well by a Gaussian distribution. Given the distribution $p(a, b, x_m^* | x_m > x_E, y_m)$ we can use the change of variable formula to determine the distribution

$p(a, y_M^*, x_m^* | x_m > x_E, y_m)$ where $y_M^* = a + bx_M$. The simple nature of the re-parametrisation means that $p(a, y_M^*, x_m^* | x_m > x_E, y_m)$ can also be determined analytically, up to a normalising constant, using density functions associated with a Gaussian distribution. The marginalised posterior distribution $p(y_M^* | x_m > x_E, y_m)$ can then be determined using numerical quadrature.

4. Numerical example

We illustrate the impact of information from an experiment yet to fail using data simulating experiments arising from a 'test method for determining the long-term compression strength for a specified period on boxes made of thermoplastics materials for non-pressure underground conveyance and storage of non-potable water'.[5] Boxes of particular designs have an observed behaviour for which the model in (1) is appropriate. The test method typically last over a period of $t_E = 5000$ hours and are designed to estimate the maximum permissible load that can be applied if the structures are required to last for $t_M = 50$ years.[a]

Table 1. Simulated data associated with 10 experiments. The unit associated with y_i and σ_i is kPa while $x_{0,i}$, x_i and τ_i represent the $\log_{10}$ of times in units of hours.

i	$t_{0,i}$	$x_{0,i}$	x_i	τ_i	y_i	σ_i
1	200	2.30	2.03	0.12	357.96	3.58
2	200	2.30	2.16	0.12	357.96	3.58
3	200	2.30	2.44	0.12	357.96	3.58
4	500	2.70	2.58	0.13	342.04	3.42
5	500	2.70	2.84	0.13	342.04	3.42
6	1000	3.00	2.81	0.15	330.00	3.30
7	1000	3.00	2.97	0.15	330.00	3.30
8	2500	3.40	3.03	0.17	314.08	3.14
9	2500	3.40	3.16	0.17	314.08	3.14
10	5000	3.70	3.70	0.18	302.04	3.02

Table 1 shows data associated with simulated 10 experiments recording time to failure for different stresses. The times $t_{0,i}$ represent a nominal estimate of the time to failure in hours and $x_{0,i} = \log_{10} t_{0,i}/\text{hr}$. If t_i is the observed time to failure in hours then $x_i = \log_{10} t_i/\text{hr}$. The stresses

[a]Parts of the 'great drain' in Rome, the Cloaca Maxima, have been operational for well over 2000 years.

y_i represent the 'measured' applied load, that is, the load setting of the instrument. The actual applied load is assumed to vary at random from the indicated load setting. The standard uncertainties τ_i and σ_i correspond to a relative uncertainty of 5 % and 1 % respectively in the log time and applied load. (Recall that the uncertainty associated with the log time relates mainly to the variation in strength in a batch of components.) The design of the experiments allows for the experiments to be conducted using three test instruments: experiments 1 to 7 can be run consecutively on one instrument, experiments 8 and 9 consecutively on a second instrument with experiment 10 running on a third instrument. The data x_i and y_i have been generated at random based on $t_{0,i}$ with the exception of x_{10} which has been assigned to be equal to $x_E = \log_{10} t_E$ where $t_E = 5000$ hrs, the duration of the experiment.

Below we evaluate information about response parameters $\boldsymbol{a} = (a, b)^\top$, the mean log time to failure x_m^* corresponding to a load $y_m^* = a + bx_m^*$ and the load y_M^* that would have an expected time to failure of t_M. The evaluation is based on three sets of information:

a) data derived from the first $m - 1$ measurements (labelled $m - 1$),
b) data derived from all m measurements, corresponding to observing a failure for load y_m^* at $x_m = x_E$, that is an observed failure just as the experiment is ending, (labelled m), and
c) the experiment ending at x_E with no failure observed corresponding to load y_m^* (labelled E).

Figure 1 graphs data points (x_i, y_i) in Table 1, along with the linear fit (solid line) defined by a_E and b_E where a_E and b_E are the means of the posterior distributions $p(a|x_m > x_E, y_m)$ and $p(b|x_m > x_y, y_m)$, respectively. Also shown are the GDR line fit to all m data points, the GDR line fit to the first $m-1$ data points, and the point $(\hat{x}_m^*, y_m)$, estimating the expected log time to failure, x_m^*, for the measured load y_m based on the first $m - 1$ points. The observed log time to failure x_m is greater than that $\hat{x}_m,^*$ predicted on the basis of the first $m - 1$ points so that the GDR line fit to all m data points is closer to (x_m, y_m) and has smaller slope than the fit based on $m - 1$ points. The fit based on scenario c), no failure yet observed at x_m, has smaller slope again, reflecting the fact that the mth experiment suggests that the material is more resilient than if a failure had occurred.

Figure 2 graphs the distributions associated with the intercept parameter a corresponding to the three sets of information a) to c) above. Also shown is the Gaussian distribution $p_{E,0}(a)$ that approximates

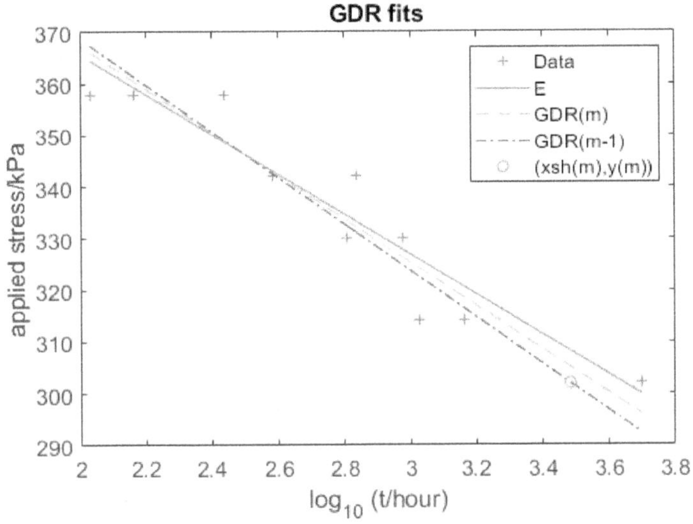

Fig. 1. Simulated data points '+' associated with $m = 10$ experiments (Table 1) along with the linear fit (solid line) defined by a_E and b_E where a_E and b_E are the means of the posterior distributions $p(a|x_m > x_E, y_m)$ and $p(b|x_m > x_y, y_m)$, respectively. Also shown are the GDR line fit to all m data points (dashed line), the GDR line fit to the first $m-1$ data points (dot-dashed line), and the point $(\hat{x}_m^*, y_m)$, ('o') estimating the expected log time to failure, x_m^*, for the measured load y_m based on the first $m-1$ points, 'o'.

$p(a|x_m > x_E, y_m)$ derived from the Gaussian approximation to the mode of $p(\boldsymbol{a}, x_m^*|x_m > x_E, y_m)$. The component of the mode of the joint distribution $p(\boldsymbol{a}, x_m^*|x_m > x_E, y_m)$ corresponding to the intercept parameter is close to but not equal to the mode of the marginalised distribution $p(a|x_m > x_E, y_m)$. The Gaussian approximation, derived using only optimisation techniques with no need to apply numerical quadrature, is a good basis for inferences about a. Figure 3 corresponds to Fig. 2 but graphs the distributions associated with the slope parameter b.

Table 2 gives the estimates of the parameters $\boldsymbol{a} = (a, b)^\top$ and their standard uncertainties derived from four calculations i) the mean and standard deviations of the posterior distributions, $p(a|x_m > x_E, y_m)$ and $p(b|x_m > x_E, y_m)$, approximating $p(\boldsymbol{a}, x_m^*|x_m > x_E, y_m)$ by a Gaussian distribution about its mode, and the mean and standard deviation of the Gaussian distributions p_{m-1} and p_m derived from a fit to the first $m-1$ points and all m data points. The uncertainties associated with p_E corresponding to no observed failure for load y_m are approximately half way

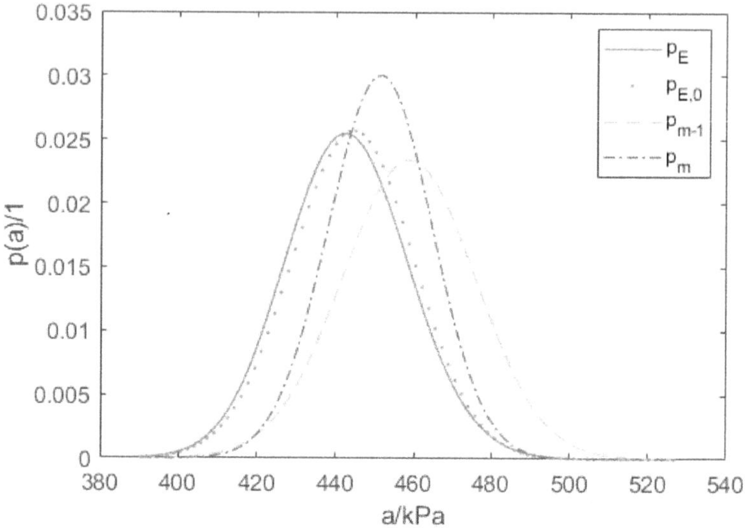

Fig. 2. Distributions associated with intercept parameter a for the data in Fig. 1. The solid line is the posterior distribution $p_E(a) = p(a|x_m > x_E, y_m)$, the dashed line, the Gaussian distribution p_{m-1} derived from a generalised distance regression fit to the first $m-1$ points, and the dash-dotted line, the Gaussian distribution p_m derived from a generalised distance regression fit to all m points. Also shown, in the dotted line, is the Gaussian distribution $p_{E,0}$ that approximates p_E derived from the Gaussian approximation to the mode of $p(a, b, x_m^*|x_m > x_E)$.

Table 2. Estimates of the parameters (first column of two) and their standard uncertainties (second column) derived from $p(a|x_m > x_E, y_m)$ and $p(b|x_m > x_E, y_m)$ (first set), maximising $p(\boldsymbol{a}, x_m^*|x_m > x_E, y_m)$ (second set), GDR fit to $m-1$ points (third set), GDR fit to m points (fourth set) derived from the data in Table 1.

	p_E		$p_{E,0}$		p_{m-1}		p_m	
a	443.09	15.69	444.41	15.52	458.69	17.06	451.29	13.27
b	-38.71	5.87	-39.20	5.80	-44.97	6.50	-41.96	4.89

between that for $m-1$ data points and m data points. We can say, loosely, that the information associated with no observed failure is worth approximately 'half a data point'.

Figure 4 graphs distributions associated with x_m^* for the data in Fig. 1 for the cases a) and c) above. The posterior distribution $p(x_m^*|x_m > x_E, y_m)$ shows the information gain from knowledge derived from the absence of failure in the mth experiment at log time x_E: low values of x_m^* that are plausible on the basis of the first $m-1$ experiments are no longer plausible.

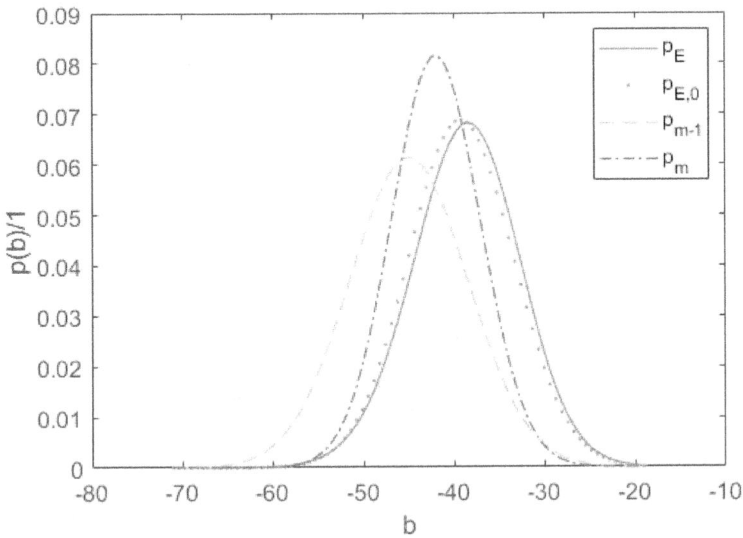

Fig. 3. As Fig. 2 but for distributions associated with slope parameter b.

The distribution does give probability mass in the region close to but less than x_E corresponding to the case that the material sampled happens to be more resilient that the batch average. The nonlinear of the propagation model $x^* = (y^* - a)/b)$ is in evidence in that the Monte Carlo distribution is somewhat different from a Gaussian distribution.

Figure 5 graphs the distributions associated with y^*_M for the data in Fig. 1 for cases a) to c) above. The posterior distribution $p(y^*_M|x_m > x_E, y_m)$ is noticeably asymmetric.

5. Concluding remarks

This paper has described the analysis of data associated with estimating time to failure for engineering components. We have described the application of generalised distance regression to making inferences based on observed failures for different load strengths and shown how information from an experiment yet to fail can be incorporated into the analysis using Bayesian approaches that can be implemented using optimisation and numerical quadrature approaches. The numerical example shows that valid inferences can based on the Gaussian approximation to the posterior distribution for the parameters of interest, obviating the need for applying numerical quadrature techniques. There are many interesting questions

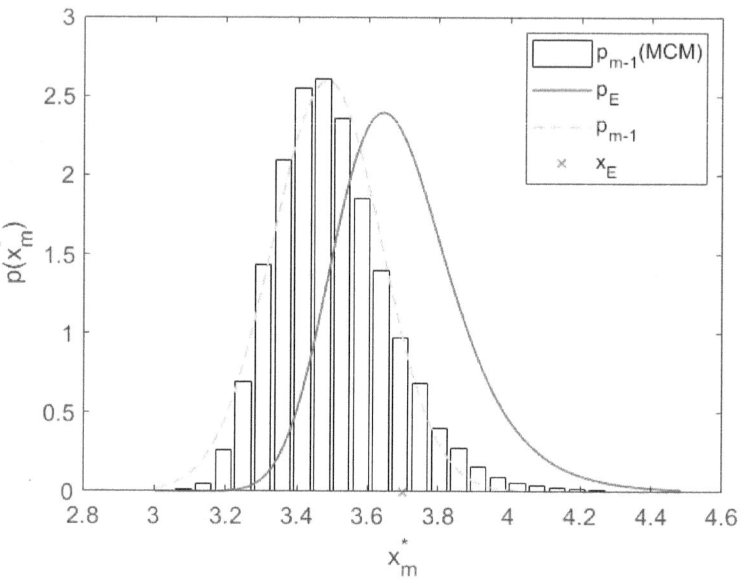

Fig. 4. Distributions associated with x^*_m for the data in Fig. 1. The solid line is the posterior distribution $p(x^*_m|x_m > x_E, y_m)$, the dashed line, the Gaussian distribution associated with $\hat{x}^*_m$ determined on the basis of the GDR fit to the first $m-1$ points as in (7) and (8), while the histogram gives distribution associated with $\hat{x}^*_m$ derived using the Monte Carlo method.

relating to the design of experiments in order to estimate time to failure. These design considerations have not been discussed in this paper but the analysis given in this paper should enable some of these design issues to be addressed.

Acknowledgements

The author is very grateful to Professor Nigel Jennet of Strategic Measures Consultancy Ltd for his inputs relating to the test method standards, typical test data, simulation software and for many insightful discussions.

This work has been supported by the UK's National Measurement Programme for Data Science.

References

1. R. M. Barker, M. G. Cox, A. B. Forbes and P. M. Harris. Software Support for Metrology Best Practice Guide no. 4: Modelling Discrete

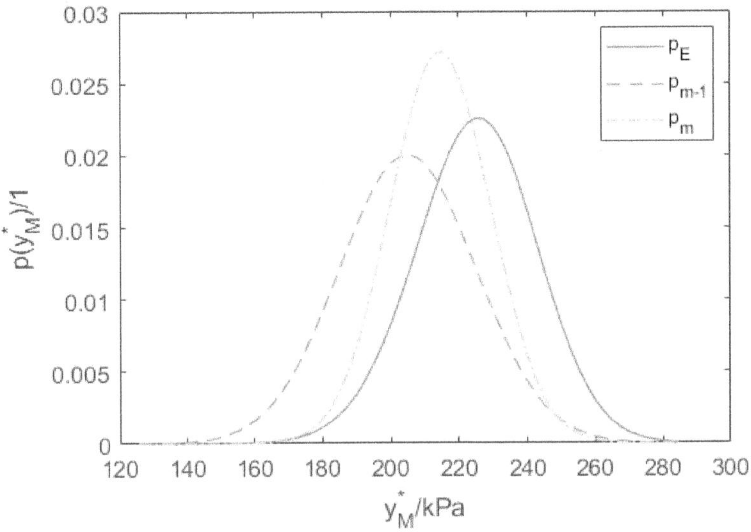

Fig. 5. Distributions associated with y_M^* for the data in Fig. 1. The solid line is the posterior distribution $p(y_M^*|x_m > x_E, y_m)$, the dashed line, the Gaussian distribution associated with $\hat{y}_M^*$ determined on the basis of the GDR fit to the first $m-1$ points, and the dash-dotted line, the Gaussian distribution associated with $\hat{y}_M^*$ determined on the basis of the GDR fit to all first m points.

Data and Experimental Data Analysis. Technical Report DEM-ES 018, National Physical Laboratory, Teddington, 2007.

2. BIPM, IEC, IFCC, ILAC, ISO, IUPAC, IUPAP and OIML. Evaluation of measurement data — Guide to the expression of uncertainty in measurement. Joint Committee for Guides in Metrology, JCGM 100:2008, 2008.
3. BIPM, IEC, IFCC, ILAC, ISO, IUPAC, IUPAP and OIML. Evaluation of measurement data — Supplement 1 to the "Guide to the expression of uncertainty in measurement" — Propagation of distributions using a Monte Carlo method. Joint Committee for Guides in Metrology, JCGM 101:2008, 2008.
4. BIPM, IEC, IFCC, ILAC, ISO, IUPAC, IUPAP and OIML. Evaluation of measurement data — Supplement 2 to the "Guide to the expression of uncertainty in measurement" — extension to any number of output quantities. Joint Committee for Guides in Metrology, JCGM 102:2011, 2011.
5. CEN. *EN17151: Plastics piping systems for non-pressure underground*

conveyance and storage of non-potable water - Test method for determination of long-term compression strength of boxes. European Centre for Standardization, Brussels, 2019.

6. M. G. Cox, A. B. Forbes, P. M. Harris and I. M. Smith. Classification and solution of regression problems for calibration. Technical Report CMSC 24/03, National Physical Laboratory, May 2003.
7. A. B. Forbes. Nonlinear least squares and Bayesian inference. In F. Pavese, M. Bär, A. B. Forbes, J.-M. Linares, C. Perruchet and N.-F. Zhang, editors, *Advanced Mathematical and Computational Tools in Metrology VIII*, pp. 103–111, Singapore, 2009. World Scientific.
8. A. B. Forbes. Parameter estimation based on least squares methods. In F. Pavese and A. B. Forbes, editors, *Data modeling for metrology and testing in measurement science*, pp. 147–176, New York, 2009. Birkhäuser-Boston.
9. A. Gelman, J. B. Carlin, H. S. Stern, D. B. Dunson, A. Vehtari and D. B. Rubin. *Bayesian Data Analysis.* Chapman & Hall/CRC, Boca Raton, Fl., third edition, 2014.
10. P. E. Gill, W. Murray and M. H. Wright. *Practical Optimization.* Academic Press, London, 1981.
11. International Organization for Standardization, Geneva. *ISO/TS 28037: Determination and use of straight-line calibration functions*, 2010.
12. K. Klauenberg, S. Martens, A. Bosnjakovic, M. G. Cox, A. M. H. van der Veen and C. Elster. The GUM perspective on straight-line errors-in-variables regression. *Measurement*, 187:110340, 2022.
13. H. S. Migon and D. Gamerman. *Statistical Inference: An Integrated Approach.* Arnold, London, 1999.

F Pavese, A Bošnjaković, S Eichstädt, A B Forbes, J Alves e Sousa (eds.):
Advanced Mathematical and Computational Tools in Metrology and Testing XIII
Series on Advances in Mathematics for Applied Sciences, Vol. 94

https://doi.org/10.1142/9789819800674_0006

A competency question driven approach to conceptual data model design for digital verification and validation

Joao Gregorio*, Moulham Alsuleman, Michael Chrubasik and Paul Duncan
Data Science Department, National Physical Laboratory, Glasgow, UK
**E-mail: joao.gregorio@npl.co.uk*

Graeme Bisland
National Manufacturing Institute Scotland, University of Strathclyde, Glasgow, UK
E-mail: graeme.bisland@strath.ac.uk

This work introduces a data-driven credibility assessment to quantify simulation quality in industrial part re-manufacturing. The framework evaluates the dependability of sources, data, and methodologies, focusing on robustness and uncertainty for data quality and simulation confidence. A conceptual data model, designed using competency questions, maps data requirements, digitalising credibility evaluation and promoting data traceability and accessibility.

Keywords: data quality; credibility assessment; simulation quality; data traceability; conceptual data model.

1. Introduction

The credibility of simulation outputs and their data is often a blocker when trying to replace experiments with purely digital systems [1]. Take, for example, Digital Twins (DTs), which are limited by the quality of the data and accuracy of the measurements used to create them and will fail to produce accurate forecasts if they are inadequate [2]. Ensuring that proper data curation exists and that all associated measurements are traceable and validated is critical for reliably replacing physical testing with simulations. As the tendency towards the more widespread adoption of computational measurements/experiments, such as modelling and simulations, as the *de facto* methods for producing data and results used in decision-making increases so does the need to address the data quality for these types of experiments [3].

National Metrology Institutes (NMIs) play an increasingly pivotal role in ensuring the traceability and quality of data used in digital systems while also providing measurement services spanning a wide-range of different areas. In recent years an increasing amount of focus has been placed on ensuring the traceability, verification, validation and robustness of data, as key aspects on ensuring the data quality of measurements [4]. The importance of the data quality in decision making would become more crucial when it is associated with high risk, whether the measurement was done physically or digitally [5–8]. And while digital systems do not have the ability to fully replace physical testbeds, digital testbeds can, if appropriately validated, replace part of the physical framework [9]. Moreover, physical testing might not be possible when ensuring the continued application of the test object is a requirement, as is the case for destructive tests.

The work presented in this paper stems from a collaboration between the National Physical Laboratory (NPL) & the National Manufacturing Institute Scotland (NMIS) to develop a data-driven credibility assessment to quantify simulation quality for the re-manufacturing process of industrial parts. This process, Digital Verification & Validation (DVV), aims to take advantage of the benefits of digital testing while mitigating its shortcomings by quantitatively assessing data quality attached to simulations. While digital testbeds can function as standalone systems to aid in verification and validation processes, they can also serve as modular building blocks for systems such as DTs and Machine Learning (ML) algorithms. Both of these are commonplace methods to simulate, test and validate systems to predict how they will behave in real-world scenarios [1, 2, 10–12]. However, they have significant data infrastructure needs to support their proper use, particularly for data to comply with FAIR principles [4].

The National Manufacturing Institute Scotland (NMIS) is currently pioneering a project, ReMake, based on the concept of a circular economy, by repairing and re-purposing old and damaged industrial parts, prolonging their lifetime. In addition to the cost saving of using repurposed parts, this process aims to reduce greenhouse emissions from manufacturer processes by 45% of their current total, moving closer to a net zero economy. However, the detriment to re-purposing old parts is that they often require extensive testing to ensure their compliance with requirements, making the entire process not cost effective for the majority of applications. A digital testbed for testing repurposed parts is being established, with the aim of avoiding the costs associated with physical tests, making the process viable.

Finite Element Analysis (FEA) simulations form the core of this testbed and can have several workflows: they could be comprised of a single run; small sets of runs to carry out simple sensitivity analysis — model bias towards slight parameter variations; or full MC simulations to evaluate uncertainties. Due to the almost exclusive digital nature of the testbed, NASA's

Standard for Models and Simulations (NASA-STD-7009A) [13] served as the basis for the development of a risk assessment framework geared towards scoring simulation credibility on eight key areas: **Verification; Validation; Input Pedigree; Results Uncertainty; Results Robustness; Use History; Model and Simulation Management (M&S Management); and Technical Review**. An initial risk assessment establishes the minimum values for each of these areas, henceforward called "credibility axis", that the re-conditioned item must achieve in order to be successfully re-deployed. It is also worth noting that this is a translation of the NASA standard to the ReMake process, thus modifications undertaken to suit the latter should be anticipated in the data model.

By using a conceptual data model that highlights key terms in the process, it is possible to map the data requirements within physical systems and digitalise the evaluation of the credibility assessment. An initial credibility matrix is generated by assessing the risk attached to the deployment of remanufactured parts and serves as the baseline for the simulation quality needed to justify the remanufacture. This work outlines the required data architecture necessary for the implementation of a fully automated credibility assessment process based on the aforementioned risk assessment method. The process ensures that all necessary data is readily available for the automated calculation of the risk factors involved. A generalised framework based on this conceptual data model also has the potential to act as a template for simulation management. It provides value to organisations that require data traceability and reliability, thereby promoting a FAIR data approach.

Additionally, the design framework presented in this work puts a strong emphasis on using tools more often associated with ontology engineering [14–16] for the data modelling process. Competency questions (CQs), validated by domain experts, are used to identify additional criteria and the data necessary for automating the calculation of credibility scores attached to priority concepts [17]. CQs were originally used in ontology model validations and our data model building follows the principles of modularity and validation [17–19].

2. Methodology

The data model should be able to capture all data entities required for the calculation of a full credibility assessment, which needs to be programmatically compared with an initial minimum standard generated from a risk assessment process. A modular approach to designing the data model was taken. Initially, it was necessary to formalise the problem domain, by establishing the structure and format of the credibility assessment, in a way that it could be programmatically interacted with. This stage was followed by developing a list of CQs, which the conceptual data model would need to be able to answer, in order to calculate the value of each credibility axis. Entities within the model are split into classes which have attached properties, similar to an ontology. Additionally, classes are translatable into data tables and properties into columns using SQL [20].

2.1. *Credibility assessment as the problem domain*

The credibility assessment, shown in Fig. 1, establishes the problem domain from which the conceptual data model can be built. Higher scores indicate higher quality metrics, while lower scores are indicative of lower quality metrics often associated with lack of traceable data.

Each credibility axis, independent of its score value, is representative of a specific data pathway. Within the scope of this study we define: **Verification** as the process of ensuring correct models/methods meet requirements [21]; **Validation** as the process of ensuring model results meet requirements, error margins [21]; **Input Pedigree** as the degree of data traceability in simulations/experiments; **Results Uncertainty** as the dispersion in model results due to input variability; **Results Robustness** as the variability in results due to methodology changes; **Use History**

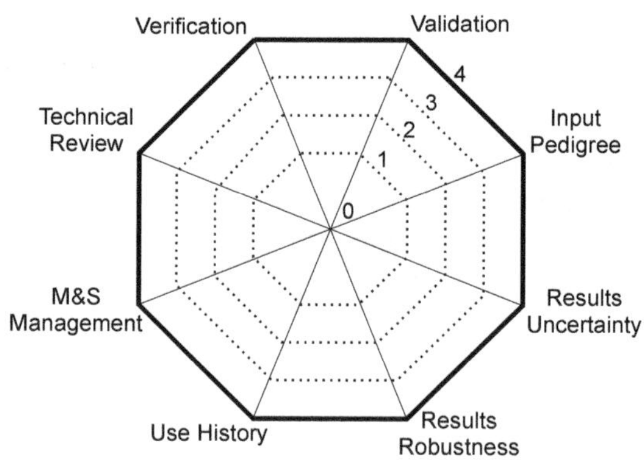

Fig. 1. Outline of the credibility assessment process. A minimum credibility score is produced via an initial risk assessment and establishes the minimum requirements. An actual credibility score is output from the credibility assessment process, using the data model designed in this study. The nature of the credibility assessment allows for an automated comparison, resulting in an response with the final decision.

as the record of past simulations/experiments and their methodologies; **M&S Management** as the assessment of simulation procedure management; **Technical Review** as the degree of documentation from processes or reviewed.

The axes themselves do not directly translate into the core concepts for the design of the conceptual data model, but instead provide insight into the data needs that the model must meet in order to calculate the score of each axis. These needs are translated into competency questions which serve as the main guide for the development of the conceptual data model.

2.2. *Competency questions for requirements capture*

Model design typically involves user interviews to capture requirements, extract data, and identify application needs, documented as user requirements. Subsequently, functional requirements, defined by the user, are specified. As our model is based on a domain expert risk assessment, requirement capture was completed in a similar manner to ontology engineering. CQs that the data model should be able to answer were captured, extracting the required classes and their attached properties.

By choosing this approach we were able to validate the model with our domain experts [22]. This was harmonised against the NASA standard [13], the credibility assessment and its requirements. As our model is based on a domain expert risk assessment, we used a form of user requirement capture more typically encountered in ontology engineering, specifically competency questions (CQs). Here we captured what kind of questions the data model should be able to answer, extracting the required classes and their attached properties. By choosing this approach we could directly validate the model with our domain experts [22]. This was harmonised against the NASA-STD-7009A standard [13] and the credibility assessment, shown in Fig. 1, and its inherent requirements.

One drawback of CQs is that they may not accurately reflect user preferences, and they are difficult to modify as they translate opinions of domain experts into logic. User stories are agile and user-centric, capturing user perspectives. They are easier to prioritise, estimate, and modify based on changing user requirements. Another drawback and distinction between the two is that CQs may not capture non-functional requirements, user interface design, and business rules. However, none of these applies to the work, making CQs preferable. An example of a CQ, for our domain, would be "What type of simulation method is used?". CQs such as these are inherently answered by model development. The existence of a "Method" class would automatically answer this CQ. Hence, output focused CQs are more relevant for model development, as credibility axes can be calculated by determining if certain classes, or their properties, exist in the model.

In addition to serving as a foundation for data model development tasks, the use of CQs enables a facilitated pathway towards the future integration of semantic models, such as the European Materials Modelling Ontology (EMMO) and the Semantic Sensor Network (SSN) ontology for material properties and sensors, respectively.

3. Results

The conceptual data model includes six core concepts, distributed over three key areas, as shown in Fig. 2. The starting point is the concept of "Item Description", which contains all the existing information of an item in the form of classes such as "Original Documentation" and "Inspection" which directly feed into the input pedigree credibility axis. The concept of "Item Description" along with that of "Infrastructure", which contains all data regarding people, qualifications, hardware and equipment, make up the entirety of the administrative network that supports the simulation and experimental frameworks.

"Simulation Input" is the most expansive of all the concepts in the model, as seen in Fig. 2, and contains multiple classes spanning the different data properties necessary to build a simulation input file. The "Input" class acts as a summary class as it pulls data from every other class within this concept to provide a single source of data for input building. Classes such as "Method" contain data properties relating to technical information such as method name, description, source, and application, and feed into the "Input" class. Each data class contains numerous data properties, such as exemplified by the "Input" class, which includes the "operatorID", "verifierID", "workstationID", "inputParametersID", "itemID", "itemXrayID", and "comments" properties. As the "Input" class is a summary class associated with the "Simulation Input" concept, all of its properties, with the exception of one, pertain to other data classes.

The "Experimental Setup" concept is optional as it is only required to achieve specific credibility axis scores, as determined by the CQs, and follows the same structure as "Simulation Input" with the "Experiment" class acting as a summary class. Both of these concepts serve to aggregate the data necessary for the application of either simulations of experiments.

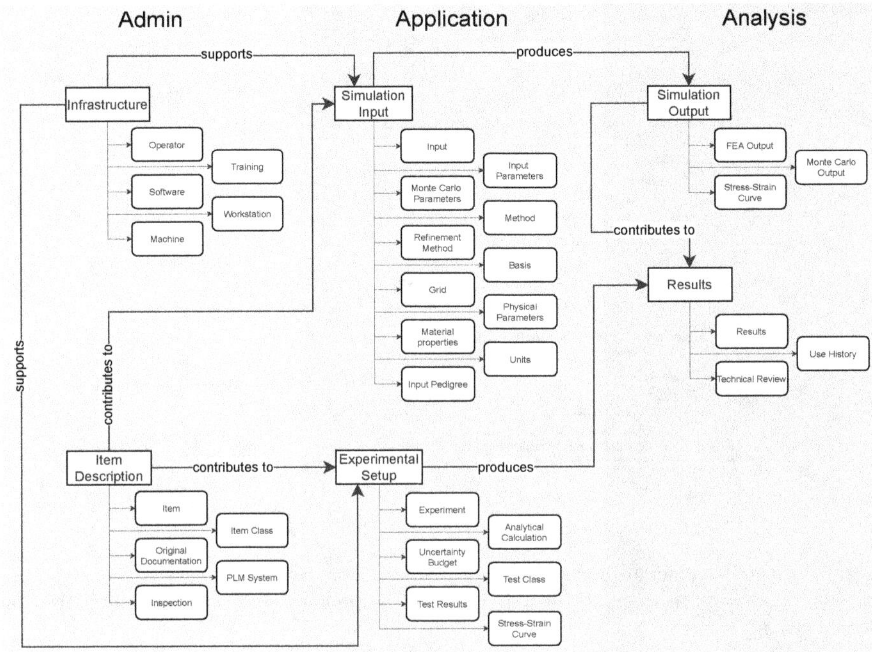

Fig. 2. The developed data model is depicted simply with rectangles identifying key areas of the ReMake testbed, and rounded shapes detailing the different data classes necessary for automating the digital validation process based on credibility assessment.

Lastly, the analysis of simulations/experiments is achieved via data stored in the "Simulation Output" and "Results" concepts. The first of these is generated via the data produced by simulations based on provided inputs, while the latter acts as a summary concept containing links to data generated by either simulations or experiments. The structure of the "Results" class also allows it to act as a Boolean summary of requirements specified by the initial risk and credibility assessments. Querying the "Results" class provides the final credibility assessment calculation, which determines if an item is suitable for re-deployment.

The process for calculating the credibility score for the Verification axis is shown in Fig. 3. It starts by querying the model to see if a simulation output exists for a specific input, indicating that the simulation is solvable. A return NULL value indicating the absence of a simulation output prompts a score of zero. If an output is retrieved by the query, then the model passes that information into the summary "Results" class, which can be queried to assess if the results match expectations based on expert opinion or other

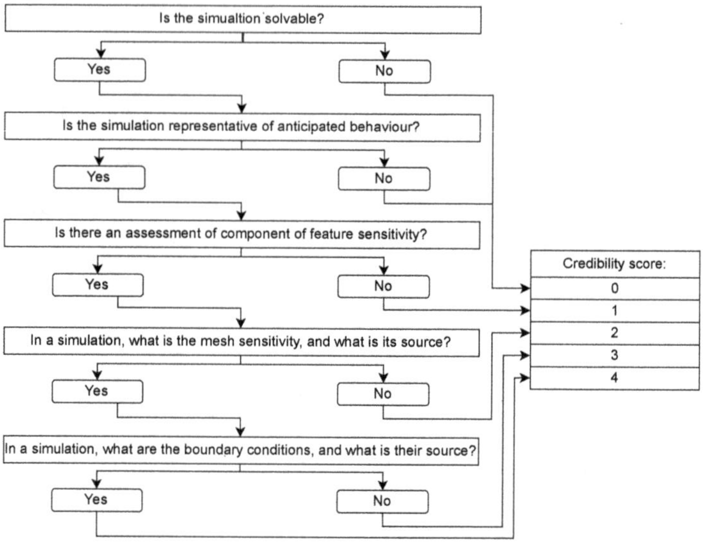

Fig. 3. Logic for calculating the verification credibility axis value based on using competency questions to query the data model. Competency questions are numbered in sequence.

experiments also contained within the same class. If this query returns a negative result, then the achieved score is also zero.

If the first two queries both return non-NULL values, then the model can be queried in regards to the sensitivity parameters used in simulations. These can be found by retrieving data from both the "Input" and "Output" classes. A NULL value returned by this query prompts a credibility score of one. If a non-NULL value is returned, then the model can be queried about the source and strictness of the sensitivity parameters used. While this data is accessible via the "Input" summary class, it originates from the "Method" and "Physical Parameters" classes. Again, a NULL value returned by the query prompts a credibility score of two, while a non-NULL value permits the continuation of the credibility score assessment.

The final query for determining the value of the credibility score for the Verification axis targets the boundary conditions used in the simulation, whether their original source documentation exists, as well as their trustworthiness. While this data can be accessed via the "Input" summary class, it originates in the "Grid" class. If a NULL value is returned, the highest and final credibility score achieved is three. If a non-NULL value is returned, then the maximum credibility score of four is achieved for the Verification axis.

4. Discussion

The data model developed in this work is built on six core concepts, compartmentalising data into discrete areas for easy traceability and findability. This gives the model a modular character, facilitating future expansion. The framework differs from NASA's as it provides guidance on meeting credibility assessment requirements. It's tailored towards the ReMake testbed, resulting in changes such as a different Technical Review credibility axis and the absence of a "people qualifications" axis.

The Technical review class indicates the depth of technical review performed on the results of a simulation or set of simulations and experiments. In addition, a "verifierID" data property exists within relevant classes like the "Input" and "Experiment" classes. This allows for a step-by-step review of the simulation and experiment procedures. The "people qualification" credibility axis was excluded from the model for simplicity, as it was not anticipated to significantly influence credibility scores for ReMake. However, the model includes administrative and infrastructure concepts, along with data classes related to people and training. This allows for the future implementation of a people qualifications credibility axis, if needed.

A competency questions-driven approach was used to create a data model that captures expert knowledge, is understandable by non-experts, and facilitates automated reporting. This approach has been instrumental in defining the essential questions that the data model should be able to answer, thereby outlining the required classes and their properties. The model's classes directly connect with CQs, identifying key missing data classes for specific credibility scores. Additionally, the model can be implemented for an application back end. However, ontology engineering and CQ formulation can be time-consuming and costly, and their use may reduce model flexibility. Despite potential drawbacks, feedback indicated the method's effectiveness, particularly due to the applied NASA framework and domain expert involvement.

The model development involved certain necessary assumptions. It was presumed that a single operator would be assigned to a specific input file, but the model allows for multiple operators per input or vice versa, due to its modular nature. Based on expert opinion, the model assumes that only MC simulations validate core FEA simulations, but other supporting simulation types can be added as new classes in the future. The "Results" class's Boolean array shape facilitates the addition of these new simulation classes, demonstrating the model's modularity. While the model that can facilitate the automation of the calculation of a credibility assessment based on a digital simulation-based testbed, the initial credibility assessment is still derived from a risk assessment and requires human input. This initial generation is not covered in this work, but it identifies an opportunity for future research to explore the digitisation of the requirements necessary for standardising this part of the process.

The foundation of our digital verification and validation model, established through CQs and expert inputs, not only serves our current needs for data modelling, but also paves the way for future enhancements trough the formalisation of an ontological representation. Future integration of existing ontologies into our data model would provide benefits in regards to improved model efficacy, allowing for a more comprehensive simulation quality framework. Harmonising ontologies such as the European Materials Modelling Ontology (EMMO) and the Semantic Sensor Network (SSN) ontology could provide a pathway for incorporating added detail, in the form of material properties and sensor descriptors respectively. This would also have the added benefit of further cementing our data model on well-established peer-reviewed knowledge. However, for the purposes of our demonstrator, the added complexity of this integration was beyond the initial scope of this work.

5. Conclusion

The aim of this study was to develop a data model for the automated calculation of a credibility assessment, with a view to increasing robustness and confidence in simulations performed to assess the capability of reconditioned industrial parts. In addition to establishing this framework, the data model presented in this study is implemented by sharing many design principles with ontology engineering. Specifically, the use of an expert-in-the-loop approach based on competency questions made it easier for non-experts to interact with the model, thereby facilitating model design and providing easier entry points. Furthermore, data concepts were selected with data compartmentalization in mind, making it easier to expand the model and carry out future development tasks. The proposed data model provides value to organisations that require data traceability and reliability, which includes a large number of National Metrology Institutes. In addition to serving as a framework for data modelling and simulation metadata, the proposed model can also serve as a template for simulation management.

Acknowledgements

This work was funded by the UK Government Department for Science, Innovation and Technology through the UK's National Measurement System. We would also like to thank our partners at the NMIS. Thanks to Frederic Brochu and Louise Wright for providing feedback on the manuscript.

References

[1] D. Jones, C. Snider, A. Nassehi, J. Yon and B. Hicks, Characterising the digital twin: A systematic literature review, *CIRP Journal of Manufacturing Science and Technology* **29**, 36 (2020).

[2] S. Boschert and R. Rosen, Digital twin — the simulation aspect, *Mechatronic futures: Challenges and solutions for mechatronic systems and their designers*, 59 (2016).

[3] Smart design: certification by analysis https://hvm.catapult.org.uk/what-we-do/programmes/certification-by-analysis/, Accessed August 12 2024.

[4] M. D. Wilkinson, M. Dumontier, I. J. Aalbersberg, G. Appleton, M. Axton, A. Baak, N. Blomberg, J.-W. Boiten, L. B. da Silva Santos, P. E. Bourne *et al.*, The fair guiding principles for scientific data management and stewardship, *Scientific data* **3**, 1 (2016).

[5] W. Jung, A review of research: an investigation of the impact of data quality on decision performance, in *Proceedings of the 2004 international symposium on information and communication technologies*, 2004.

[6] C. Cichy and S. Rass, An overview of data quality frameworks, *IEEE Access* **7**, 24634 (2019).

[7] L. Cai and Y. Zhu, The challenges of data quality and data quality assessment in the big data era, *Data science journal* **14** (2015).

[8] C. W. Fisher and B. R. Kingma, Criticality of data quality as exemplified in two disasters, *Information & Management* **39**, 109 (2001).

[9] R. Damgrave and E. Lutters, Smart industry testbed, *Procedia CIRP* **84**, 387 (2019), 29th CIRP Design Conference 2019, 08–10 May 2019, Póvoa de Varzim, Portugal.

[10] S. Haag and R. Anderl, Digital twin — proof of concept, *Manufacturing Letters* **15**, 64(02 2018).

[11] K. Alexopoulos, N. Nikolakis and G. Chryssolouris, Digital twin-driven supervised machine learning for the development of artificial intelligence applications in manufacturing, *International Journal of Computer Integrated Manufacturing* **33**, 1(04 2020).

[12] M. J. Kaur, V. P. Mishra and P. Maheshwari, The convergence of digital twin, iot, and machine learning: transforming data into action, *Digital twin technologies and smart cities*, 3 (2020).

[13] R. R. Roe, *Standard for Models and Simulations*, Tech. Rep., National Aeronautics and Space Administration (2016).

[14] L. Sebastian-Coleman, Chapter 3 - data management, models, and metadata, in *Measuring Data Quality for Ongoing Improvement*, ed. L. Sebastian-Coleman, MK Series on Business Intelligence (Morgan Kaufmann, Boston, 2013) pp. 27–37.

[15] M. West, Chapter 3 - some types and uses of data models, in *Developing High Quality Data Models*, ed. M. West (Morgan Kaufmann, Boston, 2011) pp. 23–36.

[16] C. D. Tupper, Chapter 11 - model constructs and model types, in *Data Architecture*, ed. C. D. Tupper (Morgan Kaufmann, Boston, 2011) pp. 207–221.

[17] J.-L. Hippolyte, M. Chrubasik, F. Brochu and M. Bevilacqua, A domain-agnostic ontology for unified metrology data management, *Measurement: Sensors* **18**, p. 100263 (2021).

[18] J.-L. Hippolyte, M. Romanchikova, M. Bevilacqua, P. Duncan, S. E. Hunt, F. Grasso Toro, A.-S. Piette and J. Neumann, Using ontologies to create machine-actionable datasets: Two case studies, *Metrology* **3**, 65 (2023).

[19] J.-L. Hippolyte, P. Duncan, M. Romanchikova and M. Chrubasik, Ontologies for experimental mechanics, in *Proc. 6th International Conference on Advances in Experimental Mechanics (BSSM, Oxford, UK, September 2022)*, Smarter Testing of Aerospace Structures (Oxford, UK, 2022).

[20] T. Teorey, S. Lightstone, T. Nadeau and H. Jagadish, Chapter 5 - transforming the conceptual data model to sql, in *Database Modeling and Design (Fifth Edition)*, eds. T. Teorey, S. Lightstone, T. Nadeau and H. Jagadish, The Morgan Kaufmann Series in Data Management Systems (Morgan Kaufmann, Boston, 2011) pp. 85–108, fifth edn.

[21] Verification, validation and uncertainty quantification https://www.asme.org/codes-standards/publications-information/verification-validation-uncertainty, Accessed July 6 2023.

[22] S. Lane, P. O'Raghallaigh and D. Sammon, Requirements gathering: The journey, *Journal of Decision Systems* **25**, 302 (2016).

F Pavese, A Bošnjaković, S Eichstädt, A B Forbes, J Alves e Sousa (eds.):
Advanced Mathematical and Computational Tools in Metrology and Testing XIII
Series on Advances in Mathematics for Applied Sciences, Vol. 94

https://doi.org/10.1142/9789819800674_0007

Digitalisation of units and dimensions

Blair D. Hall

Measurement Standards Laboratory of New Zealand,
Lower Hutt 5010, New Zealand
E-mail: blair.hall@measurement.govt.nz

The symbolic representation of quantities, units, and dimensions is reviewed with regard to information carried by conventional notation. The formal structure of unit systems like the International System of Units (SI) has been an efficient way for scientists to share data about the magnitudes of quantities, but it is an unsatisfactory basis for digitalisation. SI notation does not convey information about kind of quantity, which must be obtained from some other source, and the use of dimensional expressions in the SI also loses information about kinds of quantity. It is argued that digitalisation of quantities, units, and dimensions should focus on the more fundamental measurement concepts of quantity kinds and scales of measurement to provide strong foundations on which to build support for quantities, including features familiar from SI unit notation.

Keywords: digitalisation; SI unit; dimension; dimensional expression; quantity; quantity kind.

1. Introduction

Most scientists, after a cursory introduction to the topic, acquire familiarity with notions related to units of measurement through experience. To people skilled in a technical discipline, the units, dimensions, and quantities in their field become familiar entities. But is this captured in notation? Does an expert extract the information they need from notation alone or does experience and tacit knowledge play a role? We believe that contextual information is often used to disambiguate notational shortfalls. This may be an intuitive process that its scarcely noticed, allowing symbols to be regarded as loci for information that, in fact, they do not carry.

Digitalisation in metrology, as in any other domain, must seek appropriate representations for concepts that people use. These concepts must be formalised, and any need for tacit knowledge must be recognised and rendered explicit. Digital representations must be based on logical models that can be implemented with algorithms and data structures: recourse to human judgement is not an option.

This paper reviews conventional scientific notation representing units and dimensions with regard to the information carried. We conclude that the notation does not convey reliable information about kinds of quantity and dimensions. This is important information that people glean from unstructured contextual sources. Digital systems, on the other hand, require detailed and explicit information to handle data correctly.

This paper is organised as follows. Section 2 summaries the important characteristics of the concept of a physical quantity, then Sec. 3 looks at how these characteristics relate to scientific notation for expressed quantities. Section 4 summarises the rules of quantity calculus and reviews the relationship between terms in quantity equations and units of measurement. Section 5 explains the distinction between a dimension and a dimensional expression, and then Sec. 6 explains how dimensional expressions relate to unit names in the SI. Taken together, the considerations outlined in Secs. 2–6 make it clear that the SI system is not a suitable conceptual model for digital representation of measurable quantities. Section 8 outlines a more suitable approach. We discuss how these ideas can be developed further in Sec. 9 and conclude in Sec. 10.

2. Quantity

The concept of a physical quantity has a number characteristics:

- a quantity has magnitude;
- there is an intrinsic zero magnitude for each kind of quantity;
- quantities of the same kind can be compared by ratio;
- quantities of the same kind can be compared by difference;
- quantities of the same kind can be compared by order.

In addition to these characteristics, there is a sense in which individual quantities of the same kind can be combined (added or subtracted) with the result being a quantity of the same kind.[a] This means that the sense of zero magnitude is somehow equivalent to the sense of zero magnitude difference. For example, when several rods have the same length, the amount of length that must be added to one rod to make it the same length as another is zero. Not all measurable properties interpret zero in this way, so these properties

[a]The way in which addition and subtraction correspond to physical processes depends on the quantity, but a sense for addition is needed to construct a scale. It is possible for the same unit and same intrinsic zero to be associated with different non-linearly related scales [1], although this does not seem to arise in practice.

are not strictly speaking quantities. For example, it is nonsense to think that a temperature difference of 0 °C is equivalent to a temperature of 0 °C, which corresponds to the freezing point of water.

3. Unit names

Comparing a quantity, Q, by ratio to a reference for the same kind of quantity, Q_{ref}, produces a numerical magnitude[b]

$$\{Q\} = \frac{Q}{Q_{\mathrm{ref}}}. \tag{1}$$

The same quantity can be compared with different references, say $Q_{\mathrm{ref}\cdot 1}$ and $Q_{\mathrm{ref}\cdot 2}$, to obtain different ratios:

$$\{Q\}_1 = \frac{Q}{Q_{\mathrm{ref}\cdot 1}} \text{ and } \{Q\}_2 = \frac{Q}{Q_{\mathrm{ref}\cdot 2}}. \tag{2}$$

Combining these equations, and eliminating the ratio of Q to itself, leaves

$$\begin{aligned}\frac{\{Q\}_1}{\{Q\}_2} &= \frac{Q}{Q_{\mathrm{ref}\cdot 1}} \frac{Q_{\mathrm{ref}\cdot 2}}{Q} = \frac{Q_{\mathrm{ref}\cdot 2}}{Q_{\mathrm{ref}\cdot 1}} \frac{Q}{Q} \\ &= \frac{Q_{\mathrm{ref}\cdot 2}}{Q_{\mathrm{ref}\cdot 1}}.\end{aligned} \tag{3}$$

So, the ratio of the initial and final references serves as a factor that converts values expressed in terms of $Q_{\mathrm{ref}\cdot 2}$ to values expressed in terms of $Q_{\mathrm{ref}\cdot 1}$,

$$\{Q\}_1 = \{Q\}_2 \frac{Q_{\mathrm{ref}\cdot 2}}{Q_{\mathrm{ref}\cdot 1}}. \tag{4}$$

Equation (4) underpins the meaning of notation used to express quantities.

In formal treatments of quantities and units, the magnitude of a quantity is expressed as a product of two factors,

$$\{Q\}\,[Q], \tag{5}$$

a numeric value and the name of a unit for Q, denoted $[Q]$. For example, 10 kg takes this form, where 'kg' is a name for the numeric conversion factor $Q_{\mathrm{ref}\cdot 2}/Q_{\mathrm{ref}\cdot 1}$ in Eq. (4). This interpretation of unit symbols was explained by Maxwell [2]:

> The formulae at which we arrive must be such that a person of any nation, by substituting for the different symbols the numerical value of the quantities as measured by his own national units, would arrive at a true result.

[b]A term in braces, like $\{Q\}$, denotes a numeric value of magnitude for Q.

So, this notation was intended to address the needs of whosoever used the data. Each symbol represents the ratio of the initial (named) reference to the (unnamed) final reference. Different users may, of course, have different final references, and the factor will be unity if references are the same. So, this intuitive process ensures that values can be easily obtained in terms of local references. For example, 3 ft takes the value 3 when working in feet and $3 \times 0.3048 = 0.9144$ when working in metres, because a one-foot reference measures 0.3048 on a length scale calibrated in metres.

The interpretation of unit names and symbols as conversion factors, rather than labels for reference quantities, will be shown to explain why unit names are sometimes shared by quantities. This situation reflects the fact that, when units change, the same conversion factors may apply to different quantities; that is, the different quantities scale in the same way.

4. Quantity and unit calculations

It is common nowadays to express relationships between quantities in the form of quantity equations. Familiar examples include Newton's second law, $f = ma$, and Ohm's law, $R = V/I$. The terms in these equations are understood to represent quantities as described in Sec. 2, and the relations are understood to hold regardless of any choice of reference quantities.

Arithmetic operations are restricted in quantity equations: quantities of any kind can be multiplied or divided, but only terms for the same kind of quantity can be added or subtracted. These rules — called the quantity calculus — prohibit formulation of some nonsensical equations, but not all equations that are legitimate under these rules are necessarily meaningful.

The interpretation of a product or quotient of terms is straightforward [3]. A product is proportional to the magnitudes of its factors, while a quotient is proportional to magnitudes in the numerator and inversely proportional to magnitudes in the denominator. Unlike addition and subtraction, the quantity calculus does not determine the kind of quantity associated with a product or quotient; products and quotients are not somehow composed in any physical sense from their arguments. For example, there is no sense in which speed is somehow the comparison by ratio of length to time.

Terms whose numerical expressions are unaffected by the choice of reference quantities are treated as plain numbers, and are exempt from the rules of quantity calculus. They are called 'dimensionless' quantities, although 'unit-less' quantities might be more appropriate [4]. For example, a ratio of lengths does not depend on the reference chosen for length, so linear scale factors and aspect ratios are both dimensionless quantities.

A quantity equation can be evaluated when its all terms can be expressed quantitatively. In the first step of this process, each term, Q, is replaced by a value and the name of a unit, $\{Q\}\,[Q]$. In the second step, conversion factors are assigned to unit names, to effect any changes required.[c] The final step evaluates the resulting expression.

Because quantity calculus places no restrictions on the terms in products and quotients, there are no restrictions on the unit names appearing in products and quotients either. These names, being numeric variables, can be treated algebraically. For example, average speed is the ratio of distance travelled to the time elapsed, $V = L/T$. If L is measured in miles and T in hours then a unit name for V is mile-per-hour ($\mathrm{mi} \cdot \mathrm{h}^{-1}$).

Unit names representing products and quotients can be simplified algebraically too. This is where confusion can arise if a one-to-one association is assumed between a name and a corresponding reference quantity [5, 6]. Section 2 explained that the notation does not support this interpretation, so it must be remembered that the names are simply conversion factors. When simplification is possible, the decision to do so is often a matter of convention. For example, an amount of rain fall is expressed in terms of length, not in terms of volume divided by area, although the latter is physically meaningful. Conversely, fuel consumption expressed in litres-per-100 kilometres could be simplified to an expression in terms of area, but is not.

5. Dimension

Maxwell referred to terms enclosed in square brackets, like $[Q]$, both as concrete units and as dimensions. This was not a contradiction. The dimension of a quantity is the class of units for that quantity [1]. So, a term $[Q]$ refers to an unspecified unit for Q; i.e., the name of any unit in the dimension for Q. In the discipline of dimensional analysis, the square-bracket notation is associated with a class of units, but also with dimensional expressions involving products of powers of terms for classes [7].

Dimensional expressions can be obtained from a quantity equation describing one quantity in terms of others. For example, starting with the equation for kinetic energy,

$$K = \frac{1}{2} M V^2, \tag{6}$$

[c]Only quantities of the same kind may be added or subtracted. However, ratios with respect to the same references are required for numeric computation.

and substituting pairs of factors for each term, we can separate terms into a value equation for kinetic energy,

$$\{K\} = \frac{1}{2}\{M\}\{V\}^2, \tag{7}$$

and a dimensional expression,

$$[K] = [M][V]^2. \tag{8}$$

It is important to distinguish between the idea of quantity dimensions and dimensional expressions — products or powers of dimensional terms. Equations, like (8), do not define individual quantity dimensions. Rather, the equation relates to the proportionality of conversion factors: the conversion factor for kinetic energy (on the left) is equal to the conversion factor for mass multiplied by the conversion factor for speed squared (on the right).

6. Unit systems

The formal structure for unit systems proposed by Maxwell facilitates the exchange of quantity data among scientists working in different units [2]. Notation for the International System of Units (SI) is based on these ideas.

A Maxwellian system identifies a small number of base quantities, in terms of which other 'derived' quantities can be described (e.g., $V = L/T$). Dimensional expressions for derived quantities can be obtained from the defining equations, as explained above. There is one dimensional expression for each quantity kind, but several quantities can share an expression.

Unit systems can be constructed by associating base units with base quantities. For example, precursor unit systems to the SI include the metre-kilogram-second (MKS) system and the centimetre-gram-second (CGS) system, both with length, mass, and time (L, M, and T) as base quantities.

Names for derived-quantity units are generated systematically from dimensional expressions. Each term in an expression is replaced by the name, or symbol, of the corresponding base unit. So, for example, the dimensional expression for kinetic energy, $[M]\,[L]^2\,[T]^{-2}$, produces $\mathrm{kg \cdot m^2 \cdot s^{-2}}$ in the MKS system (and in the SI) and $\mathrm{g{\cdot}cm^2{\cdot}s^{-2}}$ in the CGS system. Systematic naming depends only on the dimensional expression. So, quantities with the same dimensional expressions get the same names. Well-known examples of degenerate names in the SI include the shared unit for frequency and activity, $\mathrm{s^{-1}}$, and the shared unit for torque, work, and energy, $\mathrm{kg \cdot m^2 \cdot s^{-2}}$. Perhaps because of name degeneracy, some derived units are provided with alternative special names. Examples include the joule (J), for energy, and the hertz (Hz) and becquerel (Bq), for frequency and activity, respectively. These special names can be used in compound names, such as N · m, which is a degenerate unit name for torque, work, and energy.

It is important to keep the distinction between well-defined reference quantities and unit notation in mind. The measure of a quantity must, in some sense, always involve a comparison with a standard reference of the same kind of quantity (i.e., Q_{ref} in Eq. (1)). However, derived unit names do not reliably identify reference quantities. The names are mnemonics for dimensional expressions, which describe unit conversion factors.

7. Other units and measurable properties

In addition to formal unit systems, like the MKS, CGS and SI, less formal collections of units may also be referred to as systems, such as Imperial Units and US Customary Units. These systems do not define a set of base quantities and so eschew systematic unit naming. There tends to be more diversity of names as a result, such as: the inch, foot, yard, furlong, etc., which are all Imperial units of length; or the pound, ounce, and ton, which are units of mass. Conversion factors are sometimes published by authoritative bodies, such as the National Institute of Standards and Technology in the United States [8], to convert customary units to SI units.

In many cases, customary unit names also serve as conversion factor variables. However, there are exceptions. One well-known case is temperature expressed in degrees Fahrenheit. This scale for temperature is similar to the Celsius scale for temperature in the sense that both define points representing zero at different arbitrary temperatures. So, neither scale expresses a quantity as described in Sec. 2. To convert values from one scale to another, an offset must be applied as well as a conversion factor. So, it makes no sense to associate °C or °F with conversion ratios.

There are also many measurable properties that are defined by strict protocols, which are needed to obtain satisfactory levels of reproducibility and comparability in measurement data. In such method-defined measurements, the property is not a physical quantity in the sense of Sec. 2. Nonetheless, SI units may be associated with such data, and doing so can be misleading because the characteristics of a quantity are not satisfied and the meaning attributable to data depends on the method used.

8. Digitalisation

Digitalisation aims to develop logical models that represent certain concepts of interest to a group of people. Digitalisation should provide ways for systems to manipulate data that are consistent with conceptual models and facilitate use of data. Digitalisation often produces a logical model that can be presented clearly to people and implemented algorithmically by machines. An excellent example is the real-number representations now ubiquitous in digital systems, which allow mathematical calculations to be defined by people and executed by machines.

Attempts have been made to develop software support for physical quantities and units systems since the late 1960s. Recent reviews attest to the amount of work done and the fact that no general solution has emerged [9]. Literally hundreds of open-source software libraries for units of measurement were identified in a study by Bennich and McKeever [10]. There are many reasons for the number and diversity of such projects; however, one that may not have been considered yet is that the conceptual model developers are focusing their attention on is inappropriate.

The preceding sections have explained that Maxwellian unit systems are intended to facilitate manual data processing. Unit symbols easily accompany numeric values on the page, where they may prompt a need for a unit conversion or help check equations for dimensional homogeneity. People associate familiar unit names with quantities but the names themselves do not formally encode that information. Maxwell left the task of identifying quantity kinds to people, because it was safe to assume that scientists in the 19th century would not err in this detail. However, in the 21st century, digitalisation strives for autonomy from human supervision. So, some way of tracking kinds of quantity is now needed.

The considerations presented in earlier sections strongly suggest that digitalisation of Maxwellian unit systems is not an appropriate objective. Better ways of identifying and keeping track of quantity kinds and quantity dimensions are needed to correctly represent and interpret quantity equations. Unit conversion is a secondary concern, which can be layered on top of these foundational concepts. The rest of this section outlines a fundamentally different approach to digitalisation.

8.1. *Unique digital identifiers*

When using mathematical expressions in scientific writing, a suitable notation must always be found to distinguish terms associated with different concepts. Documents usually describe notation in detail, unless familiarity with a standard notation can be assumed (e.g., SI unit symbols are standardised). In a similar way, digital systems label data with identifiers that allow algorithms to be applied. Digital identifiers are sometimes intended for internal system use only (e.g., the address of a block of data in memory may serve as an identifier); however, when data is widely shared, a 'unique digital identifier' (UID) is needed to ensure that data can be properly identified anywhere and at any time.[d]

A public collection of UIDs labelling different kinds of quantity and a collection of UIDs for scales associated with units are a basic requirement to support the digital representation of measured quantities. Identifiers from these collections are used in pairs: one quantity kind (denoted by angle brackets $\langle\cdot\rangle$) and one scale (denoted by double square brackets $[\![\cdot]\!]$). Expressions of quantities in digital systems annotate numeric data with identifiers in much the same way as written numeric data is annotated by a unit. The uniqueness and public access to the UID collections ensures that data can be freely exchanged without risk of misinterpretation.

8.2. *Disambiguating quantities*

Quantity–scale identifier pairs overcome the problem of identifying quantity kinds in traditional unit notation. For example, one identifier pair for measures of activity expressed in SI units is $\langle\text{activity}\rangle, [\![\text{s}^{-1}]\!]$ and another is $\langle\text{activity}\rangle, [\![\text{Bq}]\!]$. No ambiguity would result if either pair were used to express activity data. Similarly, $\langle\text{frequency}\rangle, [\![\text{s}^{-1}]\!]$ and $\langle\text{frequency}\rangle, [\![\text{Hz}]\!]$ are pairs that both express frequency data. Another example is the alternative pairs of identifiers for energy: $\langle\text{energy}\rangle, [\![\text{J}]\!]$, $\langle\text{energy}\rangle, [\![\text{N}\cdot\text{m}]\!]$, and $\langle\text{energy}\rangle, [\![\text{kg}\cdot\text{m}^2\cdot\text{s}^{-2}]\!]$, which would not be confused with pairs for torque: $\langle\text{torque}\rangle, [\![\text{N}\cdot\text{m}]\!]$, and $\langle\text{torque}\rangle, [\![\text{kg}\cdot\text{m}^2\cdot\text{s}^{-2}]\!]$.

Derived quantities, like area and volume, are also attributed UIDs. Some dimensionless quantities are given individual identities too. For example, plane and solid angles have special SI unit names (rad and sr) but are indistinguishable from all the other dimensionless quantities if the alternative SI unit one is used. This difficulty is overcome by identifiers $\langle\text{plane angle}\rangle$ and $\langle\text{solid angle}\rangle$. Quantities defined as a ratio of the same kind of quantity may also be identified by UIDs, such as: $\langle\text{length ratio}\rangle$, $\langle\text{voltage ratio}\rangle$, $\langle\text{reflection coefficient}\rangle$, etc.

---

[d]There are many forms of unique identifiers. One that may be familiar to readers is the digital object identifier (DOI) used to uniquely identify documents and other digital objects. Detailed information about a digital object may be indexed by its DOI.

8.3. *Conversions and dimensions*

Conversion from one expression to another depends on the initial and final UID pairs that annotate data. It is convenient to collect information about conversions in a table, where table entries are indexed by UID pairs for the initial and final expressions. For instance, the conversion factor $180/\pi$ can be stored as an entry indexed by an initial UID pair $\langle$plane angle$\rangle$, $[\![$rad$]\!]$ and a final pair $\langle$plane angle$\rangle$, $[\![{}^\circ]\!]$.

The set of entries for a particular quantity in the table embodies a representation of the dimension for that quantity. That is, all the entries that transform a quantity from one scale to another. This set includes any scales with conversion factors entered in the table. So, for example, the Imperial inch and foot would be included in the dimension of length alongside the SI metre.

9. Discussion

The approach described in the previous section supports quantity equations and numerical calculations involving quantities. Information about quantity kinds is directly available, so the rules of quantity calculus can be imposed, and when terms are added or subtracted, scales can be checked immediately for compatibility, and conversions applied if necessary.

Familiar features, like unit-calculus, which many expect from any digital unit representation, can be implemented on top of the foundations described [11]. Information about the unit system, or systems, associated with a scale can be stored and accessed using the scale UID as a key. For instance, the UID $\langle$s$\rangle$ can provide access to a dimensional expression of time in the SI. With this additional information, unit arithmetic can be implemented in the same way as in other digital libraries.

Section 7 described situations where measurable properties are not quantities and should not be manipulated as such. SI notation offers no remedy for such cases, but the approach described here can be helpful. First, the notion of scale is actually more nuanced than has been explained so far. More details are available elsewhere [12, 13] but, for example, the difficulties posed by different scales for temperature can be overcome by recognising that Celsius and Fahrenheit temperatures are expressed on a different type of scale to other SI units. In these scales, both the zero point and the size of scale divisions must be defined. For all other SI quantities, which have intrinsic zeros, only the size of scale divisions needs to be defined.

Second, more detailed labelling of measurable properties is possible. For example, a distinction can be made between the quantity thermodynamic temperature and the scale called ITS-90, which is widely used to approximate thermodynamic temperature [14]. With extra information available, systems can ensure that data processing is appropriately controlled. This is needed for method-defined measurements, where processing must have access to information about the standard methods used to obtain data.

10. Conclusion

It is not possible to reliably determine a kind of quantity from SI unit names or symbols. The notation used in Maxwellian unit systems, like the SI, is simply not intended to do this. People can infer which quantities are involved from contextual information; however, digital systems need to resolve this detail immediately and without ambiguity. So, a fundamentally different approach is proposed to support measurable quantities in digital systems. This approach presents no particular technical difficulties and resolves the problems inherent in SI notation, which have dogged attempts to provide digital support for units for decades.

A publicly accessible register of unique digital identifiers for quantity kinds and scales of measurement provides unambiguous labels for quantities and scales. Pairs of identifiers annotate numerical data to denote a particular form of expression. Conversions from one expression to another are tabulated, with entries indexed by the initial and final expression identifiers. This representation inherently captures the dimension of a quantity in the set of conversion table entries for that quantity. Additional information can be stored in the registry and accessed by the unique identifiers for quantities and scales. This enables additional features associated with unit systems to be implemented.

Acknowledgement

This work was funded by the New Zealand government. The author thanks Peter Saunders for a careful review of the manuscript.

References

[1] B. Ellis, On the nature of dimensions, *Philosophy of Science* **31**, 357 (1964).

[2] J. C. Maxwell, *A Treatise on Electricity and Magnetism* (Clarendon Press, Oxford, 1873).

[3] A. Lodge, The multiplication and division of concrete quantities, *Nature* **38**, 281 (1888).
[4] I. Johansson, Metrological thinking needs the notions of parametric quantities, units and dimensions, *Metrologia* **47**, 219 (2010).
[5] W. H. Emerson, On the Algebra of Quantities and Their Units, *Metrologia* **41**, L33 (December 2004).
[6] W. H. Emerson, On Quantity Calculus and Units of Measurement, *Metrologia* **45**, 134 (April 2008).
[7] G. I. Barenblatt, *Dimensional analysis* (Gordon and Breech, 1987).
[8] K. S. Butcher, L. D. Crown, E. J. Gentry and C. Hockert, *The International System of Units – Conversion Factors for General Use*, NIST Special Publication, Vol. 1038 (NIST, May 2006).
[9] O.-A. Salah and S. McKeever, Lack of adoption of units of measurement libraries: Survey and anecdotes, in *Proceedings of the ACM/IEEE 42nd International Conference on Software Engineering: Software Engineering in Practice*, ed. Association for Computing Machinery 2020 pp. 81–89.
[10] O. Bennich-Björkman and S. McKeever, The next 700 unit of measurement checkers, in *Proceedings of the 11th ACM SIGPLAN International Conference on Software Language Engineering*, ed. Association for Computing Machinery 2018 pp. 121–132.
[11] B. Hall, Modelling expressions of physical quantities, in *Proceedings of the 15th International Joint Conference on Knowledge Discovery, Knowledge Engineering and Knowledge Management - KEOD*, (SciTePress, 2023).
[12] B. Hall and M. Kuster, Representing quantities and units in digital systems, *Measurement: Sensors* **23**, p. 100387 (2022).
[13] B. Hall, Representing and expressing measurement data in digital systems, *SN Computer Science* **4**, p. 120 (2023).
[14] B. Hall, P. Saunders and D. R. White, Digital representation of scales and units for temperature and related quantities, *Temperature: Its Measurement and Control in Science and Industry (Proc. 10th Int. Temp. Symp.)* **10** (2023), in press.

F Pavese, A Bošnjaković, S Eichstädt, A B Forbes, J Alves e Sousa (eds.):
Advanced Mathematical and Computational Tools in Metrology and Testing XIII
Series on Advances in Mathematics for Applied Sciences, Vol. 94

https://doi.org/10.1142/9789819800674_0008

Uncertainty quantification for results of AI-based data processing: Towards more feasible algorithms

C. Q. Lauter, M. Ceberio and V. Kreinovich*

Department of Computer Science, University of Texas at El Paso
El Paso, Texas 79968, USA
**E-mail: vladik@utep.edu, www.cs.utep.edu/vladik*

O. M. Kosheleva

Department of Teacher Education, University of Texas at El Paso
El Paso, Texas 79968, USA
E-mail: olgak@utep.edu

AI techniques have been actively and successfully used in data processing. This tendency started with fuzzy techniques, now neural network techniques are actively used. With each new technique comes the need for the corresponding uncertainty quantification (UQ). In principle, for both fuzzy and neural techniques, we can use the usual UQ methods — however, these techniques often require an unrealistic amount of computation time. In this paper, we show that in both cases, we can use specific features of the corresponding techniques to drastically speed up the corresponding computations.

1. Formulation of the problem

Need for indirect measurements. In many practical situations, we are interested in the value of a quantity y that is difficult (or even impossible) to measure directly. Since we cannot measure y directly, we measure it *indirectly*, i.e.:

- we find easier-to-measure quantities $x_1, \ldots, x_n$ that are related to y by a known dependence $y = f(x_1, \ldots, x_n)$,
- we measure these quantities, and
- we use the measurement results $\widetilde{x}_1, \ldots, \widetilde{x}_n$ to estimate y as

$$\widetilde{y} = f(\widetilde{x}_1, \ldots, \widetilde{x}_n).$$

Computing $\widetilde{y}$ is an important case of what is called *data processing.*

Similar data processing is also used for prediction, to use the current values $x_1, \ldots, x_n$ of different physical quantities to compute estimates y for future values of quantities of interest.

Need for uncertainty quantification (UQ). Measurements are never 100% accurate. Each measurement result $\widetilde{x}_i$ is, in general, different from the actual (unknown) value x_i; see, e.g.,[14]. As a result, even if the dependence $y = f(x_1, \ldots, x_n)$ is exact, the estimate $\widetilde{y}$ is, in general, different from the actual value y.

From the practical viewpoint, it is important to know how big can be the difference $\Delta y \stackrel{\text{def}}{=} \widetilde{y} - y$. For example, if we are estimating the amount of water in an underground location as 100 units, then:

- if it is 100 ± 20, this is worth exploring, while
- if it is 100 ± 150 then maybe there is no water at all, so we better perform more measurements before starting the expensive drilling.

Typical types of uncertainty. To estimate Δy, we need to have information about measurement errors $\Delta x_i \stackrel{\text{def}}{=} \widetilde{x}_i - x_i$. Usually (see, e.g.,[14]):

- we either know the probability distribution; this is known as *probabilistic uncertainty*;
- or we know the upper bound Δ_i on the absolute value $|\Delta x_i|$.

Let us describe the two corresponding UQ problems.

Comment about probabilistic uncertainty. In the probabilistic cases, usually, the mean value of the measurement error is 0. Indeed, if the mean is not 0, i.e., if the measuring instrument has bias, then we can re-calibrate all the measurement results by subtracting this known bias — and thus, get the mean to be 0.

Also, usually, the measurement error is caused by many relatively small effects. It is known that the joint effect of a large number of small errors leads to an approximately normal distribution; this is known as the Central Limit Theorem (see, e.g.,[15]). It is known that to uniquely describe a normal distribution with 0 mean, it is sufficient to describe its standard deviation σ_i. Thus, in the case of probabilistic uncertainty, we usually know the standard deviation σ_i of the corresponding measurement error.

Measurement errors of different measurements are usually caused by different factor and are, therefore, independent.

Comment about the second type of uncertainty. In the second case, once we get the measurement result $\widetilde{x}_i$, all we know about the actual value x_i of the corresponding quantity is that this value is in the interval $[\widetilde{x}_i - \Delta_i, \widetilde{x}_i + \Delta_i]$. The value Δ_i is then equal to the half-width of this interval. Because of this fact, this case is known as *interval uncertainty*; see, e.g.,[5,8,9,11,14].

UQ: formulation of the problem in the case of probabilistic uncertainty. We know:

- the data processing algorithm $f(x_1, \ldots, x_n)$;
- the measurement results $\widetilde{x}_1, \ldots, \widetilde{x}_n$; and
- the standard deviations $\sigma_1, \ldots, \sigma_n$ of the corresponding measurement errors $\Delta x_i = \widetilde{x}_i - x_i$; the mean values of these errors are 0s, and different measurement errors are independent.

Based on this information, we need to compute the standard deviation σ of the resulting estimation error $\Delta y = \widetilde{y} - y$, where $\widetilde{y} = f(\widetilde{x}_1, \ldots, \widetilde{x}_n)$ and $y = f(x_1, \ldots, x_n)$.

UQ: formulation of the problem in the case of interval uncertainty. We know:

- the data processing algorithm $f(x_1, \ldots, x_n)$;
- the measurement results $\widetilde{x}_1, \ldots, \widetilde{x}_n$; and
- the upper bounds $\Delta_1, \ldots, \Delta_n$ of the absolute values of the corresponding measurement errors $\Delta x_i = \widetilde{x}_i - x_i$: $|\Delta x_i| \le \Delta_i$.

Based on this information, we need to compute the largest possible value Δ of the absolute value of the resulting estimation error $\Delta y = \widetilde{y} - y$, where $\widetilde{y} = f(\widetilde{x}_1, \ldots, \widetilde{x}_n)$ and $y = f(x_1, \ldots, x_n)$.

Possibility of linearization. Measurement errors Δx_i are usually reasonable small, in the sense that terms quadratic in terms of these errors are much smaller than the errors themselves — and can, thus be safely ignored. For example, if $\Delta x_i \approx 10\%$, then $(\Delta x_i)^2 \approx 1\%$, which is much smaller than 10%. In such situations, we can expand the dependence of Δy on Δx_i in Taylor series and keep only terms which are linear in Δx_i:

$$\begin{aligned}\Delta y &= \widetilde{y} - y = f(\widetilde{x}_1, \ldots, \widetilde{x}_n) - f(x_1, \ldots, x_n) \\ &= f(\widetilde{x}_1, \ldots, \widetilde{x}_n) - f(\widetilde{x}_1 - \Delta x_1, \ldots, \widetilde{x}_n - \Delta x_n) \\ &= f(\widetilde{x}_1, \ldots, \widetilde{x}_n) - \left(f(\widetilde{x}_1, \ldots, \widetilde{x}_n) + \sum_{i=1}^{n} f_{,x_i} \cdot (-\Delta x_i) \right),\end{aligned}$$

where $f_{,x_i}$ denotes the partial derivative of the function $f(x_1, \ldots, x_n)$ with respect to x_i:

$$f_{,x_i} \stackrel{\text{def}}{=} \frac{\partial f}{\partial x_i}_{|x_1=\widetilde{x}_1, \ldots, x_n=\widetilde{x}_n}.$$

Thus, we have

$$\Delta y = \sum_{i=1}^{n} f_{,x_i} \cdot \Delta x_i. \tag{1}$$

Comment. In the following text, we will use the above simplifying notation for partial derivatives in other cases as well, i.e., we will denote

$$a_{,b} \stackrel{\text{def}}{=} \frac{\partial a}{\partial b}.$$

Linearization helps to solve both UQ problems. If we know all the partial derivatives $f_{,x_i}$, then we can use the known formulas for solving both UQ problems:

- in the probabilistic case, we have

$$\sigma^2 = \sum_{i=1}^{n} (f_{,x_i})^2 \cdot \sigma_i^2; \tag{2}$$

- in the interval case, we have

$$\Delta = \sum_{i=1}^{n} |f_{,x_i}| \cdot \Delta_i. \tag{3}$$

How can we compute the partial derivatives? Sometimes, the data processing algorithm $f(x_1, \ldots, x_n)$ is straightforward, so partial derivatives can be computed by explicit differentiation. In most other cases, a natural idea is to use sensitivity analysis (i.e., in effect, numerical differentiation) to find the derivatives $f_{,x_i}$ of f with respect to the inputs x_i:

$$f_{,x_i} \approx \frac{f\left(\widetilde{x}_1, \ldots, \widetilde{x}_{i-1}, \widetilde{x}_i + h_i, \widetilde{x}_{i+1}, \ldots, \widetilde{x}_n\right) - \widetilde{y}}{h_i}. \tag{4}$$

In this approach, we need several calls to the function $f(x_1, \ldots, x_n)$:

- we need one call to compute the value $\widetilde{y} = f\left(\widetilde{x}_1, \ldots, \widetilde{x}_n\right)$, and
- we need n calls to compute n values

$$f\left(\widetilde{x}_1, \ldots, \widetilde{x}_{i-1}, \widetilde{x}_i + h_i, \widetilde{x}_{i+1}, \ldots, \widetilde{x}_n\right)$$

  needed to compute n derivatives.

So overall, we need $n+1$ calls to the program that computes $f(x_1, \ldots, x_n)$.

Comment. When the number of inputs n is large, this idea — of computing all the derivatives and then applying formulas (2) and (3) — is not always practical. Indeed, in some such cases, computing f often takes hours on a high-performance computer. So we do not have the luxury of calling the function $f(x_1, \ldots, x_n)$ for each of thousands of variables. In such cases, we can use Monte-Carlo (MC) techniques:

- traditional MC techniques for probabilistic uncertainty (see, e.g.,[15]), and
- Cauchy-based MC techniques for the case of interval uncertainty; see, e.g.,[7].

AI-related challenges. AI techniques are becoming ubiquitous in data processing. Their results are good, but in terms of uncertainty quantification (UQ) they lead to new challenges.

First challenge: the use of fuzzy techniques. The first challenge is that, for some inputs x_i, we only have expert estimates. These estimates are often formulated by using imprecise ("fuzzy") words from natural language. There are special techniques — known as fuzzy techniques[2,6,10,12,13,16] — for translating these estimates into numerical machine-understandable form. In this techniques, an expert information about the quantity x_i is described by assigning, to each possible value of this quantity, a degree $\mu_i(x_i) \in [0, 1]$ with which this value is possible. The corresponding function $\mu_i(x_i)$ is called a *membership function* or, alternatively, a *fuzzy set.*

This is largely equivalent to describing, for several degrees of confidence α, the interval $\mathbf{x}_i(\alpha)$ of all the values which are possible with at least this degree of confidence:

- for $\alpha > 0$, we have

$$\mathbf{x}_i(\alpha) = \{x_i : \mu_i(x_i) \geq \alpha\},$$

  and
- for $\alpha = 0$, we have

$$\mathbf{x}_i = \overline{\{x_i : \mu_i(x_i) > 0\}},$$

  where $\overline{S}$ means the closure of the set S.

In principle, there are UQ techniques for processing such fuzzy inputs.

In effect, we need to:

- select several certainty levels α, e.g., the values $\alpha = 0, 0.1, 0.2, \ldots, 0.9, 1$, and
- on each level, perform interval estimations, i.e., compute the range

$$\mathbf{y}(\alpha) = \{f(x_1, \ldots, x_n) : x_i \in \mathbf{x}_i(\alpha)\}. \quad (5)$$

The challenge is that this multiplies the UQ time — which is often already long — by a factor of 11. This often makes it not feasible.

Second challenge: the use of neural techniques. Second, data processing is now often performed by a deep neural network (NN), with millions (and even billions) of parameters; see, e.g.,[3]. In this case, each computation already requires a significant amount of computation time. So, for large number of inputs n, repeating the computations $n + 1$ times — as in the traditional UQ techniques — is not possible.

Comment: it is important to be cautious about neural data processing. It is known that in many neural networks, a small change in the inputs can cause a large change in the output; see, e.g.,[3,4]. There are known examples when, e.g., changing a few pixels can classify a panda as a vulture. Similar huge changes can be provoked when a neural network processes data.

A lot of research has been performed to minimize this undesired phenomenon. Even when this misclassification effect is still present, the resulting neural network-based system can still be very useful. For example, if we have a neural-based system that successfully detects cancer based on an X-ray more frequently than a human doctor (e.g., in 80% of the cases instead of human-doctor's 70%), this is a great application, even though in 20% of the cases, the system still makes a wrong conclusion.

However, in situations — typical in engineering — when we want guaranteed bounds on the uncertainty of the result of data processing, we cannot use neural networks exhibiting such small-change-in-inputs-causing-huge-changes-in-output effect. In such situations, we need to restrict ourselves to neural networks for which small changes of inputs lead to reasonably small changes in the output.

What we do in this paper. In this paper, we provide possible solutions to both challenges.

2. What we propose: fuzzy case

Idea. For fuzzy techniques, we can use the fact that all input fuzzy sets usually have similar shapes: e.g., they are all triangular or they are all trapezoid.

Case of triangular membership functions. A (symmetric) triangular membership function $\mu_i(x_i)$ is different from 0 on some interval $[\underline{x}_i, \overline{x}_i] = [\widetilde{x}_i - \Delta_i, \widetilde{x}_i + \Delta_i]$, in which its value is equal:

- to $\mu_i(x_i) = (\overline{x}_i - x_i)/\Delta_i$ for $x_i \in [\widetilde{x}_i, \overline{x}_i]$,
- to $\mu_i(x_i) = (x_i - \underline{x}_i)/\Delta_i$ for $x_i \in [\underline{x}_i, \widetilde{x}_i]$, and
- to 0 when $x_i \notin [\underline{x}_i, \overline{x}_i]$.

For this membership function, the half-width $\Delta_i(\alpha)$ of the α-cut interval is equal to $(1 - \alpha) \cdot \Delta_i$. By plugging in this expression into the formula (3), we conclude that

$$\Delta(\alpha) = \sum_{i=1}^{n} |f_{,x_i}| \cdot (1 - \alpha) \cdot \Delta_i = (1 - \alpha) \cdot \sum_{i=1}^{n} |f_{,x_i}| \cdot \Delta_i.$$

So, in this case, there is no need to apply the formula (3) 11 times:

- we can apply the formula (3) once, for $\alpha = 0$, when $\Delta_i(0) = \Delta_i$, and
- then, to find the values $\Delta(\alpha)$ corresponding to all other α, we can simply multiply $\Delta(0)$ by $1 - \alpha$.

Case of trapezoid membership functions. Similarly, a (symmetric) trapezoid membership function $\mu_i(x_i)$ is different from 0 on some interval $[\underline{x}_i, \overline{x}_i] = [\widetilde{x}_i - \Delta_i^{(1)} - \Delta^{(2)}, \widetilde{x}_i + \Delta_i^{(1)} + \Delta_i^{(2)}]$, in which its value is equal:

- to $\mu_i(x_i) = 1$ when $x_i \in [\widetilde{x}_i - \Delta_i^{(1)}, \widetilde{x}_i + \Delta_i^{(1)}]$,
- to $\mu_i(x_i) = (\overline{x}_i - x_i)/\Delta_i^{(2)}$ for $x_i \in [\widetilde{x}_i + \Delta_i^{(1)}, \overline{x}_i]$,
- to $\mu_i(x_i) = (x_i - \underline{x}_i)/\Delta_i^{(2)}$ for $x_i \in [\underline{x}_i, \widetilde{x}_i - \Delta_i^{(1)}]$, and
- to 0 when $x_i \notin [\underline{x}_i, \overline{x}_i]$.

For this membership function, the half-width $\Delta_i(\alpha)$ of the α-cut interval is equal to $\Delta_i^{(1)} + (1 - \alpha) \cdot \Delta_i^{(2)}$. By plugging in this expression into the formula (3), we conclude that

$$\Delta(\alpha) = \sum_{i=1}^{n} |f_{,x_i}| \cdot [\Delta_i^{(1)} + (1-\alpha) \cdot \Delta_i^{(2)}] = \sum_{i=1}^{n} |f_{,x_i}| \cdot \Delta_i^{(1)} + (1-\alpha) \cdot \sum_{i=1}^{n} |f_{,x_i}| \cdot \Delta_i^{(2)}.$$

So, in this case, there is no need to apply the formula (3) 11 times:

- we can apply the formula (3) twice, first for the values $\Delta_i^{(1)}$, and then for the values $\Delta_i^{(2)}$, resulting in the values $\Delta^{(1)}$ and $\Delta^{(2)}$;
- then, to find the values $\Delta(\alpha)$ corresponding to all other α, we can simply compute $\Delta^{(1)} + (1-\alpha)\cdot\Delta^{(2)}$.

More general case. We can have other cases in which the half-width $\Delta(\alpha)$ of the α-cut interval has the form

$$\Delta_i(\alpha) = \sum_{j=1}^{k} f_j(\alpha)\cdot\Delta_i^{(j)}, \tag{6}$$

for some $k \ll 11$. Here, the functions $f_j(\alpha)$ are common for all membership functions of this type, while the values $\Delta_i^{(j)}$ depend on the specific membership function from this family. For example:

- for triangular membership functions, we have $k = 1$ and $f_1(\alpha) = 1-\alpha$;
- for trapezoid membership functions, we have $k = 2$, $f_1(\alpha) = 1$, and $f_2(\alpha) = 1-\alpha$.

In general, substituting the expression (6) into the formula (3), we get

$$\Delta(\alpha) = \sum_{j=1}^{k} f_j(\alpha)\cdot\left(\sum_{i=1}^{n} |f_{,x_i}|\cdot\Delta_i^{(j)}\right).$$

Here too, there is no need to apply the formula (3) 11 times:

- we can apply the formula (3) k times, each time $j = 1,\ldots,k$ for the values $\Delta_i^{(j)}$, resulting in the values $\Delta^{(1)},\ldots,\Delta^{(k)}$;
- then, to find the values $\Delta(\alpha)$ corresponding to all other α, we can simply multiply compute

$$\Delta(\alpha) = \sum_{j=1}^{k} f_j(\alpha)\cdot\Delta^{(j)}.$$

Thus, we arrive at the following algorithm.

Formulation of the problem. We know:

- the data processing algorithm $f(x_1,\ldots,x_n)$;
- the measurement results $\widetilde{x}_1,\ldots,\widetilde{x}_n$; and

- for different values $\alpha \in [0, 1]$, the half-widths $\Delta_1(\alpha), \ldots, \Delta_n(\alpha)$ of the corresponding *alpha*-cut intervals, for which, for some small value k and for some known functions $f_1(\alpha), \ldots, f_k(\alpha)$, we have

$$\Delta_i(\alpha) = \sum_{j=1}^{k} f_j(\alpha) \cdot \Delta_i^{(j)},$$

for some values $\Delta_i^{(j)}$.

Based on this information, we need to compute the half-width $\Delta(\alpha)$ of the α-cut intervals for y.

Resulting algorithm.

- First, for each j from 1 to k, we compute

$$\Delta^{(j)} = \sum_{i=1}^{n} |f_{,x_i}| \cdot \Delta_i^{(j)}.$$

- Then, for each α, we compute

$$\Delta(\alpha) = \sum_{j=1}^{k} f_j(\alpha) \cdot \Delta^{(j)}.$$

Discussion. By using this algorithm, instead of 11 applications of the time-consuming formula (3), we only apply it $k \ll 11$ times:

- $k = 1$ time for triangular membership functions,
- $k = 2$ times for trapezoid membership functions, etc.

Thus, we get a significant 5–10 times speedup.

3. What we propose: neural case

Idea. For neural data processing, we can use the fact that:

- when a neural network is trained,
- it already computes a lot of partial derivatives.

Our idea us to use these partial derivatives to estimate the desired derivatives $f_{,x_i} = y_{,x_i}$. To explain our idea, let us recall how neural networks process data and how they are trained.

How neural networks process data: a brief reminder. A neural network consists of several layers.

- First, the inputs $x_1, \ldots, x_n$ are fed to the neurons from the first layer. Based on these inputs, each of these neurons produces the following output:

$$y_j = s\left(\sum_{i=1}^{n} w_{ji} \cdot x_i + w_{j0}\right), \tag{7}$$

  where w_{ji} are constants called *weights*, and $s(z)$ is nonlinear function called an *activation function*.
- The outputs of the first layer then serve as inputs for the second layer, whose neuron perform similar computations.
- The outputs of the second layer become inputs for the their layer, etc.
- In general, outputs of each layer serve as inputs for the next layer.

The output of the neuron in the last layer is then returned as the result y of data processing. This whole process is known as *forward propagation*.

How neural networks are trained: a brief reminder. Training the neural network means finding the values of all the weights for which, for each inputs for which we know the desired output $\widetilde{y}$, the value y generated by the neural network is close to this desired output.

To train the neural network, we need to select an appropriate measure of distance $J(y)$ between y and $\widetilde{y}$ — e.g., $J(y) = (y - \widetilde{y})^2$ — that the training will minimize. To minimize this measure, neural networks use gradient descent, i.e., they update each weight w by using the gradient descent formula:

$$w \mapsto w - \beta \cdot J_{,w},$$

where $J_{,w}$ denotes the partial derivatives of the objective function J with respect to the weight w. A special efficient algorithm — known as *back-propagation* — is used to compute these partial derivatives $J_{,w}$.

As intermediate results in these computations:

- we compute the values $s'\left(\sum\limits_{i=1}^{n} w_{ji} \cdot x_i + w_{j0}\right)$, where $s'(z)$, as usual, means the derivative of the function $s(z)$; we will denote these values by y'_j; and
- we compute the derivatives $J_{,y_j}$ of the objective function J with respect to all the intermediate results y_j.

Once the network is trained, we freeze the weights and use the resulting fixed-weights network to process the data.

Comment. From the computational viewpoint, backpropagation is, in effect, a particular case of a reverse differentiation algorithm well known in automatic differentiation; see, e.g.,[1,3].

Mathematical analysis of the problem. Let us show that, based on the partial derivatives computed on the backpropagation stage, we can compute the partial derivatives $f_{,x_i}$ — that, as we have mentioned, can be used to propagate uncertainty.

Indeed, as part of the usual training-related computations, the neural network computes the values $J_{,y_j}$, where y_j are outputs of the first layer.

From the general formulas relating partial derivatives, we can conclude that $J_{,y_j} = J_{,y} \cdot f_{,y_j}$. Thus, if we know the values $J_{,y_j}$, we can compute the derivatives $y_{,y_j}$ as the ratios $y_{,y_j} = J_{,y_j}/J_{,y}$.

Also, from the general formulas relating partial derivatives, we can conclude that $f_{,x_i} = \sum\limits_j y_{,y_j} \cdot y_{j,x_i}$. From the formula (7) describing the dependence of y_j on the inputs x_i, we conclude that $y_{j,x_j} = y'_j \cdot w_{ji}$. Substituting this expression for y_{j,x_j} into the above formula for $f_{,x_i}$, we conclude that

$$f_{,x_i} = \sum_j \frac{J_{,y_j}}{J_{,y}} \cdot y'_j \cdot w_{ji}. \tag{8}$$

Thus, it is reasonable to come up with the following algorithm.

Formulation of the problem: reminder. We know:

- the function $f(x_1, \ldots, x_n)$ computed by a trained neural network; and
- the measurement results $\widetilde{x}_1, \ldots, \widetilde{x}_n$.

We want to compute the partial derivatives $f_{,x_i}$ of the function $f(x_1, \ldots, x_n)$ with respect to all its inputs $x_1, \ldots, x_n$. Once we know these derivatives, we can use the formulas (2) or (3) to propagate uncertainty.

Resulting algorithm. Once we have the inputs $x_1, \ldots, x_n$, we first apply the data-processing neural network to these inputs, i.e., we apply forward propagation to the inputs — and get the result y of data processing.

Then, we pick some reasonable value $\widetilde{y}$ and form the corresponding objective function $J(y)$. For this objective function, we apply one back-propagation step, on which we compute all the partial derivatives $J_{,w}$ — but, in contrast to real back-propagation, we do not change the weights.

From the neural network, we know all the weights w_{ji}. During the backpropagation step, we have computed:

- the values $J_{,y_j}$ — the partial derivatives of the objective function with respect to the outputs y_j of the first data processing layer, and
- the values $y'_j = s'\left(\sum_{i=1}^{n} w_{ji} \cdot x_i + w_{j0}\right)$ corresponding to this layer.

Since we know the objective function $J(y)$, so we can easily compute the derivative $J_{,y}$. Then, we can compute the desired partial derivatives $f_{,x_i}$ by using the formula (8). Once we know these derivatives, we can use the formulas (2) and/or (3) to propagate uncertainty.

Discussion. As we have mentioned, for data processing algorithms $f(x_1, \ldots, x_n)$ that require a large amount of computations — in particular, for data processing performed by a neural network — the most time-consuming time-consuming part of uncertainty estimation is applying this algorithm f.

In the usual approach to uncertainty propagation, to compute the value σ and/or Δ describing the uncertainty of the result y of processing data $x_1, \ldots, x_n$, we need to apply the neural network $n+1$ times. In contrast, if we apply the proposed algorithm, we need to apply the neural network only one more time: in addition to the forward propagation step computing the result of data processing, we also perform one backpropagation step. Thus, the number of applications of the neural network decreases from $n+1$ to 2. As a result, the computation time needed to propagate uncertainty decreases in the same proportion. For example:

- when we process $n = 10$ inputs, we reduce the number of applications of a neural network from 11 to 2 — i.e., we get a 5 times speedup, similar to the speedup in the fuzzy case;
- when we process $n = 100$ inputs, we reduce the number of applications of a neural network from 101 to 2 — i.e., we get a 50 times speedup, much larger than in the fuzzy case.

Acknowledgements

This work was supported in part by the National Science Foundation grants 1623190 (A Model of Change for Preparing a New Generation for Professional Practice in Computer Science), HRD-1834620 and HRD-2034030 (CAHSI Includes), EAR-2225395 (Center for Collective Impact in Earthquake Science C-CIES), and by the AT&T Fellowship in Information Technology.

It was also supported by the program of the development of the Scientific-Educational Mathematical Center of Volga Federal District No. 075-02-2020-1478, and by a grant from the Hungarian National Research, Development and Innovation Office (NRDI).

The authors are greatly thankful to Franco Pavese for his encouragement, to all the participants of the International Conference on Advanced Mathematical Tools in Metrology and Testing AMCTM 2023 (September 26–28, 2023) for valuable discussions, and to the anonymous referees for valuable suggestions.

References

1. A. G. Baydin, B. A. Pearlmutter, A. A. Radul and J. M. Siskind, *Automatic Differentiation in Machine Learning: A Survey*, arXiv:1502.05767, 2015.
2. R. Belohlavek, J. W. Dauben and G. J. Klir, *Fuzzy Logic and Mathematics: A Historical Perspective*, Oxford University Press, New York, 2017.
3. I. Goodfellow, Y. Bengio and A. Courville, *Deep Learning*, MIT Press, Cambridge, Massachusetts, 2016.
4. I. J. Goodfellow, J. Shlens and C. Szegedy, *Explaining and Harnessing Adversarial Examples*, arXiv:1412.6572, 2014.
5. L. Jaulin, M. Kiefer, O. Didrit and E. Walter, *Applied Interval Analysis, with Examples in Parameter and State Estimation, Robust Control, and Robotics*, Springer, London, 2012.
6. G. Klir and B. Yuan, *Fuzzy Sets and Fuzzy Logic*, Prentice Hall, Upper Saddle River, New Jersey, 1995.
7. V. Kreinovich and S. Ferson, "A new Cauchy-based black-box technique for uncertainty in risk analysis", *Reliability Engineering and Systems Safety*, 2004, Vol. 85, No. 1–3, pp. 267–279.
8. B. J. Kubica, *Interval Methods for Solving Nonlinear Constraint Satisfaction, Optimization, and Similar Problems: From Inequalities Systems to Game Solutions*, Springer, Cham, Switzerland, 2019.
9. G. Mayer, *Interval Analysis and Automatic Result Verification*, de Gruyter, Berlin, 2017.
10. J. M. Mendel, *Uncertain Rule-Based Fuzzy Systems: Introduction and New Directions*, Springer, Cham, Switzerland, 2017.
11. R. E. Moore, R. B. Kearfott and M. J. Cloud, *Introduction to Interval Analysis*, SIAM, Philadelphia, 2009.

12. H. T. Nguyen, C. L. Walker and E. A. Walker, *A First Course in Fuzzy Logic*, Chapman and Hall/CRC, Boca Raton, Florida, 2019.
13. V. Novák, I. Perfilieva and J. Močkoř, *Mathematical Principles of Fuzzy Logic*, Kluwer, Boston, Dordrecht, 1999.
14. S. G. Rabinovich, *Measurement Errors and Uncertainty: Theory and Practice*, Springer Verlag, New York, 2005.
15. D. J. Sheskin, *Handbook of Parametric and Nonparametric Statistical Procedures*, Chapman and Hall/CRC, Boca Raton, Florida, 2011.
16. L. A. Zadeh, "Fuzzy sets", *Information and Control*, 1965, Vol. 8, pp. 338–353.

F Pavese, A Bošnjaković, S Eichstädt, A B Forbes, J Alves e Sousa (eds.):
Advanced Mathematical and Computational Tools in Metrology and Testing XIII
Series on Advances in Mathematics for Applied Sciences, Vol. 94

https://doi.org/10.1142/9789819800674_0009

Big Data and dataism: Some metrological reflections

F. Pavese†
Independent Scientist
Torino, Italy
E-mail: frpavese@gmail.com

Two worldwide events opened novel reflections in a highly vast scientific literature concerning a universal concept like that of “data”, namely as the results of observations, either experimental or human-mind generated: the “Big Data” and the universal use of informatics, related to each other and both in an exponential increase.

The capacity and speed of modern computers allow us to obtain such immense amounts of data, both from experimental setups or from the elaboration of algorithms, either human-built or through AI. Their handling too is necessarily operated via informatics means, considered distinct from mathematical means. Apparently, these facts have disconnected data evaluation from traditional fields, not only science but also the way to make decisions based on them.

This paper contains reflections on the properties and meaning of the data in themselves, specifically as the vehicle of information almost universally necessary to make decisions, the latter being a frame that is mixed and often prevalent in the literature about big data.

Keywords: data; Big Data; dataism; informatics; measurement science; metrology.

1. Introduction

Two worldwide events opened novel reflections in a highly vast scientific literature concerning a universal concept like that of “data”, namely as the results of observations, either experimental or human-mind generated: the “Big Data” and the universal use of informatics, related to each other and both in an exponential increase.

The capacity and speed of modern computers allowed us to obtain such immense amounts of data, both from experimental setups or from the elaboration of algorithms, either human-built or through artificial intelligence (AI). Their handling too is necessarily operated via informatics means,

---

†Former Research Director in Metrology at the National Research Council of Italy (IMGC-CNR; then INRiM from 2006).

considered distinct from mathematical means. Apparently, these facts have disconnected data evaluation from traditional fields, not only of science but also in the way to *make decisions* based on them [1].

This paper contains reflections on the properties and meaning of (not only numerical) data in themselves, as also the vehicle of information almost universally necessary to *make decisions*, a frame that is often prevalent in the literature about Big Data.

In particular, it tackles the issue of the concept of "uncertainty" in some of the new frames, necessarily related to the "quality" of the obtained information.

2. From a single datum to the Big Data

The reflections in this paper are restricted to data arising from the information acquired from the "external" world, i.e. each datum will be considered here to be the exclusive elementary component of human *knowledge* in its (numerical) form. In Sec. 3, its distinct origin is considered, to be the result of a mind reflection generating, e.g., scientific theories, the above two being the basic frames of our (limited) knowledge.

A quantitative datum, obtained from the observation of the external world, is commonly considered a "fact", meaning an *objective* piece of *information* — further, most often expressed by a number — as opposed to a mind reflection, taken as *subjective*. The interested reader is directed to a previous author's paper on the limits of this distinction [2].

Each datum is an instance of a "series", which can be of different nature and kind according to a well-established set of classifications. [3] The first kind is typically obtained "experimentally", i.e. from an "observation"/"vote" expressed most often — exclusively in the following — in numerical form. These data are treated by a discipline called "measurement science" setting "a process intended to share common ways to transmit the knowledge to the Community, not limited to a single generation of scientists and practitioners, and is intended to obtain from it the necessary consensus". [4] Different tools are used, one of the foremost being statistics.

The definitions of Big Data found in the scientific literature are various since the 1990s, [5–7] and not much more informative than the one chosen here by the Author:

> Big Data are series of data [*in the principle of both the above origins*] so much more extensive than the "historical series" that they can only be treated and analyzed by employing informatics, primarily since the

available modern ones are exceptionally performing from the point of view of speed and memory capacity.

As a consequence, Author's position — not in contrast with literature definitions [5–7] — is that they are synonyms of "a large amount of data", with the *additional* characteristics to be *so large* that they can only/also be elaborated by informatics means, informatics is a discipline that in many respects is presently considered by many of its users as distinct from measurement science. Such a distinction is certainly true as to the technicalities and the apparatuses. Additionally, it is a fact that most of the activities of experimental science are today based on computers.

In addition, Big Data apparently is a collation of *raw data, raw* implies that no specific "qualification" is associated with each datum, but only the source of the complete series is provided. A consequence, in what one can find in the literature, is, e.g., the fact that the discussion on the Big Data is limited to a mere analysis of the dataset, very seldom on their "quality" (see "dataism" later).

3. Data from models vs. empirical data

With the rapid diffusion of informatics, the use of *models* operated by computers has rapidly grown — becoming one of the preferred tools in Internet socials for "informing" people. This preference for *models* has pushed an increasing number of scientists — namely economists — to exercise their use. However, that is a field potentially risky, namely because in it the limit between science and economics/politics is almost invisible and certainly quite uncertain.

Typically, "*data* is interpreted and processed by *a priori* established models", [8] assumed to be based on them. On the other hand, the models might, conversely, also be considered an important generator of big data, especially when concerning simulation: a (generally) mathematical modelling of a phenomenon. Conversely, modelling can be directly based on existing databases (extended or not) and is customarily used for the *prediction of future behaviour* of an observed phenomenon. However, since prediction (forecast) does not usually make use of random types of mathematical representations but of deterministic functions, the future trends obtained from these functions cannot be considered independent datasets.

Instead, the big datasets of empirical data tend to even replace analytical models today. Due to that, "the predictive capabilities of such models entirely lie in data itself" [8] a form of *dataism* (see later). In this respect, such a tendency concerning *prediction* often masks a critical misunderstanding about the meaning and the use of data [9].

In fact, the expected *validity duration of a prediction* depends first of all on the observed and estimated "law(s)": in the case of mechanics, e.g., of the orbit of the sky big bodies like the Earth, a forecast can confidently be provided for extremely long periods. The same level of confidence is not available for thermodynamics in physics, basically dominating natural events, or for economics [1].

Problems may arise from the fact that no (set of) mathematical function used as a model is indefinitely "flexible", i.e. apt to "correctly" interpolate *any* cluster of data — namely the less when less are the parameters of the function(s). There is a vast literature on this subject matter, for example, to detect a "changing point" in a trend. Fitting is considered a good technique when is merely a reasonable balance between the best "copying" of behaviour with time (like when the function has to match a given profile) and a satisfactory "averaging" of the behaviour supplied by experimental values — and the more extended is the period and without "masking" of changing points.

As a consequence, usually, a sort of balance should always exist between the *numerosity* of data *already* available — plus the length of the period during which they were taken — and those collected in a future period. Only in that case the function so obtained is apt to be "safely" extrapolated and considered to provide a sufficiently good forecast — i.e., by remaining accurate without the support of any constraints, except those (purely mathematical) set by the function. Big Data could partially compensate for their more extensive number of data, but *only* when related to a short extension in time of the extrapolation.

For example, if the "safe" *observation* time is considered to be 30 years, it is hardly possible to assume any sensible extrapolation to a further period of the same length — the shorter is the reference period. In fact, the more rapidly increasing (or decreasing) of the function derivatives — or non-simple shapes of the function — the more the variation in the extrapolated period becomes problematic.

Further, a qualification that is often absent in the information supplied together with the extrapolations is the *quality of the available dataset*. Instead, the existing data *uncertainty* — whose information must always be provided as it determines *quality* — must be taken into consideration, since the *quality* of the fitting on the available data is vital for the *quality* of the subsequent extrapolation. Again, the *size* of the Big Data series *has in itself no influence* on data precision. There are even instances where the original data quality is so low that is already sufficient to make one understand that their fit would be unreliable and the extrapolation meaningless.[1]

Accordingly, the forecast almost always consists of a *trend area* (typically increasing its width with time) where the future determinations are assumed to fall within an *assigned* probability. Most often, the trend is monotonic because usually *changing points cannot be predicted*, but, in a few cases, the extrapolation may also show a change in sign of the first or/and of the second derivative — e.g., for trend causes foreseen to become exhausted, or instead be born in future.

4. Dataism vs. measurement science

The existence of Big Data induced the creation of a new frame called "dataism" by the New York Times journalist who created it in 2013, David Brooks. [10, 11] Subsequently, a discussion started, especially on Internet sites, considering it as a replacement for "humanism" or even thinking that a scientific revolution has taken place. Also in the frame of philosophy of science, the issue raised some interest regarding the concept of "measurement" when "data science" is pretended to replace measurement science [5, 12].

According to [12], dataism: (italics added)

"– perceives the entire world as a flow of data;
- believes that data provide a fair and exhaustive representation of *reality*;
- has unconditioned confidence in data and bases their everyday judgments *only on data*;
- believes that *artificial intelligence* will overcome human intellect".

It looks like a new form of "objectivism" (the belief that certain things/situations exist independently of human knowledge or perception of them), certainly contrasting (current) modern science/economic theory.

Even without reaching such an extreme, in the field of data expressed in numerical form, one can observe in the scientific/economic literature how the information about the *origin* of the data is more and more omitted. In other cases, it went lost in procedures based on informatics, even though with the consciousness of the importance, in particular, of the fact that each dataset has a

[1]Frequently, supplying the results from more than one model is preferred, as a multiplicity can be able to allow an indirect evaluation of the possible variability of the forecast. The comparison of models can certainly mitigate the risk of false extrapolations, if made with different fitting (set of) equations — and of different complexity — on the same data, but places other problems — such as the choice of the "best" model, and the fact that all are, in general, deterministic function.

quality that must remain explicit. However, what is frequently lost is the fact that *uncertainty* is associated with *each* datum *since its first record*, influencing the whole dataset. Then, in order to be reliable, each datum should have been obtained in strict compliance with the rules established by measurement science.

5. Metrological analysis of Big Data

The notable and peculiar issue in the Big Data literature is that, in the discussion about the set, terms are used based on several disciplines, e.g. those intended for decision-taking more than that of measurement science.

Another peculiar characteristic, found in the elaborations of Big Data, was that, basically, only the *dispersion* of the values with respect to some kind of smoothing or fit is reported to characterise the "quality" of the set — and/or to discuss the role of dataset quality in respect to *their* use [13].

However, dataism should not mean that the Big Data analysis could start from the recorded numerical values, omitting first the analysis of their *quality* and assigning to each group or set of them a *degree of confidence* — and an analysis of their uncertainty components bringing to the overall quality. Increasing the number, or the frequency, of observations, cannot in itself be sufficient for mitigating uncertainty. The increase in the number of observations or their acquisition frequency does not ensure an increase in the *quality* of the collected information.

Therefore, dataism neither can mean an overestimation of the importance of basing science (and suggestions) on data, partially including their use in the economic discipline: they are to be treated as *partial* knowledge, though often the most reliable available but often also far from being univocally meaningful. They cannot replace theory [13].

In fact, the information content of data, the only important one, should be evaluated in terms of its "quality", a property that is *not intrinsic* in any data but that *has to be verified.* According to [8], not specifically referring to Big Data, (italics added):

"The dimensions of *information quality* include (and is valid for any discipline):

1. *Consistency*: the condition that data is within the assumed value domain and is not duplicated;
2. *Availability*: the fraction of time that data is made available by the system that stores it;
3. *Currency and timeliness*: the degree to which data is updated and readily available for use, respectively;

4. *Specificity*: a condition related to the quantity of syntactical information: stating, e.g., that a length is in the interval (10.5 ± 0.1) m is more specific, and therefore of better quality, than stating that it is in the interval (10 ± 1) m; when referring to measurement results, specificity is also called *precision*;
5. *Trueness*: a condition related to the faithfulness of semantic information: were, e.g., 10.55 m the value of a length provided by the best independent method, stating that the length is 10.50 m is truer, and therefore of better quality, than stating that it is 10.40 m as synthesized in terms of *accuracy* in metrology" (for the definition of the metrological terms used above, see [3, 14]).

In the Big-Data literature, *almost all the above scientific terms*, decorating the meaning of the terms "uncertainty" and "quality" and still commonly used in it though often related only to the decisions to be taken, *are lacking*. This fact is sufficient in itself to qualify its present metrological quality.

6. Example of a possible "dataism-syndrome" in big data analyses and use

Measurement science is *multidisciplinary* so many of its scientific evaluations can be performed irrespective of their specific kind or origin.

One field where immense data collations exist today is the one concerning Earth parameters, especially in the current period where the climate has become a dominant and critical activity of vast Communities of scientists. They may be considered Big Data. It mainly involves the disciplines of physics and chemistry — meteorology and environment being part of them — but many of the relevant observed effects then need decisions of a political and economic nature. An example is taken in the following from a field that the author recently has already partially explored from his metrologist's point of view. It is the Global Mean Surface Temperature (GMST) parameter — chosen here for temperature metrology being one of the specific professional expertise of the Author [15, 16].

The GMST is probably the most popular climate *parameter* that is commonly used to support decisions recommended by International Organizations, namely the IPCC, HadCRUT, etc. It is computed from the data obtained by the meteorological stations of the WMO network, amounting to millions, whose distribution on the Earth's surface is however quite non-homogeneous [17]. The resulting database can be considered a Big Data one consisting of the numerical values of surface air temperature recorded according to the WMO protocol for the meteorological stations.

In the following, only a few peculiarities are summarised, being already more extensively analysed and discussed in [18]. To the Author, they look like an example of symptoms of dataism applied to Big Data treatment.

The only direct statistical information that is supplied in the literature for Big Data is obtained from the *fitting* of datasets with a suitable function to obtain the trend for a period: the obtained information basically consists of the *dispersion* of the data values (evaluated via the standard deviation or similar parameters). However, that is *not the uncertainty* of the dataset, but only one of the *several uncertainty components* of the collected data arising from several data manipulations: sampling, interpolation where data are lacking, normalization, adaptation, smoothing, homogenization, etc.

It is scholarly known that, with sufficient and competent statistical methods, it is possible to get an estimate of a set of sparse values having a resulting much lower dispersion and higher coherence/consistency. However, each of the required many manipulations (not only methodological/computational, though today computer-based, requiring *choices*) and assumptions, add components to the final uncertainty: each and all contribute to the *overall* uncertainty.

In this case, one has even billions of data, produced by temperature sensors (of medium-low quality, of both of the contact or radiation types) affected by an original uncertainty and all placed in *different and distant locations* on the Earth. Nobody is metrologically able to mitigate or reduce their *original* uncertainty, arising from *calibration/traceability* (two pillar procedures of measuring science and of WMO) and from the measuring-system uncertainties.

Having studied the literature and examined some of those databases [14], no evidence of application of the basic metrological requirements was found, *namely being supported by a published Budget of Uncertainty*, applied to the immense work done by the hundreds of authors involved in that frame.

As an example of the published treatment of this Big Data set, in Fig. 1, taken from [1], the GMST record from 1850 is reported. Its trend is not a simple one even not considering the clear effect of the II World War, and became an effective increase basically only after ≈1970 (incidentally, with the human population increasing from 2 to 8 billion since the year ≈1950). The published value of the standard deviation (s.d.) for a confidence level of 97.5% for the 1970–2021 period fit is ±0.12 °C: the Author of this paper got, for the same data and again for the 97.5% confidence level, a s.d. of ±0.27 °C. For the GMST values computed for the IX century, 1850–1900, the reported uncertainty is 0.5–0.7 °C while the Author's fit of the same data in the same period and for the GMST change of –0.32 °C, brings to ±0.36 °C. These discrepancies are facts whose origin remains unknown but make the reliability of the original data

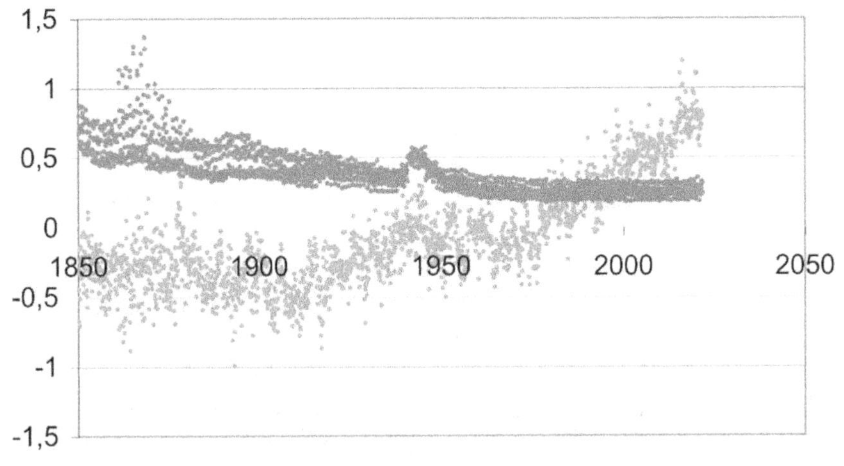

Fig. 1. GMST *vs* increased behaviour (lower set of dots) between the year 1850 and 2021 according to HadCRUT (database downloaded and analyzed by the Author). [20] The higher set of dots reports the upper and lower values of the standard deviation interval for a confidence level of 97.5%. Notice the clear effect of the World War II.

treatment insufficient. Note that the above reported s.d. values only mean *reproducibility*, *not* the (quite smaller) *precision*, of the GMST trend.

A similar attitude concerning the *reliability of the published computations and forecasts* was found concerning other climate parameters [18]: the interested reader can find the Author's findings about the *melting of the ice* on the Earth's surface on [15], and a discussion about the future increase of the mean water level in the oceans on this book in [18], Figs. 4(a)–4(c).

7. Final remarks

The massive increase of data brought to the very extended databases called Big Data forced the users to concentrate on the informatics methods concerning not only the storage and retrieval but also the complementing of the traditional tools used in science, for qualifying and making use of the information that they supply by means of totally informatics ones. Actually, most of the literature on Big Data comes just from the informatics discipline and from Institutions dedicated to it, instead of metrological Institutions.

That feature favoured the expansion of the use of terms of the latter discipline, in most cases replacing the traditional measurement science. The critical issue put in evidence in this paper is that, in too many cases, the scientific terms used in the latter discipline, which define the way a measurement can be conducted and then analyzed in a scientifically reliable way, are ignored, or at best omitted.

That way of dealing with data does not only involve and affect the *credibility* of the analyses performed on the data, which is, in most cases, the main reason for the analyses. Decisions following the analyses should obviously be as much as possible based on credible information, but the latter is maximised only by increasing the reliability of the considered data [18].

Data do not fit a "purpose" but a need, also in the case of Big Data. The need is not to prove a position but to reach an understanding of a situation or the reason for the need. They are supposed to *describe*, *support* or *disclose* a situation or a scenario needing further scientific exploration.

Dataism assumes that data directly *originate from reality*: this is a *too strong* assumption, as political inference can instead be dominant with respect to "neutral" technical aspects. Trust in data cannot turn into a kind of "faith" in data: faith is not science, nor can "true" value be taken as a reference.

The above is a too frequent misunderstanding that arises from blind confidence in the data as *objective facts* — when not even replacing scientific procedures as dataism does. However, a lack of compliance with the rules of science cannot be mitigated *by* the numerosity of the available data — often in a short time.

Data uncertainty is not ignored, but the discussion on it, in general, omits resorting to the most important fundamental parameters of measurement science. In a word, it is enough to summarize the situation with the *fact* of the total absence of a published quantitative "uncertainty budget" [3].

8. Conclusions

The present Big-Data frame has no reason in itself for derogating from the current context/foundation of science. However, it could make easier false manipulations if not correctly and strictly handled according to scientific standards. In that respect, big data does *not* deserve a higher authority with respect to previous critically-formed datasets of more limited extensions. They are a mere technicality. *Big* does not mean necessarily *better or more powerful* with respect to *qualified* information.

References

1. F. Pavese, Decisions at the time of Big Data, Global Journal of Science: Frontier Research (A) 23 (7) 1–7 (2023) Print ISSN: 0975-5896.
2. F. Pavese, On the degree of objectivity of uncertainty evaluation in metrology and testing, *Measurement* 2009 **42** 1297–1303.
3. *Vocabulaire international de métrologie* – Concepts fondamentaux et généraux et termes associés, 3rd edition (VIM3), 2012, BIPM, Sèvres, France.
4. F. Pavese: "Comment to: L. Mari 'Is our understanding of measurement evolving?' ", *Acta Imeko* ISSN: 2221-870X, December 2021, **10** (4), 209–213, *Acta Imeko* 2022 **11** (2) 1–3.
5. Hilbert, Martin; López, Priscila (2011). "The World's Technological Capacity to Store, Communicate, and Compute Information". Science. **332** (6025): 60–65. Bibcode: 2011Sci...332...60H. doi:10.1126/science.1200970.
6. Wikipedia, Big Data. Visited on December 14, 2023.
7. A. De Mauro, M. Greco, M. Grimaldi (2016), A Formal Definition of Big Data Based on its Essential Features, Library Review, **65** (3) pp. 122–135, doi: 10.1108/LR-06-2015-0061.
8. L. Mari, D. Petri, The Metrological Culture in the Context of Big Data: Managing Data-Driven Decision Confidence, IEEE Instrumentation and Measurement Magazine **20** (5), 4–20, 2017, doi:10.1109/MIM.2017.8036688.
9. F. Pavese, Uncertainty in the case of lack of knowledge: the case of data extrapolation, with an example on climate forecast models, Ukrainian Metrological Journal 2022 (3).
10. D. Brooks, New York Times Newspaper, Article 2013.
11. Y.N. Harari, Homo Deus: *A Brief History of Tomorrow*. London, UK: Vintage, Penguin Random House, 2016.
12. D. Petri, Big Data, Dataism and Measurement IEEE Instrumentation & Measurement Magazine May 2020, 1094-6969/20.
13. C. Anderson, The end of theory: the data deluge makes the scientific method obsolete, Wired, 2008. https://www.wired.com/2008/06/pb-theory.
14. ISO 5725-1:1994, *Accuracy (trueness and precision) of measurement methods and results* – Part 1: General principles and definitions, International Organization for Standardization, 1994.

15. F. Pavese, Graphic method for retrieval of quantitative data from computer-mapped qualitative information, with a NASA video as an example, 2020, ESIN **13** 655–662.
16. F. Pavese, The choice of cartographic system on the calculation of Earth's surface parameters from maps, namely of the GMST, 2023, ESIN 14, online 18 August 2023, https://doi.org/10.1007/s12145-023-01061-0.
17. https://databasin.org/maps/new/#datasets=de8f7f71e3334eba863ff6003484364f.
18. F. Pavese, From uncertain time-series data can one infer an extrapolated deterministic trend? Taking uncertainty correctly into account in (measurement) science, this book (AMCTM XIII), pp. 1–23.
19. Byung-chul Han, Im Schwarm. Ansichten des Digitalen, book, Matthes & Seitz, 2013.
20. C.P. Morice, J. J. Kennedy, N. A. Rayner, J. P. Winn, E. Hogan, R. E. Killick, R. J. H. Dunn, T. J. Osborn, P. D. Jones, I. R. Simpson, An Updated Assessment of Near-Surface Temperature Change From 1850: The HadCRUT5 Data Set, J. of Geophysical Research: Atmospheres **126** (3), 2021. e2019JD032361. doi: https://doi.org/10.1029/2019JD032361.

F Pavese, A Bošnjaković, S Eichstädt, A B Forbes, J Alves e Sousa (eds.):
Advanced Mathematical and Computational Tools in Metrology and Testing XIII
Series on Advances in Mathematics for Applied Sciences, Vol. 94

https://doi.org/10.1142/9789819800674_0010

Conceptual model for measurement service provision in support of a digital quality management system

R. M. White
National Research Council,
Ottawa, ON, K1A 0R6, Canada
E-mail: ryan.white@nrc-cnrc.gc.ca
https://orcid.org/0000-0003-3589-5900

Measurement service provision is an established process of measurement quality management systems, which includes the exchange of information between service providers and consumers. It is an informative use case for modeling digital systems for quality management. This paper describes a conceptual model of quality system processes that govern traceable measurement services. The model identifies metrology concepts, measurement processes, and their relationship with measurement data. The model demonstrates applicability for machine-interpretable digital calibration certificates.

Keywords: quality management system; digital transformation; metrology; measurement system; metrological traceability.

1. Introduction

The International Committee for Weights and Measures (CIPM) intends to develop and adopt a digital framework for the International System of Units (SI) [1]. The framework comprises digital services based on metadata representing measurement data. The metadata describes measurement values, units, and uncertainty. The framework may facilitate the exchange of calibration information between digital systems.

This paper looks beyond the scope of digitizing calibration information to the description of a conceptual model of metrologically traceable measurement service provision. This model identifies metrology concepts and their relationships with measurement data by representing quality processes that govern traceable measurement services. The description of those processes is a prerequisite to digitalization [2]. Digital systems that support the quality infrastructure described in Sec. 2 should be capable of managing the variety of processes that govern the quality of measurement* [3].

Digitalization efforts thus far rely on current practices that summarize details of complex processes in human-readable documentation to define requirements. This overlooks the prerequisite to have a description of processes that produce and consume measurement information. The measurement service provision process relies on contextual information describing measurands and measured quantities that give meaning to measurement data and descriptions of measurement standards, measurement procedures, and qualified technical staff to demonstrate competence. These are some of the details that may not be evident when relying on current practices to define digitalization requirements.

Robust digitalization requires the meaning of data items to be correct and unambiguous. Digital systems ability to manage and exchange data depends on automatic availability and accessibility [4]. Robust data use in digital systems requires correct and complete contextual knowledge of the underlying processes. The purpose of this paper is to model the measurement service process and identify the metrology concepts that give context to measurement information and describe quality processes as a prerequisite to digitalization in metrology.

This paper is as follows: Section 2 describes the international quality infrastructure. Section 3 presents a conceptual model for measurement service provision. Section 4 reviews existing metadata models. Section 5 discusses their application to the digital calibration certificate. This paper concludes with a summary and future work.

2. International quality infrastructure

The SI is a common system of measurement adopted globally. National Metrology Institutes (NMI) realize SI units and NMIs disseminate the SI through national and international networks [5]. These networks comprise the metrology pillar of the international quality infrastructure (IQI). The dissemination of the SI is made practicable by international mutual recognition of calibration and measurement capabilities (CMC) where the International Bureau of Weights and Measures (BIPM) provides the overall coordination of the arrangement under the authority of the CIPM [6].

NMI calibration and measurement capabilities (CMCs) are outcomes of the

*Ashby's law of requisite variety states that a system's capacity as a regulator cannot exceed its capacity as a channel for variety. A necessary, though not sufficient, condition for a regulator to exert anticipatory control by responding to its current input.

arrangement which are documented in the Key Comparison Database (KCDB) managed by the BIPM and made publicly available on the web [7]. In a similar manner, national accreditation bodies publish calibration and testing laboratory scopes of accreditation (SoAs). CMCs and SoAs are quality records demonstrating laboratory competence that conform to ISO/IEC 17025 [8], whereas calibration, testing, and analysis certificates record evidence of measurement quality. In certain cases, laboratories may reference their CMC or SoA in their certificates for quality assurance. These records allow for the quality infrastructure to function and measurement standards to flow from a national metrology institute down to the measurement consumer with evidence of metrological traceability and measurement quality. Digital transformation in metrology intends to exchange these records digitally, and the success of this transformation relies on the ability of digital systems to manage the quality processes as well.

3. Conceptual model for measurement service provision

Bharti *et al.* discussed shortcomings in measurement system models, in particular, existing models lack key metrological concepts that arise in the measurement process. The authors concluded that software models should represent the measurement system, the data flow of measurement results, and measurement uncertainty [11]. However, a model for measurement service provision requires additional metrological concepts to describe the quality system processes that govern the traceable measurement. Moreover, to model the measurement service provision process, the process should be modelled from different perspectives.

Three perspectives are considered, referred to as model views, in order to represent the differences that arise in measurement service process and identify the metrological concepts. The initial view defines the scope and relationship of quality management systems and describes the service provision process as a whole. The second and third views define complementary views from the consumer and provider perspective.

Figure 1 is a view of the conceptual model for the Digital Quality Management System (D-QMS) that represents the scope of measurement capabilities and their relationship. The model represents two entities: the measurement service provider and the measurement consumer. These entities have many-to-many relationships, representing organizations in the IQI described in Sec. 2. They also have different processes related to the

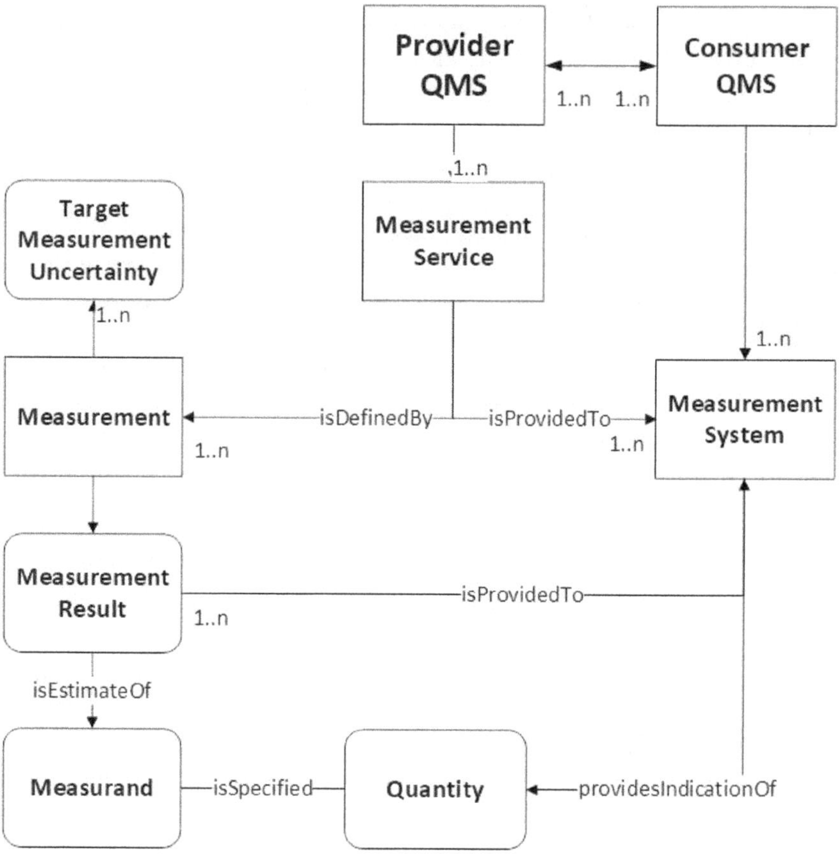

Fig. 1. Conceptual model for the digital quality management system. This view describes the scope and relationship of provider and consumer quality management systems.

measurement result that is the outcome of the service provision process. These different processes are views of the same D-QMS conceptual model.

The scope of a provider's QMS is the collection [1..n] of measurement services provided to consumers, and the scope of the consumer's QMS is the collection [1..n] of measurement systems. This represents consumers of measurement services that are governed by quality processes who intend to carry out measurements also governed by quality processes. This implies that the measurement consumer should use calibrated measurement systems in their processes, and that each consumer measurement system has [1..n] corresponding traceable measurement service(s) for calibration. In addition to the many-to-many relationship representing the IQI, measurement systems and measurement services have a many-to-many relationship that is established in the measurement process. When measurement uncertainty is considered in the data flow[†] between systems, metrological traceability is a property of the measurement process.

Traceable measurement services must specify the provider's target measurement uncertainty to establish a relationship between the provider's measurement service and consumer's measurement system when it falls below an acceptable threshold for the consumer's measurand. The degree of clarity and correctness of the measurand also affects how this relationship is determined because measurement "presupposes a description of the quantity commensurate with the intended use of a measurement result" [9, 2.1]. The measurand [9, 2.3 and 2.6] specification requires knowledge of the measured quantity, and the state of the measurement system(s) and measurement conditions.

Once the relationship is established, the service provider performs measurement and attributes quantity values to a measurand together with measurement uncertainty, exchanging the measurement result to that consumer so they can assess the quality of their measurement system. The indication provided by the consumer's measurement system does not constitute a measurement until the service provider's measurement result is part of their measurement procedure, resulting in distinct measurement stages governed by quality processes.

The consumer view of the service provision process (Fig. 2) describes the request for service and evaluation of the outcome. The measurement consumer should specify the measurand(s) corresponding to their measurement system and the acceptable upper threshold of measurement uncertainty. The measurand specification communicates the intent of the consumer. As represented in Fig. 1, the provider specifies the measurand which they attribute to the measured quantity value. The provider's specification of the measurand informs the consumer of the use of the measurement result during their measurement process. The consumer may also use this information to assess whether the result is fit for purpose. Once assessed that the measurement result meets the acceptable threshold of measurement uncertainty (*i.e.*, quality), the indication of

---

[†]Metrological traceability may also be defined as the property of the measurement process wherein there is an unbroken chain of measurement stages each producing a measurement result and contributing to measurement uncertainty [10]. Each measurement stage may be described with a measurement model that involves several input quantities. Each input quantity may be related to a measurement standard that has associated measurement uncertainties.

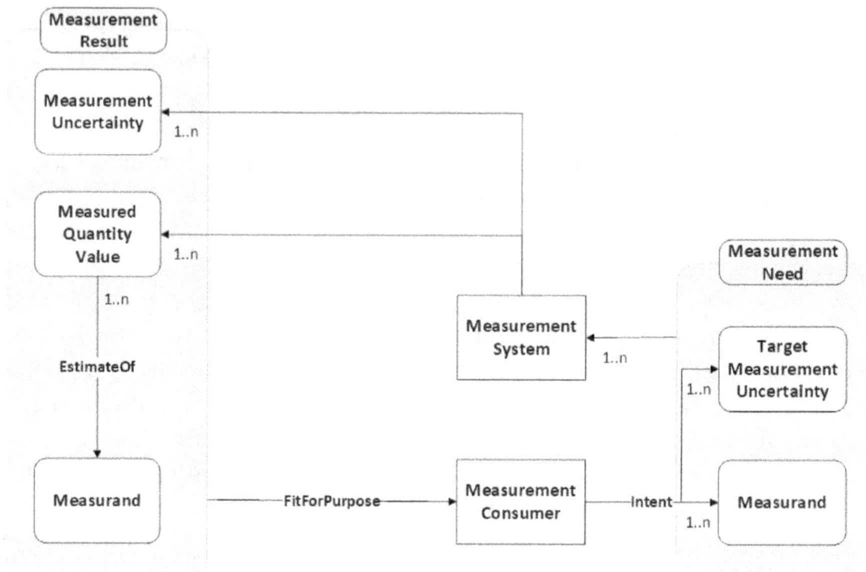

Fig. 2. Consumer view of measurement service provision describes the consumer's use of the measurement result.

the measurement system with the measurement result provides an estimate of the consumer's measurand.

Figure 3 represents the calibration process [9, 2.39 and 6.11]. It is an operation carried out under specified conditions in two stages. In the first stage, the operation establishes a relationship between the quantity values provided by measurement standards and indications of the measurement system. The measurement standard(s) quantity values and the indications of the measurement system have associated measurement uncertainties. In the second stage, authorized technical staff carry out measurement procedure(s) based on information from the first stage. They use the information from the first stage to establish a relation for obtaining a measurement result from an indication of the measurement system. Each procedure results in a set of measured quantity values experimentally obtained which the provider can reasonably attribute to a measurand together with measurement uncertainty.

The measurement processes should be described from the various quality system perspectives to model metrology concepts and their relationship with measurement data. The D-QMS model represents complimentary ways to conceptualize a measurement result and the process of service provision.

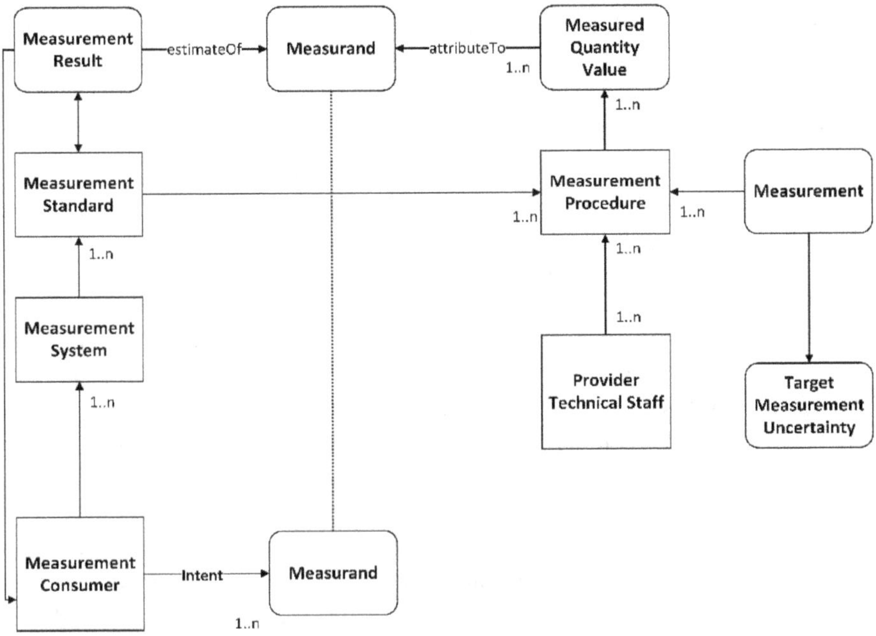

Fig. 3. The service provider's view of measurement service provision describes the calibration process.

4. Metrology metadata models

The M-layer [12] addresses ambiguity in digital systems when interpreting an expressed quantity. This is needed to address limitations in measurement units, such as the SI units, which do not identify the quantity. This is not generally a problem for humans who can infer the expressed quantity from contextual or domain knowledge. However, the use of traditional semantics may lead to incorrect inference by digital systems. The semantics of digital systems should identify the quantity independent of the unit. The M-layer is a tripartite representation,

$$Q = \{Q\}[Q]\langle Q\rangle \tag{1}$$

composed of a numerical component, $\{Q\}$, the scale $[Q]$, and the expressed aspect $<Q>$ of the quantity.[‡] The aspect and measurement scale are contextual

[‡]General notion of quantity where the scale for the expression is not specified. For example, the mass is an aspect of an object that can be expressed on different ways (scales). Kind of quantity (VIM 1.2) is an aspect common to mutually compatible quantities.

information for digital systems. Digital systems use this information for validation, rather than inference, of the expressed quantity's digital representation.

The M-layer handles measurement data when associated with aspects and measurement scales. The metadata is enough for expressing measurement data since the aspect is the most general information needed to disambiguate an expressed quantity. However, the M-layer does not describe details of the measured quantity as that requires specifying the measurand.

The measurand is a key model concept and appears throughout the service provision process. As the model describes, the measurand establishes a relation between consumer intent and provider capability, and as part of the calibration process the provider attributes the measurement result to a measurand. The NCSL International Committee 141 has an XML model for the measurand [13]. The committee developed a taxonomy of measurands which establishes a hierarchy of related quantities encountered in measurement laboratories. The most relevant elements of the model are the measurand name, *i.e.* the taxon, and the quantity that estimates the measurand. The model parameterizes the measurement system state and measurement conditions.

As mentioned previously in Sec. 2, the BIPM maintains the KCDB where details of NMI CMCs are documented. As part of the digital services that comprise the SI Digital Framework, the BIPM developed the KCDB API (Application Programming Interface) to provide machine-readable and machine-accessible measurement information. An important feature of the database is that each CMC has a persistent and unique identifier which serves as a digital reference. The identifier provides access to the details of a particular CMC, including the target measurement uncertainty and the status of NMI capability and may relate to one or more measurands. The measurands may clarify the description of each CMC and give context to the measurement process. The CMC identifier, measurand taxon, and M-layer metadata may improve the machine-interpretability of calibration information.

5. Constructing machine interpretable measurement results

Metrological metadata, when referenced in quality records and measurement results relates outputs of metrological processes. The M-layer, measurand taxonomy, and the KCDB are FAIR [14] (Findable, Accessible, Interoperable, Reusable) metadata that improve machine interpretable of quality records. Digital agents may access increasing levels of detail when necessary, and human-readable records are easy to generate. This supports current requirements for quality records and digitalization in metrology.

The Digital Calibration Certificate (DCC) specification [15] is an exchange format for calibration information. The XML data structure has two primary components: "Administrative Data" and "Measurement Results". The latter is the relevant part of the DCC that arises in the measurement process. The DCC intends to standardize the structure and syntax for exchanging a measurement result with an XML document. However, the DCC does not specify controlled vocabulary or standardized metadata for describing key metrological concepts that arise in the measurement process. As a result, the meaning of data items corresponding to the measurement result may be ambiguous.

The DCC is adapting to address challenges that arise in particular use cases to facilitate wider adoption. For instance, the DCC accommodates reference types which allow for domain terminology in the XML document [16], although these terminologies would require, at a minimum, domain standardization. The measurand taxonomy and M-layer differ from free form textual terms in an XML document because they have the properties of FAIR metadata. They are machine-accessible, have unique identifiers, descriptive metadata, and express measurement information independent of any data exchange format. The D-QMS model describes the required information to communicate the measurement result. These requirements are illustrated in Fig. 4 as an example for a gauge block calibration. The DCC XML includes metadata describing the provider's measurement capability, the measurand, and the measured output quantity. These metadata map to concepts described in the D-QMS model and summarize information of the service provision process based on current practices.

6. Conclusion

Models are an inherent part of software design. To have digital support for quality, models should describe metrological processes. Despite the simplicity of the D-QMS conceptual model, it provides requirements for the measurement result in digital quality records because it is model of metrological processes. Machine-readable formats of current documentary requirements rely on humans to interpret the information. This approach to software modeling hinders the digital transformation process. Current practices vary across disciplines and within networks of the quality infrastructure, limiting the standardization process. The technical details and documentary practices of quality records should be modelled conceptually to validate and verify their use in digital

```
<dcc:Result>
    <dcc:Capability>
        <cmc:Service name="Linear dimensions" id="kcdb-sc/L/DimMet-2">
        <cmc:Quantity name="End standards" id="kcdb-sc/L/DimMet-2.2.1">
        <cmc:id type="KCDB">
        <cmc:id value="https://si-digital-framework.org/kcdb-cmc/SIM-L-CA-
00000AV5-1">
    </dcc:Capability>
    <dcc:Measurand>
        <mtc:Taxon name="Measure.Length.EndStandard.GaugeBlock">
            <mtc:Parameter name="Length" mlayer:Aspect="mlayer:aspect::AS3">
        <mtc:Parameter name="Temperature" mlayer:Aspect="mlayer:aspect::AS6">
            <mtc:Result name="Length">
                <mlayer:Aspect label="Length" id="mlayer:aspect::AS3"/>
            </mtc:Result>
        </mtc:Taxon>
    </dcc:Measurand>
    <dcc:QuantityValue>
        <mlayer:Expression>
            <mlayer:Value value=1.2345>
            <mlayer:Scale label="ra_si_m" id="mlayer:scale::SC2"/>
            <mlayer:Scale external_ref="https://bipm.org/sirf/meter"/>
            <mlayer:Aspect label="Length" id="mlayer:aspect::AS3"/>
        </mlayer:Expression>
        <dcc:Uncertainty>
            <dcc:Type type="Expanded"/>
            <dcc:Coverage k=2/>
            <mlayer:Expression>
                <mlayer:Value value=0.2>
                <mlayer:Scale label="ra_si_m" id="mlayer:scale::SC2"/>
                <mlayer:Aspect label="Length" id="mlayer:aspect::AS3"/>
            </mlayer:Expression>
        </dcc:Uncertainty>
    </dcc:QuantityValue>
</dcc:Result>
```


Fig. 4. Example DCC XML document with contextual metadata based on the D-QMS conceptual model.

systems. The clarity, correctness and increasing level of descriptive metadata for concepts, processes and data will impact the ability to harmonize current metrological details and practices across the IQI. A common conceptual model can facilitate harmonization efforts and relate existing information that supports traceability (*i.e.*, CMCs, SoAs, instrument specifications, standard calibration procedures and guidelines, or regulatory standards).

A common conceptual model for quality management is a tool for documenting requirements. Models provide a common language to communicate requirements amongst stakeholders, users, and developers. They are amenable to changes over time and provide views of the same process or concept that are relevant to a particular role. Metrology organizations and stakeholders can utilize a common quality system model to ensure digitalization meets international, national, and organizational needs. Future work aims to relate the D-QMS conceptual model to technical views representing quality records.

Acknowledgements

This work is funded by the Government of Canada. The author thanks B. D. Hall, M. J. Kuster and A. Oldershaw for their contributions and support of this work.

References

1. "On the global digital transformation and the International System of Units," *Resolution 2 of the 27th CGPM,* November 2022, [online] Available: https://www.bipm.org/en/cgpm-2022/resolution-2, doi:10.59161/CGPM2022RES2E.
2. B. Hall, S. Cui and K. Yamazawa, "The APMP-DXFG: A focus group on digital transfrmation in metrology," 2022, doi: 10.21014/tc6-2022.005.
3. W. R. Ashby, "Requisite variety and its implications for the control of," *Cybernetica*, Vol. 1, no. 2, 1958.
4. J. Davies, S. Harris, C. Crichton, A. Shukla and J. Gibbons, "Metadata Standards for Semantic Interoperability in Electronic Government," in *International Conference on Theory and Practice of Electronic Governance*, Cairo, 2008.
5. BIPM, Le Système international d'unités/The International System of Units ('The SI Brochure'), Ninth ed., Bureau international des poids et mesures, 2019.
6. CIPM, "Mutual recognition of national measurement standards and of calibration and measurement certificates issued by national metrology institutes (CIPM MRA)," 1999. [Online]. Available: www.bipm.org/documents/20126/43742162/CIPM-MRA-2003.pdf/. [Accessed November 2023].
7. BIPM, "Key Comparison Database (KCDB)," Bureau international des poids et mesures, [Online]. Available: https://www.bipm.org/kcdb/. [Accessed 2023].
8. ISO/IEC, "ISO/IEC 17025:2017 General requirements for the competence of testing and calibration laboratories," International Standards Organization and International Electrotechnical Commission, Geneva, 2017.
9. BIPM, IEC, IFCC, ILAC, ISO, IUPAC, IUPAP and OIML. International vocabulary of metrology | Basic and general concepts and associated terms (VIM). Joint Committee for Guides in Metrology, JCGM 200:2012. (3rd edition).

10. B. Hall and R. White, "The notion of metrological traceability in digital systems," in *IMEKO TC6 International Conference on Metrology and Digital Transformation*, Berlin, 2022.
11. P. Bharti, Q. Yang, A. B. Forbes and Y. Koucha, "UML knowledge model for measurement process including uncertainty of measurement," *Int. J. of Metrology and Quality Engineering*, Vol. 12, p. 10, 2021.
12. B. Hall and M. Kuster, "Metrological support for quantities and units in digital systems," *Measurement: Sensors*, Vol. 18, p. 100102, 2021.
13. M. Kuster, "A guide to standardizing digital calibration and measurement capabilities," *Metrologist*, Vol. 15, no. 1, pp. 36–47, January 2022.
14. M. D. Wilkinson, M. Dumontier, I. J. Aalbersberg *et al.*, "The FAIR Guiding Principles for scientific data management and stewardship," *Scientific Data*, Vol. 3, no. 1, p. 160018, March 2016.
15. F. Härtig, J. Hornig, T. Wiedenhöfer and S. G. Hackel, "The Digital Calibration Certificate," *PTB-Mitteilungen*, no. 127, p. 7, 2017.
16. Members of the Technical Committee Mass and Weighing Instruments of the DKD, Physikalisch-Technische Bundesanstalt (PTB), ISNI: 0000-0002-1140-5252, "Instructions on how to use the DCC schema to create a digital calibration certificate for weights: Expert Report DKD-E 7-2," Physikalisch-Technische Bundesanstalt (PTB), Braunschweig and Berlin, 2022.

F Pavese, A Bošnjaković, S Eichstädt, A B Forbes, J Alves e Sousa (eds.):
Advanced Mathematical and Computational Tools in Metrology and Testing XIII
Series on Advances in Mathematics for Applied Sciences, Vol. 94

https://doi.org/10.1142/9789819800674_0011

Uncertainty analysis for the estimation of the mean from correlated data: Small and large samples

R. Willink[1,2]

[1] *Measurement Standards Laboratory of New Zealand, Lower Hutt, New Zealand*
[2] *University of Otago (Wellington), New Zealand*
E-mail: robin.willink@otago.ac.nz

This paper presents a procedure for measuring a quantity when there is serially correlated data. In this situation, the statistical method of determining the standard uncertainty that assumes independence is inaccurate and, usually, anti-conservative. The procedure presented here involves calculating (a) an alternative estimate of the standard deviation of the sample mean for use as the standard uncertainty and (b) an associated number of degrees of freedom. Simulations for large and small samples show the procedure to substantially improve on the method that assumes independence, with the amount of improvement increasing with sample size.

Keywords: time-series; standard error; serial correlation; t-distribution.

1. Introduction

The classical procedure for the evaluation of measurement uncertainty when there are n independent estimates $\{x_1, \ldots, x_n\}$ of the true value of a measurand is well known. The best combined estimate is the sample mean $\bar{x} \equiv \sum_{j=1}^{n} x_j/n$, and the square of the corresponding figure of standard uncertainty is the best estimate of the variance of the random sample mean, $\bar{X}$. Let σ^2 be the variance of the population of potential individual estimates. So — because the variance of $\bar{X}$ is σ^2/n when the estimates are independent — the standard uncertainty of $\bar{x}$ is taken to be $s/\sqrt{n}$, where

$$s^2 \equiv \sum_{j=1}^{n} (x_j - \bar{x})^2/(n-1) \tag{1}$$

is the unbiased estimate of σ^2. Furthermore, if the population is well-modelled as normal then the appropriate 95% expanded uncertainty interval for the measurand has limits $\bar{x} \pm t_{n-1,0.975} s/\sqrt{n}$, where $t_{n-1,0.975}$ is the 0.975 quantile of the t-distribution with $n-1$ degrees of freedom. These results are in accordance with standard statistical theory and with metrological recommendations, e.g., Ref. 1, Secs. 4.2 and G.4.

The material above will be familiar. However, there is no well-known general procedure for the situation where the estimates are created in a procedure with dependence between the n errors, in particular where the errors in successive measurements are correlated. In such a situation, serial correlation is likely to be positive, in which case the variance of the sample mean will be greater that σ^2/n, and $s/\sqrt{n}$ will tend to be too small. Also, there will be a smaller effective number of observations contributing to the repeatability of s^2, meaning that the figure chosen for an effective number of degrees of freedom will be smaller than $n-1$.

Neither the statistical literature nor the metrological guidelines describe an appropriate procedure for use when serial dependence is suspected and the sample size is small or moderate. So the purpose of this chapter is to present such a procedure. Our analysis here is most similar to that of Zięba and Ramza,[2] but it is more general in a number of ways, namely: (i) the expression for the standard uncertainty is derived without involving the assumption of stationarity, (ii) a number of degrees of freedom is calculated (using the theory of stationary processes and the assumption of Gaussianity), (iii) both parts of the method are evaluated by extensive simulations for both small and large samples. The results are indicative of very good performance in many practical problems, with failure only in problems where no solution is possible.

1.1. *Notation and language*

Standard statistical practice is to denote a random variable that represents a procedure with an unpredictable element using a capital letter, e.g., X, and to indicate the numerical result obtained using the corresponding lower case letter, e.g., x. For example, the n individual measurement processes are represented by the random variables $X_1, \ldots, X_n$ and the n outcomes (or 'realizations') of these variables are the observed data $x_1, \ldots, x_n$. Similarly, the random sample mean is $\bar{X} \equiv \sum_{j=1}^{n} X_j/n$ and its outcome or realization is $\bar{x}$. Probability distributions, expected values, standard deviations and variances are properties of random variables such as X_j, not of constants such as x_j. Thus, for example, when there is no dependence between the n measurements, the covariance between X_j and X_i for all $i \neq j$ is zero. Being a random variable, X_j or $\bar{X}$ is called an *estimator* of the value of the measurand. The value taken, x_j or $\bar{x}$, is called the *estimate.*

1.2. *Defining the measurand*

A stochastic process is an ordered collection of random variables $\{X_1, X_2, \ldots\}$. The process has mean μ if $\mathcal{E}[X_j] = \mu$ for all j, and it has variance σ^2 if $\text{var}[X_j] = \sigma^2$ for all j. Figure 1 shows parallel segments of three realizations of a single process with mean $\mu = 2$ and variance $\sigma^2 = 1$.

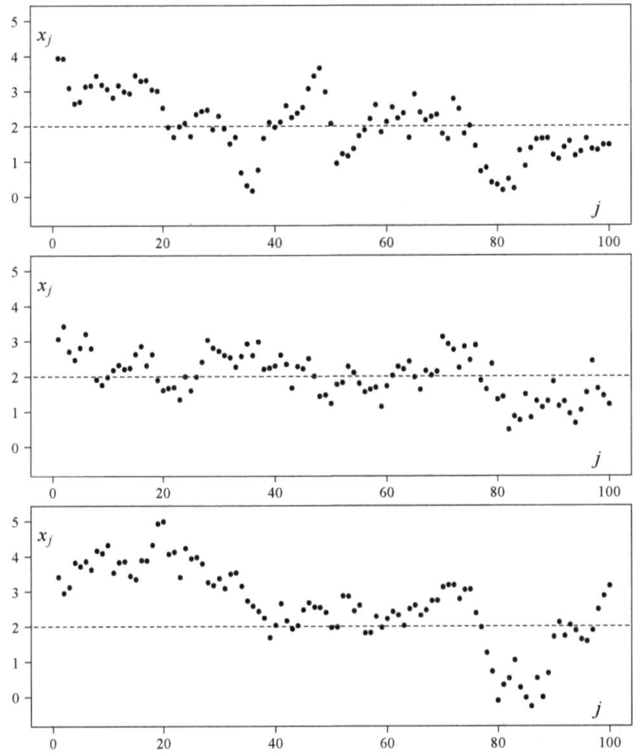

Fig. 1. Segments of three realizations (sample functions) of a certain stochastic process with mean $\mu = 2$ and variance $\sigma^2 = 1$.

A random process is said to be "ergodic in the mean" if — for every realization — the temporal mean $\theta \equiv \lim_{n=\infty} \frac{1}{n} \sum_{j=1}^{n} x_j$ exists and is equal to μ, and it is said to be "ergodic in the variance" if the temporal variance exists is equal to σ^2. Arguably, the quantity of interest for the metrologist will be θ. However, we shall assume ergodicity, so that the measurand can be taken to be μ. Then we can used the statistician's theory of random processes to answer the metrologist's practical problem.

2. Standard uncertainty $u(\bar{x})$

The expected value of the sample mean $\bar{X}$ is $\mathcal{E}[\bar{X}] = \mu$. The variance of $\bar{X}$ is easily shown to be given by the expression

$$\sigma_{\bar{X}}^2 \equiv \text{var}[\bar{X}] = \frac{\lambda_0}{n^2} + \frac{2}{n^2}\sum_{k=1}^{n-1}\lambda_k \qquad \lambda_k \equiv \sum_{j=1}^{n-k}\mathcal{E}[(X_j - \mu)(X_{j+k} - \mu)].$$

The appropriate standard uncertainty is the square root of our best estimate of $\sigma_{\bar{X}}^2$, and this will be obtained by taking our estimate of λ_k to be

$$\ell_k \equiv \frac{n}{n-1}\sum_{j=1}^{n-k}(x_j - \bar{x})(x_{j+k} - \bar{x}). \tag{2}$$

(So $\ell_0/n = s^2$.) In practice, each covariance λ_k will be non-negative unless there is a periodicity. Therefore, we consider ℓ_k to be noise if $k \geq k'$ and $\ell_{k'} < 0$. In this way, the contributions retained have lags $0, \ldots, m$ where

$$m \equiv \min\{k : \ell_k \geq 0 \cap \ell_{k+1} < 0\}. \tag{3}$$

This corresponds to the 'first transit zero' (FTZ) step of Zięba and Ramza.[2] In effect, each ℓ_k for $k > m$ is being treated as if it were zero. So, our final estimate of $\sigma_{\bar{X}}^2$ is

$$s_{\bar{X}}^2 \equiv \frac{s^2}{n} + \frac{2}{n^2}\sum_{k=1}^{m}\ell_k, \tag{4}$$

and the standard uncertainty to be associated with $\bar{x}$ as an estimate of μ is

$$u(\bar{x}) \equiv s_{\bar{X}}. \tag{5}$$

In this way, the standard uncertainty of measurement can be calculated from Eqs. (1)–(5). The analysis has not required an assumption of the normality of the distribution or the stationarity of the process.

3. Degrees of freedom ν

The analysis for the calculation of an appropriate number of degrees of freedom is more complicated. It is based on the theory of normal, stationary processes. We consider a theoretical effective number of degrees of freedom, $\mathring{\nu}$, associated with the estimator of the process variance, and we estimate $\mathring{\nu}$ by a figure ν obtained from the data.

The figure $s_{\bar{X}}^2$ is the outcome of the corresponding random variable $S_{\bar{X}}^2$, and the figure $\mathring{\nu}$ is to describe the relative variability $S_{\bar{X}}^2$ in an appropriate inverse way. The result we obtain is built on the fact that a random variable with the chi-square distribution with $\mathring{\nu}$ degrees of freedom has mean $\mathring{\nu}$ and variance $2\mathring{\nu}$. Therefore, the theoretical number of degrees of freedom when sampling independently from a normal distribution is

$$\mathring{\nu} \equiv \frac{2\mathcal{E}[S^2_{\bar{X}}]^2}{\text{var}[S^2_{\bar{X}}]},$$

and we wish to approximate this as well as possible.

The expected value of $S^2_{\bar{X}}$ is $\sigma^2_{\bar{X}}$ and the variance of $S^2_{\bar{X}}$ is

$$\text{var}[S^2_{\bar{X}}] = \frac{1}{n^4}\left(\text{cov}[L_0, L_0] + 2\sum_{k=1}^{m}\text{cov}[L_0, L_k] + 4\sum_{k=1}^{m}\sum_{h=1}^{m}\text{cov}[L_k, L_h]\right)$$

where

$$L_k \equiv \tfrac{n}{n-1}\textstyle\sum_{j=1}^{n-k}(X_j - \bar{X})(X_{j+k} - \bar{X}),$$

i.e., L_k is the random variable realized in ℓ_k. So, we wish to estimate $\text{cov}[L_k, L_h]$, which is the covariance of L_k and L_h.

For a stationary process, the covariance at lag k is

$$\gamma_k \equiv \mathcal{E}[(X_j - \mu)(X_{j+k} - \mu)]$$

and its usual estimator is

$$\hat{\gamma}_k \equiv \tfrac{1}{n}\textstyle\sum_{j=1}^{n-k}(X_j - \bar{X})(X_{j+k} - \bar{X}).$$

By Theorem 6.2.2 of Ref. 3 or Eq. (7.3.13) of Ref. 4,

$$\begin{aligned}\lim_{n\to\infty} n\,\text{cov}[\hat{\gamma}_p, \hat{\gamma}_q] &= \sum_{h=-\infty}^{\infty}\gamma_h\gamma_{h+(p-q)} + \sum_{h=-\infty}^{\infty}\gamma_h\gamma_{h+p+q} \qquad p \ge q \ge 0\\ &= \beta_{p-q} + \beta_{p+q} \qquad \beta_k \equiv \sum_{h=0}^{k}\gamma_h\gamma_{k-h} + 2\sum_{h=1}^{\infty}\gamma_h\gamma_{h+k}.\end{aligned}$$

Imagine our series is stationary. Then $L_k = n^2\hat{\gamma}_k/(n-1)$ and

$$\text{cov}[L_p, L_q] \approx \frac{n^3}{(n-1)^2}\left(\beta_{p-q} + \beta_{p+q}\right) \qquad p \ge q \ge 0,$$

and we would wish to estimate β_k. The autocovariances γ_k are considered to be zero beyond lag m, so we would define

$$\mathring{\ell}_k \equiv \begin{cases}\ell_k & 0 \le k \le m\\ 0 & m+1 \le k \le 3m\end{cases} \tag{6}$$

and then estimate β_k by $(n-1)^2 a_k/n^4$ where

$$a_k \equiv \sum_{h=0}^{k}\mathring{\ell}_h\mathring{\ell}_{k-h} + 2\sum_{h=1}^{m}\mathring{\ell}_h\mathring{\ell}_{h+k} \qquad k = 0, \ldots, 2m. \tag{7}$$

These ideas can be implemented even though the process might not be stationary. Therefore, our estimate of $\mathrm{var}[S^2_{\bar{X}}]$ is

$$w \equiv \frac{1}{n^5}\left(2a_0 + 4\sum_{k=1}^{m} a_k + 4\sum_{k=1}^{m}\sum_{h=1}^{m} a_{|k-h|} + 4\sum_{k=1}^{m}\sum_{h=1}^{m} a_{k+h}\right).$$

Exact manipulation gives the simpler expression

$$w = \frac{4}{n^5}\left(\left(m+\tfrac{1}{2}\right)a_0 - \sum_{k=1}^{m} a_k + \sum_{k=1}^{2m}(2m-k+1)a_k\right). \tag{8}$$

Subsequently, we calculate our estimate of $\mathring{\nu}$ to be

$$\nu \equiv \frac{2u^4(\bar{x})}{w} - 2. \tag{9}$$

When $m = 0$, Eq. (9) evaluates to $\nu = n - 2$. The possibility that $\nu < 0$ is discussed in Sec. 6.

This analysis gives a number of degrees of freedom ν to associate with the standard uncertainty $u(\bar{x})$. If μ is the final measurand then the realization of an approximate $100(1-\alpha)\%$ confidence interval for μ is

$$\left[\bar{x} - t_{\nu,1-\alpha/2}\,u(\bar{x}),\ \bar{x} + t_{\nu,1-\alpha/2}\,u(\bar{x})\right]. \tag{10}$$

If μ is not the final measurand then ν contributes to the final interval as appropriate (Ref. 1, G.4).

4. Performance assessment

So, the standard uncertainty and the interval in the measurement of μ can be calculated using Eqs. (1)–(10). The performance of the procedure and the performance of the classical method of Sec. 1 are now assessed against two principles: (i) $u^2(\bar{x})$ is to be an unbiased estimate of the variance of $\bar{X}$ (for use in the propagation of variance along a chain of measurements), and (ii) the interval should cover the measurand with the implied frequency, e.g., 95%. An evaluation of performance then requires the simulation of realistic random processes. We consider three types of process, each of which has mean μ and variance σ^2 and each of which is Gaussian, being derived from a sequence of independent standard normal random variables $\{Z_j\}$.

- Autoregressive process of order 1 (AR(1)) with correlation ρ,

  $$X_j = \mu + \rho(X_{j-1} - \mu) + \sigma\sqrt{1-\rho^2}\,Z_j \qquad j = 1, 2, \ldots \qquad X_0 \sim \mathrm{N}(\mu, \sigma^2).$$

  This was simulated for $\rho = -0.3, -0.1, 0, 0.1, 0.3, 0.6, 0.9, 0.99, 0.999$. When $\rho = 0$, there is independence. (The sample segments in Fig. 1 are from an AR(1) process with $\rho = 0.9$.)

- Moving average process of order q (MA(q)),

  $$X_j = \mu + \sigma(a_0 Z_j + a_1 Z_{j-1} + \cdots + a_q Z_{j-q}) \qquad j = 1, 2, \ldots \qquad \textstyle\sum_{k=0}^{q} a_k^2 = 1.$$

  This was simulated for $q = 3$ and $q = 9$, with $a_k = 1/\sqrt{(q+1)}$.
- Time-dependent correlations (TDC)

  $$X_j = \mu + \rho_j(X_{j-1} - \mu) + \sigma\sqrt{1 - \rho_j^2}\, Z_j \qquad \rho_j \equiv 1/\sqrt{j} \qquad X_0 \sim \mathrm{N}(\mu, \sigma^2).$$

  This process is non-stationary, but it is ergodic in the mean and the variance. It was simulated to obtain the segment of data $\{x_1, \ldots, x_n\}$.

For each process just described, 10^4 simulations were carried out for $n = 10, 100, 1000$ for the unadjusted method of Sec. 1 and for the proposed procedure. Without loss of generality, we set $\mu = 0$ and $\sigma = 1$. Define $r \equiv u^2(\bar{x})/\sigma_{\bar{X}}^2$. We would like $u^2(\bar{x})$ to be unbiased, so the mean value of r observed in simulations, $\bar{r}$, is to be close to 1. Also, we would like the interval with limits $\bar{x} \pm t_{\nu,(1+p)/2} u(\bar{x})$ to cover μ on $100p\%$ of occasions.

Table 1 gives $\bar{r}$ and the coverage proportions p_{95} and p_{99} for $p = 0.95$ and $p = 0.99$ for the unadjusted method. Table 2 gives the results for the proposed procedure, along with the median and minimum values of ν, and the frequency with which Eq. (9) gave a negative value of ν.

Table 1. Measures of performance when there is no adjustment for correlation.

	$n = 10$			$n = 100$			$n = 1000$		
Process	$\bar{r}$	p_{95}	p_{99}	$\bar{r}$	p_{95}	p_{99}	$\bar{r}$	p_{95}	p_{99}
MA(3)	0.21	0.64	0.79	0.25	0.67	0.80	0.25	0.67	0.80
MA(9)	0.06	0.36	0.49	0.09	0.44	0.56	0.10	0.46	0.58
TDC	0.08	0.79	0.91	0.01	0.90	0.97	0.00	0.93	0.98
AR(1) $\rho = -0.3$	1.82	0.99	1.00	1.85	0.99	1.00	1.86	0.99	1.00
AR(1) $\rho = -0.1$	1.22	0.97	0.99	1.22	0.97	1.00	1.22	0.97	0.99
AR(1) $\rho = 0$	0.99	0.95	0.99	1.00	0.95	0.99	1.00	0.95	0.99
AR(1) $\rho = 0.1$	0.82	0.93	0.98	0.82	0.92	0.98	0.82	0.92	0.98
AR(1) $\rho = 0.3$	0.53	0.87	0.96	0.54	0.85	0.94	0.54	0.85	0.94
AR(1) $\rho = 0.6$	0.24	0.69	0.84	0.25	0.67	0.80	0.25	0.66	0.80
AR(1) $\rho = 0.9$	0.04	0.32	0.44	0.05	0.33	0.43	0.05	0.34	0.44
AR(1) $\rho = 0.99$	0.00	0.11	0.15	0.00	0.09	0.11	0.00	0.10	0.13
AR(1) $\rho = 0.999$	0.00	0.03	0.05	0.00	0.03	0.04	0.00	0.03	0.04

A number of observations can be made. (1) The proposed procedure improves on the unadjusted method when there is positive serial correlation, which is a realistic type of correlation. However, as might be expected, neither method would be satisfactory when the sample size is small and the degree of correlation is high; see Sec. 6. (2) The performance of the procedure improves with n, but this is not the case with the unadjusted method. (3) When the data are independent, the method is slightly conservative.

This is to be expected, because the procedure does not make use of the independence. When $m = 0$, the method gives $u(\bar{x}) = s/\sqrt{n}$ but $\nu = n-2$. A loss in performance is incurred when the first sample covariance is non-negative, so $m \geq 1$. (4) In the unrealistic situation where the first serial correlation is negative, the procedure is conservative. Underlying negative correlations, which reduce the variability of $\bar{X}$, lead to sample correlations that are counted as zero, so the procedure overestimates $\sigma^2_{\bar{X}}$.

Table 2. Measures of performance with the proposed method of Eqs. (1)–(10).

Process	$n = 10$			$n = 100$			$n = 1000$		
	$\bar{r}$	p_{95}	p_{99}	$\bar{r}$	p_{95}	p_{99}	$\bar{r}$	p_{95}	p_{99}
MA(3)	0.50	0.87	0.92	1.06	0.95	0.99	1.06	0.95	0.99
MA(9)	0.15	0.75	0.83	0.92	0.93	0.97	1.07	0.95	0.99
TDC	0.63	0.88	0.93	1.09	0.96	0.99	1.01	0.95	0.99
AR(1) $\rho=-0.3$	1.90	0.98	0.99	1.85	0.99	1.00	1.86	0.99	1.00
AR(1) $\rho=-0.1$	1.36	0.96	0.98	1.26	0.97	1.00	1.22	0.97	0.99
AR(1) $\rho=0$	1.16	0.95	0.98	1.13	0.96	0.99	1.05	0.95	0.99
AR(1) $\rho=0.1$	1.02	0.94	0.97	1.09	0.96	0.99	1.04	0.95	0.99
AR(1) $\rho=0.3$	0.77	0.91	0.95	1.07	0.95	0.99	1.04	0.95	0.99
AR(1) $\rho=0.6$	0.46	0.85	0.91	0.99	0.94	0.99	1.05	0.95	0.99
AR(1) $\rho=0.9$	0.11	0.70	0.80	0.63	0.87	0.93	1.03	0.95	0.99
AR(1) $\rho=0.99$	0.01	0.55	0.68	0.09	0.62	0.71	0.60	0.87	0.93
AR(1) $\rho=0.999$	0.00	0.39	0.57	0.01	0.44	0.55	0.09	0.61	0.70
Process	ν_{med}	ν_{min}	Pr	ν_{med}	ν_{min}	Pr	ν_{med}	ν_{min}	Pr
MA(3)	1.1	-0.9	0.04	19.0	-0.3	0.00	215	9	0.00
MA(9)	1.0	-0.9	0.06	6.1	-0.5	0.01	70	3	0.00
TDC	2.7	-0.8	0.04	25.5	1.0	0.00	236	33	0.00
AR(1) $\rho=-0.3$	8.0	-0.6	0.01	98.0	10.9	0.00	998	998	0.00
AR(1) $\rho=-0.1$	8.0	-0.9	0.01	98.0	4.2	0.00	998	107	0.00
AR(1) $\rho=0$	8.0	-0.6	0.02	98.0	2.3	0.00	998	41	0.00
AR(1) $\rho=0.1$	8.0	-1.0	0.02	35.8	2.1	0.00	237	27	0.00
AR(1) $\rho=0.3$	3.1	-0.9	0.03	24.0	0.2	0.00	173	18	0.00
AR(1) $\rho=0.6$	2.3	-0.9	0.04	10.5	-0.3	0.00	72	5	0.00
AR(1) $\rho=0.9$	1.1	-0.7	0.06	2.8	-0.7	0.04	18	0	0.00
AR(1) $\rho=0.99$	1.0	-0.9	0.06	0.9	-0.7	0.17	3	-1	0.04
AR(1) $\rho=0.999$	1.0	-0.9	0.06	0.6	-0.8	0.22	1	-1	0.19

$\text{Pr} \equiv \text{Pr}(\nu < 0)$

5. The estimator of autocovariance

In the literature, $\hat{\gamma}_k$ is the preferred estimator of the autocovariance γ_k, seemingly because a matrix formed from the $\hat{\gamma}_k$ figures is known to be non-negative definite, e.g. Refs. 3,4. However, the random variable $\gamma^\dagger_k \equiv n\hat{\gamma}_k/(n-k)$ is a more natural estimator of γ_k, and there is no reason why it might not be superior for our purpose. Now $L_0 = nS^2$ and $L_k = n(n-k)\gamma^\dagger_k/(n-1)$, so our estimator of $\sigma^2_{\bar{X}}$ can be written as

$$S^2_{\bar{X}} = \tfrac{S^2}{n}\left(1 + 2\textstyle\sum_{k=1}^{M} \tfrac{(n-k)}{n} \tfrac{\gamma^\dagger_k}{\gamma^\dagger_0}\right),$$

while the variance of the sample mean with a stationary series with variance σ^2 and kth autocovariance γ_k is, (e.g., Eq. (3.6.1) of Ref. 5),

$$\sigma_{\bar{X}}^2 = \frac{\sigma^2}{n}\left(1 + 2\sum_{k=1}^{n-1}\frac{(n-k)}{n}\frac{\gamma_k}{\gamma_0}\right).$$

We see that our method uses $\gamma_k^\dagger$, not $\hat{\gamma}_k$, as an estimator of γ_k.

Zięba and Ramza[2] describe how using the FTZ principle improved their results considerably. And our implementation of the FTZ concept means that the weighted sum of the $\gamma_k^\dagger$ estimators is always positive, so there is no need for the guarantee of 'positive definiteness' provided by the use of $\hat{\gamma}_k$. We infer that using the natural estimator $\gamma_k^\dagger$ with the FTZ principle also makes a considerable difference, especially with small samples sizes.

6. A fundamental limitation

Consider observing the dataset of Fig. 2, where $\mu = 8$ and there is an underlying component with a wavelength of 250. The extent of our sampling has been insufficient to expose a long-range dependence, and there is almost no information in the segment of data about μ. Thus, unless we happen to sample such a process over a region where it crosses or approaches μ more than once, there is no reason why $\bar{x}$ should be close to μ and no reason why we must have reliable information about the scale of the error.

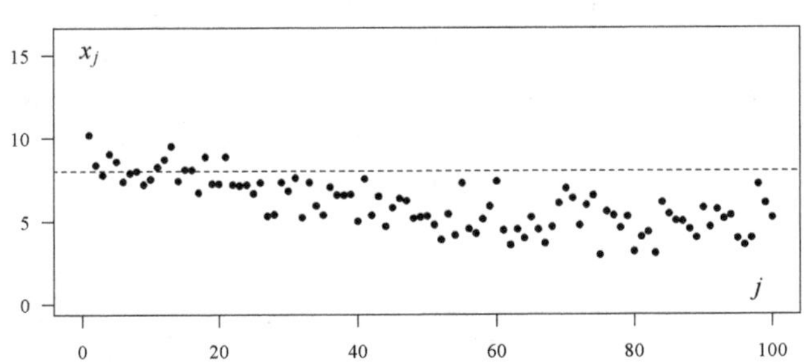

Fig. 2. Segment of realization of a stochastic process with long-range dependence.

Hypothetically, there will always be the possibility of a long-range dependence that would only become apparent with an extremely large sample. Hence, the statement of measurement uncertainty derived by *any* data-based method can be completely unreliable. This appears to be a fundamental limitation in the analysis of time-series data. Unless we can assume that there is no very-long-range dependence, we cannot estimate μ

from the dataset alone. So, the measurement of the mean when we admit the possibility of serial correlation is, to some extent, an act of faith.

The implication is that we can never properly conclude from the data that we are in a situation where performance is adequate. However, the converse does not hold: the data might conceivably provide some indication that the estimation of μ from that sample would be unreliable. We suggest that a negative value for ν in Eq. (9) would be evidence of inadequate fluctuation about the mean during the period of observation. Rather than being a nuisance, a negative value of ν would be a warning that the method should not be applied and that the sample is not adequate for its purpose.

7. Conclusion

The problem of estimating the mean of a signal from a segment of serially correlated data is an important variation on the archetypal problem where the data are independent. The sample mean is still an appropriate estimator, but the analysis must take into account the correlation when representing its variance by the squared standard uncertainty. A straightforward procedure has been given for the evaluation of the standard uncertainty and for the calculation of an associated number of degrees of freedom.

The method is based on the realistic supposition that all correlations between the data are positive. When there are negative correlations, the procedure is conservative and, hence, is statistically valid. The procedure performs well. It is applicable with samples of all sizes and, for the calculation of the standard uncertainty, it does not require the signal to be stationary or Gaussian. A more relevant condition is the condition of ergodicity, which relates to the identity of the measurand.

References

1. BIPM, IEC, IFCC, ISO, IUPAC, IUPAP and OIML, *Guide to the Expression of Uncertainty in Measurement* (1995).
2. A. Zięba and P. Ramza, Standard Deviation of the Mean of Autocorrelated Observations Estimated with the Use of the Autocorrelation Function Estimated From the Data, *Metrology and Measurement Systems* **XVIII**, 529–542 (2011).
3. W. A. Fuller, *Introduction to Statistical Time Series* (Wiley, 1976).
4. P. J. Brockwell and R. A. Davis, *Time Series: Theory and Methods* (Springer-Verlag, 1987).
5. P. Diggle, *Time Series: A Biostatistical Introduction* (Oxford, 1990).

F Pavese, A Bošnjaković, S Eichstädt, A B Forbes, J Alves e Sousa (eds.):
Advanced Mathematical and Computational Tools in Metrology and Testing XIII
Series on Advances in Mathematics for Applied Sciences, Vol. 94
 (pp. 143–150)
https://doi.org/10.1142/9789819800674_0012

Locally best linear unbiased estimation of regression curves specified by nonlinear constraints on the model parameters

Anna Charvátová Campbell[1,a], Radek Šlesinger[1,b], Petr Klapetek[1,c], Martina Chvosteková[2,d], Laura Hajzoková[2,*], Viktor Witkovský[2,e] and Gejza Wimmer[3]

[1] *Department of Primary Nanometrology and Technical Length, Czech Metrology Institute, Okružní 31, Brno 638 00, Czech Republic*
E-mail: [a]*acharvatovacampbell@cmi.cz*, [b]*rslesinger@cmi.cz*, [c]*pklapetek@cmi.cz*

[2] *Institute of Measurement Science of the Slovak Academy of Sciences, Dúbravská cesta 9, 841 04 Bratislava, Slovak Republic*
E-mail: [d]*martina.chvostekova@savba.sk*, [*]*laura.hajzokova@savba.sk*, [e]*witkovsky@savba.sk*

[3] *Mathematical Institute of the Slovak Academy of Sciences, Štefánikova 49, 814 73 Bratislava, Slovak Republic*
E-mail: wimmer@mat.savba.sk

This paper focuses on parameter estimation within the Errors-in-Variables (EIV) model with nonlinear constraints and the determination of associated uncertainties. This is motivated by a collaborative project in nanoscale material characterization, particularly the assessment of material properties like hardness and elasticity through instrumented indentation testing. Here we confront the challenge of precisely fitting curves specified by nonlinear constraints of the parameters to measurement results which come with associated uncertainties, while simultaneously managing their correlations. Such challenges are prevalent in numerous statistical and metrological applications. Our proposed approach is founded on iterative linearizations of the EIV model with nonlinear parameter constraints specified in implicit form, employing the Locally Best Linear Unbiased Estimation (LBLUE) method. LBLUE is a locally optimal technique known for providing an efficient and robust solution, particularly when dealing with weakly nonlinear models. We refer to this approach as the Optimum Estimate of Function Parameters by Iterated Linearization (OEFPIL). This method has been effectively implemented in various computing environments, including R and C, and is here introduced within the context of MATLAB, complete with illustrative examples. Compared to alternative estimation approaches in EIV models with nonlinear constraints, OEFPIL stands out for its simplicity, versatility, practicality, and computational efficiency, making it an excellent choice for applications in metrology and beyond.

Keywords: Errors-in-Variables model; nonlinear constraints on the model parameters; locally best linear unbiased estimation; uncertainty.

1. Introduction

We investigate a linear regression model for direct measurements while incorporating nonlinear parameter constraints. Our approach extends the concept initially introduced in [1], where the linear regression model integrates type II linear constraints on model parameters. In our approach, we expand this idea to allow for the inclusion of intricate nonlinear restrictions on model parameters, represented implicitly. This framework enriches the linear regression model by accommodating both linear and nonlinear constraints, all within the context of a general covariance matrix, rendering it adaptable to a wide range of measurement scenarios.

Our research is firmly grounded in the bilateral project titled *Efficient Computation Methods for Nanoscale Material Characterization*. This project places significant emphasis on the precise assessment of mechanical properties in materials, particularly hardness and elasticity, achieved through *Instrumented Indentation Testing* (IIT). This assessment presents us with two significant challenges. The first challenge involves the need to fit nonlinear curves to load (F) and depth (h) sensor data, as proposed by Oliver and Pharr in [2], with the aim of determining material hardness and elastic modulus. This challenge is compounded by the requirement for precise measurements at the millinewton and nanometer scale, introducing sensor noise and necessitating specialized methods for accurate analysis. The second challenge revolves around deriving parameters associated with the indenter contact area function, which is essential for evaluating the area of contact between the indenter and the sample. Meeting these challenges requires fitting an appropriate function specified by parameter constraints while also addressing correlations in the data. To overcome these intricate challenges, our research relies on the *Locally Best Linear Unbiased Estimation* (LBLUE) method, a locally optimal technique that offers robust solutions for modeling with nonlinear constraints on the parameters.

This approach holds significant promise for applications across various domains, particularly in the context of EIV regression with weakly nonlinear models. These applications encompass comparative straight-line and polynomial calibration [3–5], primarily aimed at addressing measurement errors in both variables. Furthermore, it has demonstrated its effectiveness in ellipse fitting [6], especially for demodulating quadrature homodyne interferometer signals. As previously mentioned, this approach is rigorously tested in IIT, where it leverages the analysis of unloading curves, following the work of Oliver and Pharr. Additionally, we will illustrate how this approach can be beneficially employed to fit specific conic curves, including circles, spheres, ellipses, hyperbolas, and parabolas. It is also a valuable tool for fitting surfaces of three-dimensional objects specified by parameter constraints, as described in [7] and [8].

The approach's capability to handle general covariance structures, along with its versatility, practicality, and computational efficiency, positions it as a compelling alternative to methods such as (weighted) total least squares regression or orthogonal least squares regression. For more in-depth information, you can refer to relevant works such as [9–11].

A suitable model for the above-mentioned purposes can be symbolically described as a simple linear regression model of direct measurements:

$$\mathbf{X} = \boldsymbol{\mu} + \boldsymbol{\varepsilon}, \quad \& \quad \mathbf{f}(\boldsymbol{\mu}, \boldsymbol{\beta}) = \mathbf{0}. \tag{1}$$

This model includes nonlinear constraints, which define the manifold in the parameter space, and are specified implicitly. In this context, $\boldsymbol{\mu}$ represents the unknown expectation of the observable random vector $\mathbf{X}$, which models the direct measurements (i.e., $E(\mathbf{X}) = \boldsymbol{\mu}$), $\boldsymbol{\varepsilon}$ represents the unobservable measurement errors with $E(\boldsymbol{\varepsilon}) = \mathbf{0}$ and $cov(\boldsymbol{\varepsilon}) = \mathbf{U}$. The unknown parameter $\boldsymbol{\beta}$ corresponds to the primary parameter of interest, representing the properties of the fitted curve, such as the intercept and slope of the straight-line calibration function or the three parameters of the fitted Oliver-Pharr function. The covariance matrix $cov(\mathbf{X}) = \mathbf{U}$ represents the known uncertainty matrix associated with the vector of all direct measurements.

An efficient computation method, OEFPIL (*Optimum Estimate of Function Parameters by Iterated Linearization*), addresses these challenges, see [11, 12]. It transforms the nonlinear regression problem into a linear model by iteratively linearizing the constraints through Taylor expansion and using linear model theory to improve parameter estimation.

2. Theoretical background

Here we recall the *Best Linear Unbiased Estimation* (BLUE) of parameters in a linear regression model with type II linear constraints, as discussed in [1]. We start with $\mathbf{X} = (X_1, \ldots, X_m)^\top$, a random vector representing observable measurements. Our goal is to estimate the parameter vector $\boldsymbol{\mu}$. Additionally, we have another p-dimensional parameter vector, denoted as $\boldsymbol{\beta}$, which cannot be directly measured but holds primary importance in our analysis. In this context, the parameter constraints, originally expressed as $\mathbf{f}(\boldsymbol{\mu}, \boldsymbol{\beta}) = \mathbf{0}$, can be simplified to the following linear constraints:

$$\mathbf{B}_1\boldsymbol{\mu} + \mathbf{B}_2\boldsymbol{\beta} + \mathbf{b} = \mathbf{0}. \tag{2}$$

Here, $\mathbf{B}_1$, $\mathbf{B}_2$, and $\mathbf{b}$ are known matrices and a vector of dimensions $q\times m$, $q\times p$, and $q\times 1$, respectively. The conditions specify that $rank(\mathbf{B}_1,\mathbf{B}_2)=q$, $rank(\mathbf{B}_2)=p$, and $p\leq q\leq m+p$. This defines our linear regression model with type II linear restrictions on its parameters:

$$\mathbf{X}=\boldsymbol{\mu}+\boldsymbol{\varepsilon}, \quad \& \quad \mathbf{B}_1\boldsymbol{\mu}+\mathbf{B}_2\boldsymbol{\beta}+\mathbf{b}=\mathbf{0}, \tag{3}$$

with a known positive definite covariance matrix $\mathbf{U}$, $cov(\mathbf{X})=cov(\boldsymbol{\varepsilon})=\mathbf{U}$.

To estimate the parameters $(\boldsymbol{\mu}^\top,\boldsymbol{\beta}^\top)^\top$ in our model, we employ BLUE. According to the *Gauss-Markov Theorem*, which relies on given assumptions, the optimal minimum variance (best) linear unbiased estimator is defined to minimize the following expression:

$$(\hat{\boldsymbol{\mu}}^\top,\hat{\boldsymbol{\beta}}^\top)^\top=\arg\min_{\boldsymbol{\mu},\boldsymbol{\beta}}\left\{(\mathbf{X}-\boldsymbol{\mu})^\top\mathbf{U}^{-1}(\mathbf{X}-\boldsymbol{\mu})-2\boldsymbol{\lambda}^\top(\mathbf{B}_1\boldsymbol{\mu}+\mathbf{B}_2\boldsymbol{\beta}+\mathbf{b})\right\}, \tag{4}$$

where $\boldsymbol{\lambda}$ is a vector of Lagrange multipliers. Solving this yields:

$$\begin{pmatrix}\mathbf{B}_1\mathbf{U}\mathbf{B}_1^\top & \mathbf{B}_2\\ \mathbf{B}_2^\top & \mathbf{0}\end{pmatrix}\begin{pmatrix}\hat{\boldsymbol{\lambda}}\\ \hat{\boldsymbol{\beta}}\end{pmatrix}=\begin{pmatrix}-\mathbf{B}_1\mathbf{X}-\mathbf{b}\\ \mathbf{0}\end{pmatrix}, \quad \hat{\boldsymbol{\mu}}=\mathbf{X}+\mathbf{B}_1^\top\hat{\boldsymbol{\lambda}}. \tag{5}$$

Simplifying further, we have:

$$\begin{pmatrix}\hat{\boldsymbol{\mu}}\\ \hat{\boldsymbol{\beta}}\end{pmatrix}=-\begin{pmatrix}\mathbf{U}\mathbf{B}_1^\top\mathbf{Q}_{11}\\ \mathbf{Q}_{21}\end{pmatrix}\mathbf{b}+\begin{pmatrix}\mathbf{I}-\mathbf{U}\mathbf{B}_1^\top\mathbf{Q}_{11}\mathbf{B}_1\\ -\mathbf{Q}_{21}\mathbf{B}_1\end{pmatrix}\mathbf{X}, \tag{6}$$

where $\mathbf{Q}_{11}$ and $\mathbf{Q}_{21}$ are blocks of the matrix $\mathbf{Q}$ defined as:

$$\mathbf{Q}=\begin{pmatrix}\mathbf{Q}_{11} & \mathbf{Q}_{12}\\ \mathbf{Q}_{21} & \mathbf{Q}_{22}\end{pmatrix}=\begin{pmatrix}\mathbf{B}_1\mathbf{U}\mathbf{B}_1^\top & \mathbf{B}_2\\ \mathbf{B}_2^\top & \mathbf{0}\end{pmatrix}^{-1}. \tag{7}$$

The covariance matrix of the estimator $(\hat{\boldsymbol{\mu}}^\top,\hat{\boldsymbol{\beta}}^\top)^\top$, and in particular, of $\hat{\boldsymbol{\beta}}$, is given by:

$$cov\begin{pmatrix}\hat{\boldsymbol{\mu}}\\ \hat{\boldsymbol{\beta}}\end{pmatrix}=\begin{pmatrix}\mathbf{U}-\mathbf{U}\mathbf{B}_1^\top\mathbf{Q}_{11}\mathbf{B}_1\mathbf{U} & -\mathbf{U}\mathbf{B}_1^\top\mathbf{Q}_{12}\\ -\mathbf{Q}_{21}\mathbf{B}_1\mathbf{U} & -\mathbf{Q}_{22}\end{pmatrix}, \quad cov\left(\hat{\boldsymbol{\beta}}\right)=-\mathbf{Q}_{22}. \tag{8}$$

2.1. *The LBLUE in EIV model with nonlinear constraints*

Now, let us delve into the EIV model with nonlinear constraints on the parameters (1). Our approach relies on iterative linearizations of the model. In each iteration step, we linearize the nonlinear model at specific parameter values $\boldsymbol{\mu}_0$ and $\boldsymbol{\beta}_0$, and then apply the BLUE procedure locally at these estimates from the previous iteration step:

$$\Delta\mathbf{X}=\Delta\boldsymbol{\mu}+\boldsymbol{\varepsilon} \quad \& \quad \mathbf{B}_1^0\Delta\boldsymbol{\mu}+\mathbf{B}_2^0\Delta\boldsymbol{\beta}+\mathbf{b}^0=\mathbf{0}. \tag{9}$$

Table 1. Examples of nonlinear parameter constraints in 2-variate EIV models.

Nonlinear Constraints: $\mathbf{f}(\boldsymbol{\mu}, \boldsymbol{\nu}, \boldsymbol{\beta}) = \mathbf{0}$	Usage/Fitting
$\beta_1 + \beta_2\boldsymbol{\mu} - \boldsymbol{\nu} = \mathbf{0}$	Straight-line calibration function
$\beta_1\boldsymbol{\mu}^{-1} + \beta_2\boldsymbol{\mu}^{-\frac{1}{2}} + \beta_3 + \beta_4\boldsymbol{\mu}^{\frac{1}{2}} + \beta_5\boldsymbol{\mu} - \boldsymbol{\nu} = \mathbf{0}$	Fractional polynomial function
$\beta_1(\boldsymbol{\mu} - \beta_2)^{\beta_3} - \boldsymbol{\nu} = \mathbf{0}$	Oliver-Pharr function
$\boldsymbol{\mu}^2 + \beta_1\boldsymbol{\nu}^2 + \beta_2\boldsymbol{\mu}\boldsymbol{\nu} + \beta_3\boldsymbol{\mu} + \beta_4\boldsymbol{\nu} + \beta_5 = \mathbf{0}$	Ellipse function

Here, $\Delta\mathbf{X}$ represents $\mathbf{X}$ after subtracting $\boldsymbol{\mu}_0$, $\Delta\boldsymbol{\mu}$ represents $\boldsymbol{\mu}$ after subtracting $\boldsymbol{\mu}_0$, $\Delta\boldsymbol{\beta}$ represents $\boldsymbol{\beta}$ after subtracting $\boldsymbol{\beta}_0$, $\mathbf{B}_1^0$ is the partial derivative of $\mathbf{f}(\boldsymbol{\mu}, \boldsymbol{\beta})$ with respect to $\boldsymbol{\mu}^\top$ evaluated at $\boldsymbol{\mu}_0$ and $\boldsymbol{\beta}_0$, $\mathbf{B}_2^0$ is the partial derivative of $\mathbf{f}(\boldsymbol{\mu}, \boldsymbol{\beta})$ with respect to $\boldsymbol{\beta}^\top$ evaluated at $\boldsymbol{\mu}_0$ and $\boldsymbol{\beta}_0$, and $\mathbf{b}^0$ is $\mathbf{f}(\boldsymbol{\mu}_0, \boldsymbol{\beta}_0)$. The final estimate is determined when a specified convergence criterion, such as a relative change in the model parameter estimates, is met within a pre-defined tolerance.

Table 1 presents selected nonlinear parameter constraints in 2-variate EIV models, using vectors $(\mathbf{X}^\top, \mathbf{Y}^\top)^\top$ and $(\boldsymbol{\mu}^\top, \boldsymbol{\nu}^\top)^\top$ instead of $\mathbf{X}$ and $\boldsymbol{\mu}$.

$$\begin{pmatrix} \mathbf{X} \\ \mathbf{Y} \end{pmatrix} = \begin{pmatrix} \boldsymbol{\mu} \\ \boldsymbol{\nu} \end{pmatrix} + \begin{pmatrix} \boldsymbol{\varepsilon}_X \\ \boldsymbol{\varepsilon}_Y \end{pmatrix} \quad \& \quad \mathbf{f}(\boldsymbol{\mu}, \boldsymbol{\nu}, \boldsymbol{\beta}) = \mathbf{0}. \tag{10}$$

Extending the 2-variate model (10) to n-variate EIV models specified by nonlinear parameter constraints, such as the 3-variate EIV model used for sphere or ellipsoid fitting, is a straightforward process.

3. The OEFPIL algorithm

The *Optimum Estimate of Function Parameters by Iterated Linearization* (OEFPIL) algorithm is a powerful yet straightforward approach specifically designed to tackle complex EIV modeling challenges. In the EIV models under consideration, nonlinear constraints are imposed on their parameters, introducing complexity into the estimation process.

Generally, OEFPIL may not ensure convergence and provide a universally optimal solution, but it offers an easy-to-implement approach to navigate this complexity. It achieves this by employing iterative linearization of the EIV model, resulting in an efficient, robust, and locally optimal estimation process. This approach is particularly effective for EIV models exhibiting weak nonlinearity, where nonlinearity is not highly pronounced, and local approximations using linear models are suitable. However, it's essential to note that a clear criterion to assess the weak nonlinearity or *linearizability* of the nonlinear model is not yet directly available, and further research in this direction is still needed.

Here, we provide a concise overview of the OEFPIL algorithm (a simplified version), emphasizing its essential components and demonstrating how it can be effectively applied to address the intricate challenges posed by these EIV models. The following steps are applied in OEFPIL in each iteration of the iterative procedure:

(1) In each iteration step, evaluate the matrices $\mathbf{B}_1^0$, $\mathbf{B}_2^0$, and $\mathbf{b}^0$ as specified earlier. Notice that the following conditions should be fulfilled in order to obtain an identifiable model: $rank(\mathbf{B}_1^0, \mathbf{B}_2^0) = q$, $rank(\mathbf{B}_2^0) = p$, and $p \leq q \leq m+p$. The necessary derivatives can be calculated using either numerical differentiation or precomputed analytical derivatives.
(2) Calculate the Cholesky decomposition of $\mathbf{M} = \mathbf{B}_1^0 \mathbf{U} \mathbf{B}_1^{0\top} = \mathbf{L}_M \mathbf{L}_M^\top$.
(3) Calculate $\mathbf{E}$ by solving $\mathbf{L}_M \mathbf{E} = \mathbf{B}_2^0$ (using a triangular solve procedure if available).
(4) Calculate a 'thin' (or 'econ') SVD of $\mathbf{E} = \mathbf{U}_E \mathbf{S}_E \mathbf{V}_E^\top$.
(5) Calculate $\mathbf{F} = \mathbf{V}_E \mathbf{S}_E^{-1}$ (divide columns of $\mathbf{V}_E$ by their respective singular values).
(6) Calculate $\mathbf{G}$ by solving $\mathbf{L}_M^\top \mathbf{G} = \mathbf{U}_E$ (using a triangular transpose solve procedure if available).
(7) Calculate $\mathbf{Q}_{21} = \mathbf{F}\mathbf{G}^\top$.
(8) Calculate $\mathbf{Q}_{11} = (\mathbf{L}_M^\top)^{-1} \mathbf{L}_M^{-1} - \mathbf{G}\mathbf{G}^\top$ (use procedures for calculating the inverse of a symmetric positive matrix from a Cholesky factor).
(9) Calculate $\mathbf{Q}_{22} = -\mathbf{F}\mathbf{F}^\top$.

These steps form the core of the OEFPIL algorithm, providing a structured approach to estimating parameters in EIV models with nonlinear constraints. Finally, we get the resulting estimates,

$$\begin{aligned} \widehat{\Delta\boldsymbol{\mu}} &= (\mathbf{X} - \boldsymbol{\mu}_*) - \mathbf{U}\mathbf{B}_1^{*\top} \mathbf{Q}_{11}^* \left(\mathbf{B}_1^* \left(\mathbf{X} - \boldsymbol{\mu}_*\right) + \mathbf{b}^*\right), \\ \widehat{\Delta\boldsymbol{\beta}} &= -\mathbf{Q}_{12}^* \left(\mathbf{B}_1^* \left(\mathbf{X} - \boldsymbol{\mu}_*\right) - \mathbf{b}^*\right), \end{aligned} \tag{11}$$

where we denote the approximated vectors and matrices in the final iteration (when convergence is achieved) as $\boldsymbol{\mu}_*$, $\boldsymbol{\beta}_*$, $\mathbf{B}_1^*$, $\mathbf{B}_2^*$, $\mathbf{b}^*$, $\mathbf{Q}_{11}^*$, $\mathbf{Q_{12}}^*$, and $\mathbf{Q}_{22}^*$. Therefore, we set the final LBLUE estimators as follows:

$$\hat{\boldsymbol{\mu}} = \boldsymbol{\mu}_* + \widehat{\Delta\boldsymbol{\mu}}, \quad \hat{\boldsymbol{\beta}} = \boldsymbol{\beta}_* + \widehat{\Delta\boldsymbol{\beta}}, \quad cov(\hat{\boldsymbol{\beta}}) = -\mathbf{Q}_{22}^*. \tag{12}$$

It is important to note that when the algorithm converges, we essentially obtain $\hat{\boldsymbol{\mu}} \equiv \boldsymbol{\mu}_*$ and $\hat{\boldsymbol{\beta}} \equiv \boldsymbol{\beta}_*$. Finally, it's worth noting that in the case of an n-variate model, the m-dimensional vectors $\mathbf{X}$ and $\boldsymbol{\mu}$ are partitioned into n equally-sized $(\frac{m}{n})$-dimensional subvectors, denoted as $\mathbf{X}^\top = (\mathbf{X}_1^\top, \ldots, \mathbf{X}_n^\top)$ and $\boldsymbol{\mu}^\top = (\boldsymbol{\mu}_1^\top, \ldots, \boldsymbol{\mu}_n^\top)$.

3.1. *The OEFPIL implementation and examples*

OEFPIL has been thoughtfully implemented across various programming environments, providing the flexibility to cater to the diverse needs of users and a wide array of applications. These implementations are freely available and can be easily located using standard web search tools.

The first implementation of OEFPIL was developed as an R package, accessible on both CRAN (https://CRAN.R-project.org/package=OEFPIL) and GitHub (https://github.com/OEFPIL/OEFPIL/), offering compatibility with the R language. Additionally, for users who prefer working with the C language, the latest developments and an efficient implementation based on advanced libraries and algorithms is available in the form of an OEFPIL library, accessible on GitLab (https://gitlab.com/cmi6014/oefpil/).

For users looking for a MATLAB version of OEFPIL, a new implementation has been introduced, and it is readily accessible on GitHub (https://github.com/witkovsky/OEFPIL-MATLAB) including both the 2-variate and n-variate versions, accompanied by a set of illustrative examples that cover EIV models specified with nonlinear constraints, as detailed in Table 1, see https://github.com/witkovsky/OEFPIL-MATLAB/tree/main/Examples. This MATLAB version is made freely available, facilitating its accessibility for researchers interested in comparisons, further enhancements, and practitioners seeking efficient solutions for complex EIV modeling challenges.

4. Conclusions

In this paper, we tackled computational challenges in nonlinear curve fitting and function calibration, especially in nonlinear EIV models with complex data uncertainties and potential correlations. We employed the LBLUE method, implemented as OEFPIL algorithm in R and C. Additionally, we introduced a new MATLAB version of OEFPIL, accompanied by a set of illustrative examples that cover diverse EIV models specified with nonlinear constraints. Our ongoing research focuses on weakly nonlinear models and characterizing regions suitable for linearization, advancing the field of computational metrology with robust solutions.

Acknowledgements

The research was supported by the Czech Ministry of Education, Youth and Sports and the Slovak Research and Development Agency through the Inter-Excellence II program, project LUASK22008/SK-CZ-RD-21-0109.

References

[1] Kubáček, L. (1988). *Foundations of Estimation Theory.* (Elsevier).

[2] Oliver, W. C., Pharr, G. M. (1992). An improved technique for determining hardness and elastic modulus using load and displacement sensing indentation experiments. *Journal of Materials Research*, 7(6), 1564–1583.

[3] ISO/TS 28037:2010, *Determination and Use of Straight-Line Calibration Functions.* (International Standards Organization, Geneva, 2010).

[4] Cox, M., Harris, P. (2015). Polynomial calibration functions revisited: numerical and statistical issues. In *AMCTM X*, 9–16.

[5] Witkovský, V., Wimmer, G. (2021). Polycal-MATLAB algorithm for comparative polynomial calibration and its applications. In *AMCTM XII*, 501–512.

[6] Köning, R., Wimmer, G., Witkovský, V. (2014). Ellipse fitting by nonlinear constraints to demodulate quadrature homodyne interferometer signals and to determine the statistical uncertainty of the interferometric phase. *Measurement Science and Technology*, 25(11), 115001.

[7] Ahn, S. J., Rauh, W., Warnecke, H. J. (2001). Least-squares orthogonal distances fitting of circle, sphere, ellipse, hyperbola, and parabola. *Pattern Recognition*, 34(12), 2283–2303.

[8] Turner, D., Anderson, I., Mason, J., and Cox, M. (1999). An algorithm for fitting an ellipsoid to data. *Technical Report*, National Physical Laboratory, UK.

[9] Golub, G. H., Van Loan, C. F. (1980). An analysis of the total least squares problem. *SIAM Journal on Numerical Analysis*, 17(6), 883–893.

[10] Malengo, A., Pennecchi, F. (2013). A weighted total least-squares algorithm for any fitting model with correlated variables. *Metrologia*, 50(6), 654.

[11] Šlesinger, R., Charvátová Campbell, A., Geršlová, Z., Šindlář V., Wimmer G. (2023). OEFPIL: New method and software tool for fitting nonlinear functions to correlated data with errors in variables. In *MEASUREMENT 2023*, Smolenice, Slovakia, May 29–31, 2023, 126–129.

[12] Charvátová Campbell, A., Geršlová, Z., Šindlář, V., Šlesinger, R., Wimmer, G. (2024). New framework for nanoindentation curve fitting and measurement uncertainty estimation. *Precision Engineering*, 85, 166–173.

F Pavese, A Bošnjaković, S Eichstädt, A B Forbes, J Alves e Sousa (eds.):
Advanced Mathematical and Computational Tools in Metrology and Testing XIII
Series on Advances in Mathematics for Applied Sciences, Vol. 94
 (pp. 151–158)
https://doi.org/10.1142/9789819800674_0013

New prioritization of QFDs' engineering characteristics through the Law of Comparative Judgment

F. Franceschini and D. A. Maisano†

Politecnico di Torino, Dept. of Management and Production Engineering
Corso Duca degli Abruzzi 24, Turin, 10129, Italy
†E-mail: domenico.maisano@polito.it

The *House of Quality* (HoQ) is the key module of *Quality Function Deployment* (QFD), which aims to support the design of new products or services by translating customer requirements (CRs) into engineering characteristics (ECs). The *independent scoring method* is the most widely used technique for the prioritisation of ECs; despite this, it has two conceptual shortcomings: (i) arbitrary numerical conversion of (ordinal) relationships between CRs and ECs (i.e., *null*, *weak*, *medium*, and *high*), and (ii) "promotion" of these relationships from an *ordinal* to a *cardinal* scale. This paper introduces a novel procedure for prioritizing ECs, inspired by the Thurstone's *Law of Comparative Judgment* (LCJ), which overcomes the conceptual shortcomings of the ISM while maintaining its flexibility and ease of implementation. A realistic application example illustrates the potential of the proposed procedure.

Keywords: house of quality; engineering-characteristic prioritization; law of comparative judgment; scale promotion.

1. Introduction

Quality Function Deployment (QFD) is a powerful methodology for designing new products and services according to *customer requirements* (CRs) (Franceschini, 2001). The *House of Quality* (HoQ) is the initial QFD module, translating CRs into *engineering characteristics* (ECs) [1]. The core of the HoQ is the *relationship matrix*, which establishes the relationships between ECs and CRs, through symbols indicating the intensity of the relationship: ∅ → *None*, Δ → *Low*, ○ → *Medium*, ● → *High*).

Figure 1 exemplifies a relationship matrix for the design of a climbing safety harness. The eight CRs shown in the rows on the left have been associated with weights, reported on the right (i.e., w_{CR_i} values, being $\sum_{\forall i} w_{\mathrm{CR}_i} = 1$), which describe their relevance to the final customer [2]. The columns of the relationship matrix refer to the ECs (i.e., EC_1, EC_2, ...) associated with the CRs. The relationship between ECs and CRs is expressed through the aforementioned symbols.

Rel. intensity	Symbol	r_{ij}
None	∅	0
Low	△	1
Medium	○	3
High	●	9

		EC1 Meets safety standards	EC2 Harness weight	EC3 Webbing strength	EC4 No. of colours	EC5 No. of sizes	EC6 Padding thickness	EC7 No. of buckles	EC8 No. of gear loops	w_{CR_i}
Easy to put on	CR1					○		●		7%
Comfortable when hanging	CR2					○	●	○		20%
Fits over different clothes	CR3					○	○	●		3%
Accessible gear loops	CR4								●	12%
Does not restrict movement	CR5		○			○	●	○		27%
Lightweight	CR6		●	○			○	△	△	10%
Safe	CR7	●	○	●						14%
Attractive	CR8		△		●		△	△		7%
	w_{EC_j}	1.26	2.20	1.56	0.63	1.71	4.69	2.48	1.18	

Fig. 1. Relationship matrix and EC prioritization using the ISM.

One of the objectives of the HoQ is to prioritize the ECs in order to identify the CRs that have the greatest impact on the final customer, so that more attention can be paid to them in the design process. In general, more attention should be given to the ECs that are related to multiple CRs, with relatively high weights, and high intensity relationships. The prioritization technique traditionally used is the *independent scoring method* (ISM), i.e., a weighted sum of coefficients obtained by numerical conversion of the symbols in the relationship matrix, using the w_{CR_i} values as weights. The resulting weight of the *j*th EC is determined by:

$$w_{\mathrm{EC}_j} = \sum_{\forall i}\left(r_{ij} \cdot w_{\mathrm{CR}_i}\right), \tag{1}$$

where r_{ij} is the numerical coefficient corresponding to the relationship between EC_j and CR_i, according to the numerical (exponential) conversion $\emptyset \rightarrow 0$, $\Delta \rightarrow 1$, $\bigcirc \rightarrow 3$, $\bullet \rightarrow 9$ (cf. top left of Fig. 1).

Although ISM is simple and practical, it has some shortcomings. First, the conversion of the relationship-matrix symbols (i.e., *ordinal* quantities) into numerical coefficients is arbitrary. Secondly, the aggregation of these numerical coefficients through a weighted sum improperly "promotes" them from ordinal to *cardinal* quantities [3]. These shortcomings make the solution provided by ISM conceptually imperfect [4, 5].

The objective of this contribution is to develop a new technique for prioritizing ECs that is alternative to ISM and overcomes its weaknesses. The new procedure will be based on Thurstone's *Law of Comparative Judgment* (LCJ), which is a well-established, conceptually rigorous, and effective statistical technique [6].

2. Methodology

This section briefly revisits the LCJ, contextualizing it within the framework of the QFD's HoQ. In general, LCJ aims to prioritize a set of objects (i.e., ECs) based on an attribute (i.e., their impact on the final customer, from the perspective of the individual CRs). Each CR — which corresponds to a specific row of the relationship matrix — can be translated into paired-comparison relationships among potential pairs of ECs. This process takes place in two stages. In the first stage, a ranking of ECs is identified for each CR, based on their impact on the CR itself. Let us illustrate this with an example related to CR_1 (cf. the first row of the relationship matrix in Fig. 1). Based on the symbols denoting intensities of relationship with the ECs, a ranking consisting of four preference blocks (in round brackets) with mutual *strict preference* relationships ("$\succ$") can be constructed:

$$(EC_{\bullet} \sim EC_7) \succ (EC_{\circ} \sim EC_5) \succ EC_{\triangle} \succ (EC_{\emptyset} \sim EC_1 \sim EC_2 \sim EC_3 \sim EC_4 \sim EC_6 \sim EC_8). \quad (2)$$

The block ($EC_{\bullet} \sim EC_7$), including ECs with high-intensity relationship is followed by the one ($EC_{\circ} \sim EC_5$) including ECs with medium-intensity relationship, then the one ($EC_{\triangle}$) including ECs with weak-intensity relationship, and finally the one ($EC_{\emptyset} \sim EC_1 \sim EC_2 \sim EC_3 \sim EC_4 \sim EC_6 \sim EC_8$) including ECs with no relationship. This ranking also includes four special (or *dummy*) objects to account for the relationship intensity: $EC_{\emptyset} \rightarrow$ *Absence of impact*, $EC_{\triangle} \rightarrow$ *Low impact*, $EC_{\circ} \rightarrow$ *Medium impact*, and $EC_{\bullet} \rightarrow$ *Maximum-possible impact.*

Table 1. Paired-comparison relationships between ECs, from the perspective of their impact on CR_1.

	EC_1	EC_2	EC_3	EC_4	EC_5	EC_6	EC_7	EC_8	$EC_{\emptyset}$	$EC_{\triangle}$	$EC_{\circ}$	$EC_{\bullet}$
EC_1	-											
EC_2	$\sim$	-										
EC_3	$\sim$	$\sim$	-									
EC_4	$\sim$	$\sim$	$\sim$	-								
EC_5	$\prec$	$\prec$	$\prec$	$\prec$	-							
EC_6	$\sim$	$\sim$	$\sim$	$\sim$	$\succ$	-						
EC_7	$\prec$	$\prec$	$\prec$	$\prec$	$\prec$	$\prec$	-					
EC_8	$\sim$	$\sim$	$\sim$	$\sim$	$\succ$	$\sim$	$\succ$	-				
$EC_{\emptyset}$	$\sim$	$\sim$	$\sim$	$\sim$	$\succ$	$\sim$	$\succ$	$\sim$	-			
$EC_{\triangle}$	$\prec$	$\prec$	$\prec$	$\prec$	$\succ$	$\prec$	$\succ$	$\prec$	$\prec$	-		
$EC_{\circ}$	$\prec$	$\prec$	$\prec$	$\prec$	$\sim$	$\prec$	$\succ$	$\prec$	$\prec$	$\prec$	-	
$EC_{\bullet}$	$\prec$	$\prec$	$\prec$	$\prec$	$\prec$	$\prec$	$\sim$	$\prec$	$\prec$	$\prec$	$\prec$	-

In the second step, each ranking is translated into paired-comparison relationships, as exemplified in Table 1, with reference to CR_1 [3]. These paired comparisons are based on the intensity of the relationship with the CR of interest, utilizing relationships of *strict preference* (e.g., "$EC_j \succ EC_k$" and "$EC_j \prec EC_k$") as well as *indifference* (e.g., "$EC_j \sim EC_k$").

The same construction can be extended to the other CRs (i.e., CR_2 to CR_8), resulting in eight total sets of paired-comparison relationships (omitted for the sake of brevity). For each specific *jk*th paired comparison of ECs (i.e., EC_j versus EC_k), the eight available relationships can be aggregated on the basis of the weights of the CRs associated with them. With reference to each *i*th CR and each *jk*th paired comparison, the following three *binary parameters* are defined:

$$\begin{aligned} &c_{i,jk}^{(\succ)} \text{ being equal to 1 if } EC_j \succ EC_k \text{ (otherwise 0)},\\ &c_{i,jk}^{(\sim)} \text{ being equal to 1 if } EC_j \sim EC_k \text{ (otherwise 0)},\\ &c_{i,jk}^{(\prec)} \text{ being equal to 1 if } EC_j \prec EC_k \text{ (otherwise 0)}.\end{aligned} \quad (3)$$

It can be seen that these parameters are mutually exclusive, and the complementarity relationship holds: $c_{i,jk}^{(\succ)} + c_{i,jk}^{(\sim)} + c_{i,jk}^{(\prec)} = 1$. A general parameter expressing the degree of preference of EC_j over EC_k, from the perspective of the *i*th CR, can be defined as:

$$c_{i,jk} = 1 \cdot c_{i,jk}^{(\succ)} + ½ \cdot c_{i,jk}^{(\sim)} + 0 \cdot c_{i,jk}^{(\prec)}, \in \{0, ½, 1\}. \quad (4)$$

Finally, the $c_{i,jk}$ parameters are utilised in the following weighted sum, which allows to determine the p_{jk} values:

$$p_{jk} = \sum_{\forall i}(c_{i,jk} \cdot w_{CR_i}), \in [0, 1]. \quad (5)$$

These p_{jk} values represent the weighted portion of CRs on which EC_j has a greater impact than EC_k.[a] By construction, the complementary relationship $p_{jk} = 1 - p_{jk}$ applies. All p_{jk} values, calculated for all possible *jk*th paired comparisons, are aggregated into a matrix of proportions (***P***), where symmetrical elements with respect to the main diagonal complement each other.

Let us briefly digress to better understand the practical meaning of the p_{jk} values. The LCJ is based on several postulates, including the representation of the impact of ECs on the final customer using a latent one-dimensional scale.

[a]In the traditional LCJ procedure, p_{jk} proportions are determined based on the judgments on the objects in question by a panel of experts, without the weighting introduced in this adaptation.

Specifically, the impact of a specific EC (such as EC_j) can be treated as a normally distributed variable, with an unknown mean and variance, reflecting its variability across different CR perspectives: $\mathrm{EC}_j \sim \mathbb{N}\left(\mu_{\mathrm{EC}_j}, \sigma^2_{\mathrm{EC}_j}\right)$. Additionally, the LCJ assumes equal variances for ECs ($\sigma^2_{\mathrm{EC}_1} = \sigma^2_{\mathrm{EC}_2} = \ldots = \sigma^2$), and equal intercorrelations between pairs of ECs (in the form of Pearson coefficients $\rho_{\mathrm{EC}_j,\mathrm{EC}_k} = \rho, \forall j, k$). The difference between the impact of EC_j and EC_k can thus, by extension, be represented as a normally distributed variable with the following parameters: $(\mathrm{EC}_j - \mathrm{EC}_k) \sim \mathbb{N}\left[\left(\mu_{\mathrm{EC}_j} - \mu_{\mathrm{EC}_k}\right), 2 \cdot \sigma^2 \cdot (1-\rho)\right]$ (Thurstone, 1928). Now, focusing on the probability:

$$P\left[(\mathrm{EC}_j - \mathrm{EC}_k) > 0\right] = P(\mathrm{EC}_j > \mathrm{EC}_k), \tag{6}$$

it represents the portion of the CRs where EC_j has more impact than EC_k and precisely corresponds to the quantity p_{jk} that we determined earlier. Therefore, this implies that p_{jk} can be expressed as a function of the unknown values μ_{EC_j} and μ_{EC_k} (demonstration omitted).

Next, the p_{jk} values are transformed into z_{jk} values using the relationship:

$$z_{jk} = \Phi^{-1}(1 - p_{jk}), \tag{7}$$

$\Phi^{-1}(\cdot)$ being the inverse of the cumulative distribution function of the standard normal distribution, and z_{jk} being a unit normal deviate. In general, p_{jk} values of 1.00 and 0.00 would correspond to z_{jk} values of $\pm\infty$; a simplified approach to address this issue is to associate p_{jk} values ≤ 0.00135 with $z_{jk} = \Phi^{-1}(1 - 0.00135) = 3$ and p_{jk} values ≥ 0.99865 with $z_{jk} = \Phi^{-1}(1 - 0.99865) = -3$. More sophisticated solutions to deal with this issue have been proposed [3].

Next, the z_{jk} values related to the possible paired comparisons are reported into a matrix $\boldsymbol{Z}$. The relationship $z_{kj} = -z_{jk}$ holds, being unit normal deviates related to complementary cumulative probabilities. It can be demonstrated that the unknown mean values of the impact of individual ECs (μ_{EC_j}) can be determined — up to a positive scale factor and an additive constant — by summing column by column the values contained in the $\boldsymbol{Z}$-matrix. In other words, the LCJ allows for to be positioned of ECs on an *interval* scale (referred to as x) with an arbitrary zero point and unit.

Drawing inspiration from the methodology developed in [3], the interval scaling resulting from the LCJ can be "anchored", using two of the previously introduced dummy objects: $EC_{\emptyset}$, which represents the absence of a relationship, and $EC_{\bullet}$, which represents the maximum-possible degree of intensity of a relationship with the CR of interest. The (interval) scale x is then transformed into a new one (y), defined within the conventional range [0, 10], through the following linear transformation:

$$\frac{y_k - 0}{10 - 0} = \frac{x_k - x_{\emptyset}}{x_{\bullet} - x_{\emptyset}} \rightarrow y_k = 10 \cdot \frac{x_k - x_{\emptyset}}{x_{\bullet} - x_{\emptyset}}, \tag{8}$$

where:

$x_{\emptyset}$ and $x_{\bullet}$ are the scale values of $EC_{\emptyset}$ and $EC_{\bullet}$, resulting from LCJ;

x_k is the scale value of a generic kth EC, resulting from LCJ;

y_k is the transformed scale value within the conventional range [0, 10].

The introduction of $EC_{\emptyset}$ and $EC_{\bullet}$ allows to anchor the scale x into a new scale (y) with a conventional unit and a zero point (which corresponds to the absence of the attribute); it is therefore admissible to consider y as a *ratio* scale and use the y-values of the ECs as w_{EC_k} values (cf. Sec. 1).

Figure 2 summarises the results of the application of the proposed procedure to the HoQ in Fig. 1.

3. Concluding remarks

This research presented an innovative procedure for prioritizing ECs within the HoQ framework, building upon Thurstone's LCJ. This procedure offers several notable advantages, including its ability to adapt to the traditional relationship matrix without the need for specialized response modes. It prioritizes ECs using a *ratio* scaling approach, anchored by the inclusion of dummy ECs. In addition, it is easy to implement, automatable, flexible, and adaptable to other response modes than the traditional one. The effectiveness and robustness of the new EC prioritization is ensured by the LCJ itself, which is a well-established technique that has been used and tested in multiple contexts [3].

In the test case, the new procedure closely aligns with ISM, demonstrating a high correlation (a comparison of the w_{EC_k} values resulting from the two methods yields $R^2 \approx 0.8602$) and producing EC rankings that are nearly identical, with only a minor rank reversal between EC_1 and EC_8:

$$\begin{aligned} &\text{ISM: } EC_6 > EC_7 > EC_2 > EC_5 > EC_3 > EC_1 > EC_8 > EC_4. \\ &\text{New procedure: } EC_6 > EC_7 > EC_2 > EC_5 > EC_3 > EC_8 > EC_1 > EC_4. \end{aligned} \tag{9}$$

While preliminary tests showed some agreement between the proposed procedure and the ISM, we acknowledge the possibility of divergent results in specific scenarios, that will be subject to further investigation. Looking ahead, the proposed procedure represents a valuable addition to the QFD toolkit, expanding the analytical possibilities. Despite the inherent limitations of LCJ [3], it is planned to explore its adaptation for situations with uncertain and/or incomplete formulation of relationships. Additionally, it is planned to conduct comprehensive comparative analyses with alternative ISM methodologies from existing QFD literature [7].

(a) Matrix ***P***

	EC_1	EC_2	EC_3	EC_4	EC_5	EC_6	EC_7	EC_8	$EC_\emptyset$	$EC_\triangle$	$EC_\circ$	$EC_\bullet$
EC_1	0.500	0.350	0.450	0.535	0.285	0.235	0.200	0.460	0.570	0.140	0.140	0.070
EC_2	0.650	0.500	0.650	0.720	0.505	0.370	0.470	0.730	0.790	0.545	0.305	0.050
EC_3	0.550	0.350	0.500	0.585	0.335	0.285	0.300	0.560	0.620	0.240	0.190	0.070
EC_4	0.465	0.280	0.415	0.500	0.250	0.235	0.200	0.425	0.535	0.070	0.070	0.035
EC_5	0.715	0.495	0.665	0.750	0.500	0.215	0.365	0.675	0.785	0.570	0.285	0.000*
EC_6	0.765	0.630	0.715	0.765	0.785	0.500	0.735	0.775	0.835	0.635	0.535	0.235
EC_7	0.800	0.530	0.700	0.800	0.635	0.265	0.500	0.760	0.870	0.655	0.335	0.050
EC_8	0.540	0.270	0.440	0.575	0.325	0.225	0.240	0.500	0.610	0.170	0.120	0.060
$EC_\emptyset$	0.430	0.210	0.380	0.465	0.215	0.165	0.130	0.390	0.500	0.000*	0.000*	0.000*
$EC_\triangle$	0.860	0.455	0.760	0.930	0.430	0.365	0.345	0.830	1.000*	0.500	0.000*	0.000*
$EC_\circ$	0.860	0.695	0.810	0.930	0.715	0.465	0.665	0.880	1.000*	1.000*	0.500	0.000*
$EC_\bullet$	0.930	0.950	0.930	0.965	1.000*	0.765	0.950	0.940	1.000*	1.000*	1.000*	0.500

(b) Matrix ***Z***

	EC_1	EC_2	EC_3	EC_4	EC_5	EC_6	EC_7	EC_8	$EC_\emptyset$	$EC_\triangle$	$EC_\circ$	$EC_\bullet$
EC_1	0.000	0.385	0.126	-0.088	0.568	0.722	0.842	0.100	-0.176	1.080	1.080	1.476
EC_2	-0.385	0.000	-0.385	-0.583	-0.013	0.332	0.075	-0.613	-0.806	-0.113	0.510	1.645
EC_3	-0.126	0.385	0.000	-0.215	0.426	0.568	0.524	-0.151	-0.305	0.706	0.878	1.476
EC_4	0.088	0.583	0.215	0.000	0.674	0.722	0.842	0.189	-0.088	1.476	1.476	1.812
EC_5	-0.568	0.013	-0.426	-0.674	0.000	0.789	0.345	-0.454	-0.789	-0.176	0.568	3.000
EC_6	-0.722	-0.332	-0.568	-0.722	-0.789	0.000	-0.628	-0.755	-0.974	-0.345	-0.088	0.722
EC_7	-0.842	-0.075	-0.524	-0.842	-0.345	0.628	0.000	-0.706	-1.126	-0.399	0.426	1.645
EC_8	-0.100	0.613	0.151	-0.189	0.454	0.755	0.706	0.000	-0.279	0.954	1.175	1.555
$EC_\emptyset$	0.176	0.806	0.305	0.088	0.789	0.974	1.126	0.279	0.000	3.000	3.000	3.000
$EC_\triangle$	-1.080	0.113	-0.706	-1.476	0.176	0.345	0.399	-0.954	-3.000	0.000	3.000	3.000
$EC_\circ$	-1.080	-0.510	-0.878	-1.476	-0.568	0.088	-0.426	-1.175	-3.000	-3.000	0.000	3.000
$EC_\bullet$	-1.476	-1.645	-1.476	-1.812	-3.000	-0.722	-1.645	-1.555	-3.000	-3.000	-3.000	0.000

(c) Scaling

	EC_1	EC_2	EC_3	EC_4	EC_5	EC_6	EC_7	EC_8	$EC_\emptyset$	$EC_\triangle$	$EC_\circ$	$EC_\bullet$
$x_k = \Sigma_k$	-6.116	0.336	-4.167	-7.989	-1.627	5.202	2.161	-5.794	-13.545	0.183	9.025	22.330
$w_{\mathbf{EC}_k} = y_k$	2.1	3.9	2.6	1.5	3.3	5.2	4.4	2.2	0.0	3.8	6.3	10.0

Fig. 2. (a) ***P*** matrix, (b) ***Z*** matrix, and (c) scaling resulting from the application of the proposed procedure to the test case. $x_k = \Sigma_k$ represents the summation of the values reported in the kth column of the ***Z*** matrix and corresponds to the *interval* scale value resulting from LCJ; the y_k values relate to the *ratio* scaling downstream of anchoring in Eq. (8).

References

1. Akao, Y. (1994). Development history of quality function deployment. The Customer Driven Approach to Quality Planning and Deployment, 339, 90.
2. Franceschini, F. (2001). Advanced quality function deployment. CRC Press.

3. Franceschini, F., Maisano, D.A., Mastrogiacomo, L. (2022). Rankings and Decisions in Engineering. Springer International Publishing, Cham, Switzerland.
4. Lyman, D., 1990. Deployment normalization. Transactions from the Second Symposium on Quality Function Deployment, pp. 307–315.
5. Huang, J., Mao, L.X., Liu, H.C., Song, M.S. (2022). Quality function deployment improvement: A bibliometric analysis and literature review. Quality & Quantity, 56(3), 1347–1366.
6. Thurstone, LL. (1927). A law of comparative judgments. Psychological Review, 34(4): 273.
7. Zare Mehrjerdi, Y. (2010). Quality function deployment and its extensions. International Journal of Quality & Reliability Management, 27(6), 616–640.

F Pavese, A Bošnjaković, S Eichstädt, A B Forbes, J Alves e Sousa (eds.):
Advanced Mathematical and Computational Tools in Metrology and Testing XIII
Series on Advances in Mathematics for Applied Sciences, Vol. 94

https://doi.org/10.1142/9789819800674_0014

Using minimum covariance determinant for simultaneous noise removal and robust planar surface fitting in coordinate metrology

E. Ghanbary Kalajahi† and M. Mahboubkhah††
Department of Mechanical Engineering, University of Tabriz
Tabriz, Iran
E-mail: †E-ghanbary@tabrizu.ac.ir, ††mahboobkhah@tabrizu.ac.ir

A. Barari
Department of Mechanical and Manufacturing Engineering, University of Ontario Institute of Technology, Oshawa, ON, Canada
E-mail: Ahmad.barari@uoit.ca

Integrated inspection system (IIS) enriches digital metrology by decreasing uncertainty due to closed-loop implementation of the cyber-components of digital metrology including planning the measurement of points, fitting the best Substitute Geometry (SGE) and evaluating the actual geometry of the surface. Optical devices are favored over contact metrology devices as they offer faster, fuller, and non-invasive measurement, but they are also susceptible to noise and outliers that may affect the accuracy and quality of the inspection. Principal Component Analysis (PCA) has been presenting high efficiency in attaining transformation matrix of the large number of discrete points given by the optical devices based on the point's distribution; however, it is not robust to noise. Noise removal prior to SGE provides pure information feedbacked to the other cyber-components of IIS, and finally correctly identifies the actual geometry of the inspected surface. In this paper, in order to obtain the robust principal component of the unorganized point cloud of the planar surfaces, a subset of the data, which is least affected by noise, is found with Minimum Covariance Determinant (MCD), and then PCA is conducted on the obtained MCD-subset. Due to the MCD's robustness, the noises can be detected by their large robust distances based on the MCD location and scatter matrix. The results of implementation on two case studies show the efficiency of MCD in simultaneous noise removal and robust surface fitting. Variety of inspection tools can utilize the methodology for point cloud filtration.

Keywords: point cloud; filtration; denoising; robust fitting; Industry 4.0; integrated inspection system; digital metrology.

1. Introduction

Coordinate metrology is the measurement of a limited number of 3-D points from the surface to estimate the *actual* manufactured surface geometry. It consists of three cyber-components: point measurement planning (PMP), by which the location and the number of the measured points are planned; SGE, by which the best substitute geometry to the point cloud (PC) is fitted by selecting an appropriate fitting criteria; deviation zone evaluation (DZE), by which the actual geometry of the surface is obtained by evaluating the deviation of the points from the substitute geometry. Closed-loop implementation of these steps in Integrated Inspection Systems (IIS), results in improving the inspection quality. Therefore, feedbacking accurate information from each cyber-component to the other components is the key point [1, 2].

Acquiring the points from the surface, at first an appropriate coordinate system should be defined for the unorganized PC to fit the best geometry (SGE). Principal Component Analysis (PCA), by which a number of correlated variables transform in to a smaller number of uncorrelated and orthogonal variables presents sufficient result for SGE [3]. PCA gives an orthogonal matrix which represents the main distribution of the points and the direction of each distribution obtained based on the covariance matrix of the data i.e. eigenvectors. Having eigenvectors and mean center of the data, transformation matrix of the data points for SGE implementation is obtained. According to the nature of the optical devices, the obtained PCs are highly prone to noise [4]. Since the PCA is conducted based on all the points' information, the result of SGE is influenced by these measurement errors. Mesh denoising is the most conventional approach of the literatures in PC filtration and a few research directly deal with raw PC denoising [5]. Hence, it is essential to remove noises directly from the unorganized PC before the fitting process, so that much robust geometry would be fitted.

Among a few research on PC filtration, some of them have considered neighborhood-based approaches. Digne *et al.* [6], and Zhang *et al.* [7] used bilateral filtering, which helps to smooth out noisy areas by applying a weighted average of neighboring points based on spatial distance and intensity similarity. In [8], based on the position information of the neighbors, a filter kernel for each point is built by extending the concept of guided image filtering approach to PC filtration, which later been improved by employing points' normal as guidance signal [9]. Optimization-based point cloud filtration approaches such as moving least squares (MLS) [10, 11], involve nonlinear problems and are computationally expensive. The probability model of the PC is obtained by kernel based clustering approach in [12], and then smoothing the point cloud is done by moving the point to the location with most probability based on mean shift clustering approach.

Minimum Covariance Determinant (MCD) estimator, which has been introduced by Rousseeuw [13, 14] as a highly robust statistical method, is used in [15] for real-time outlier detection, and in [16] for robust local planar surface fitting. MCD is an affine equivariant and highly robust estimator of multivariate location and scatter which is utilized to find a subset of data points that has minimum determinant of the covariance matrix [17]; The smallest eigenvalue of the covariance matrix of a planar PC measures the variance of the orthogonal distances from the plane. Since, the determinant of the covariance matrix is the product of the eigenvalues and is sensitive to the smallest one, it can be interpreted that a subset with minimum covariance determinant is a subset that has the smallest variance of the orthogonal distances, and so is the closest subset to the plane. The mean and covariance matrix of MCD subset can be used as robust location and scatter matrix.

This paper deals with utilizing MCD for simultaneous geometry fitting and noise removal of unorganized PC of planar surfaces, which are the widely used surfaces in industry. MCD scatter matrix is used to extract the principal components of the data robustly. Moreover, due to the MCD's robustness, noises can be detected by their large robust distances based on the MCD estimators. Section 2 will discuss the fitting process, and the concept of MCD is explained in Sec. 3 with the original figures created for this paper. Implementation of the case studies will be discussed in Sec. 4, and Sec. 5 includes the conclusion.

2. Fitting plane in PCA coordinate frame

Applying PCA on the unorganized PC ($\mathbf{P} = [x_i\ y_i\ z_i]$, i = 1,2,3,...,n) captured in the scanner coordinate frame, the transformation matrix (**T**) is obtained by the mean of the points ($\boldsymbol{\mu}$) and the obtained eigenvectors. In Eq. (1), **Cov**, **V** and $\boldsymbol{\lambda}$ are the covariance matrix of **P**, the rotation matrix that consists of the obtained eigenvectors and the vector of eigenvalues, respectively. Hence, the coordinate of the points in PCA frame can be obtained as $\mathbf{P}_{\mathbf{PCA}}$ with Eq. (2).

$$\mathbf{Cov} \times \mathbf{V} = \mathbf{V} \times \boldsymbol{\lambda}. \tag{1}$$

$$\begin{cases} [\mathbf{P}_{\mathbf{PCA}} \quad 1] = [\mathbf{P} \quad 1] \times \mathbf{T}, \\ \mathbf{T} = \begin{bmatrix} \mathbf{V}_{3\times3} & \boldsymbol{\mu}_{3\times1} \\ 0 & 1 \end{bmatrix}_{4\times4}. \end{cases} \tag{2}$$

Each of three eigenvalues shows the density of the points along its corresponding eigenvector that declares one of the axes of PCA coordinate frame. The axis with the smallest respective eigenvalue declares Z axis after fitting (Z_{PCA}) since the density of the points along the thickness of the plane is the smallest, and the new X axis corresponds to the biggest eigenvalue. Figure 1 shows the fitting procedure in the PCA coordinate frame.

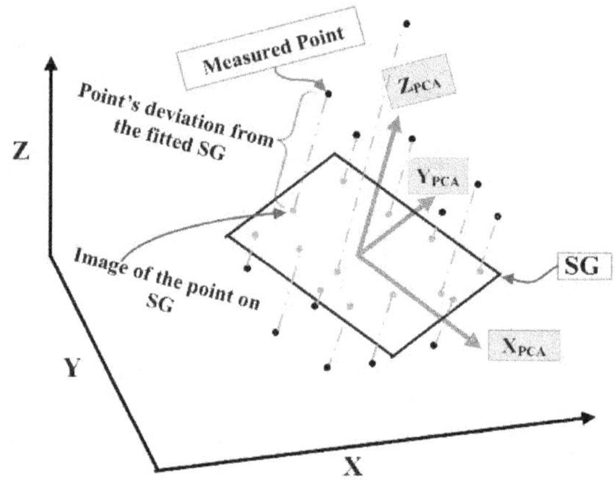

Fig. 1. Fitting procedure in PCA coordinate frame.

3. MCD for robust SGE and denoising

The aim of using MCD is to investigate through all n number of data points in matrix $\mathbf{P}_{n\times 3}$ in order to find the subset of size h ($n/2 \leq h \leq n$), such that the determinant of its covariance matrix is the smallest among all other possible subset with the same h size. Once MCD-subset is found, robust SGE can be conducted with the transformation matrix obtained by substituting the MCD location and scatter matrix, in Eq. (2). Besides, the Mahalanobis distance (Eq. (3)) as the robust distance of each point $\mathbf{p}_i$, is calculated, and the points with robust distance higher than a defined threshold are detected as noise and removed.

$$RD_i = \sqrt{(\mathbf{p}_i - \boldsymbol{\mu}_{\mathbf{MCD}})^t \, \mathbf{Cov}_{\mathbf{MCD}}^{-1} (\mathbf{p}_i - \boldsymbol{\mu}_{\mathbf{MCD}})}. \tag{3}$$

The Mahalanobis distance is a weighted sum of the squares of the distances along all the principal axes, where the weights are inversely proportional to the eigenvalues of the covariance matrix. This is in contrast with the approach of the Least squares orthogonal distance regression (LSODR) [18], which is the most conventional method used to fit the best plane to the measured data points. The LSODR method (in other words Total Least Squares (TLS) [19]), uses the orthogonal distance. Orthogonal distance is the distance measured along the normal vector of the plane, which is the same as the eigenvector corresponding to the smallest eigenvalue of the covariance matrix. Hence, the orthogonal distance is sensitive to outliers and noise, while the Mahalanobis is more robust. Another advantage of the MCD algorithm is that it is invariant to rotations and translations of the data, as is the LSODR method. This property is important for applications in coordinate metrology, where the data may be measured in different reference frames or orientations. The original MCD method requires the investigation on all the possible *h* subsets out of *n* i.e. $\binom{n}{h}$, which is highly computational and is not a practical approach. The solution is FAST-MCD algorithm [14]. In the following, the main steps and concepts of FAST-MCD algorithm will be discussed with an original figure created for this paper.

3.1 *FAST-MCD fundamental*

If $\mathbf{P}_{n\times m}$ is the data points, in which n is the number of the observations and m is the number of the variables (here $m = 3$), FAST-MCD algorithm is conducted in four main steps, starting with the random selection of many $m+1$ subsets since in this case the probability of taking the noise-free subsets is much higher than selecting many random *h* subsets:

- Step 1- Randomly select 500 numbers $(m+1)$-subsets from the data.
- Step 2- Make initial *h*-subsets from each of the previous subsets:
  - 2.1- Find covariance matrix and mean of each subset of step 1.
  - 2.2- For *all* data points, calculate robust distances with respect to this covariance matrix and mean (Eq. (3)).
  - 2.3- Consider *h* data points with the smallest robust distances as the initial *h* subsets ($\mathbf{H}_1$ matrix in Fig. 2).
- Step 3- Run C-step algorithm for each $\mathbf{H}_1$ to obtain the final *h* subsets ($\mathbf{H}_2$ matrix). Figure 2 illustrates the C-step algorithm which is the major concept of the FAST-MCD. Running the procedures similar to steps 2.1 to 2.3 on $\mathbf{H}_1$, new *h* subset is substituted for the $\mathbf{H}_1$. According to [14], the determinant of the covariance matrix of the new subset is always lower than the older. Finally, $\mathbf{H}_2$ is obtained by continuing this iteration until the determinant changes no longer or becomes zero.
- Step 4- Select the $\mathbf{H}_2$ with the smallest determinant as the ultimate MCD-subset.

In fact, the core component of the fast MCD is C-step algorithm, which is a descent procedure that updates the subset of points with minimum covariance determinant by replacing the points with the smallest distances with the points with largest distances. Although this descent procedure guarantees stopping the algorithm in finite steps, it may not lead to global solution; however, choosing many starting points would improve the solution quality. It would be useful to mention that the fast MCD algorithm is embedded in Matlab software, as *robustcov.m* function.

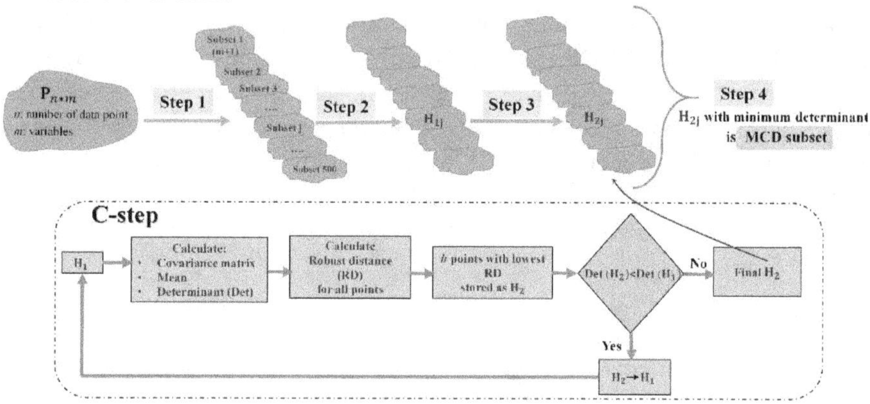

Fig. 2. FAST-MCD fundamental.

4. Implementation and discussion

The PC of a planar surface with 33000 data points is shown in Fig. 3 in which 9% of the points are the randomly added noises. Fitting the best substitute plane (SGE) by applying PCA to this noisy point cloud, the standard deviation of the deviations of all the points from the fitted plane (as flatness criteria) is 0.034 mm. By implementing MCD on this PC, and then performing PCA on the obtained MCD-subset, standard deviation from the new fitted plane is decreased to 0.003 mm that is 91% reduction. Moreover, the points which have large robust distances are detected as noise. Figure 4 displays the data before and after noise removal. Besides, according to Table 1 (surface number 1), 2545 points out of 3000 noisy points are removed by the proposed method (TP = 2545), which means noise removal process is about 85% successful; also, no healthy points is dismissed as noise (FP = 0), and 455 noisy points are not removed (FN = 455). It is worth noting that the mean and the maximum of the deviations of FN points are 0.0077 mm and 0.0172 mm, respectively; this demonstrates that the effect of the remained noisy points in the final filtered PC is trivial.

The PC with local noisy points is displayed in Fig. 5. The information of this case study is also included in Table 1 (surface number 2). Figure 6 illustrates the PC before and after removing the detected noise. The standard deviation of the deviations of all the points from the fitted plane is 0.0151 mm and 0.003 mm, before and after using the proposed procedure, respectively, which has been improved 80%. Moreover, 601 points out of all 660 noisy points, or in other words 91% of all noisy points are removed. TP and FP for this case study are 601 and 0, respectively, which again demonstrates the efficiency of the method in noise removing and keeping healthy data in the filtered PCs. The mean and the maximum of the deviations of 59 noisy points that are not detected through the procedure are 0.0081 mm and 0.0137 mm, respectively, which again declares trivial effect of the remaining noisy points in the final filtered PC.

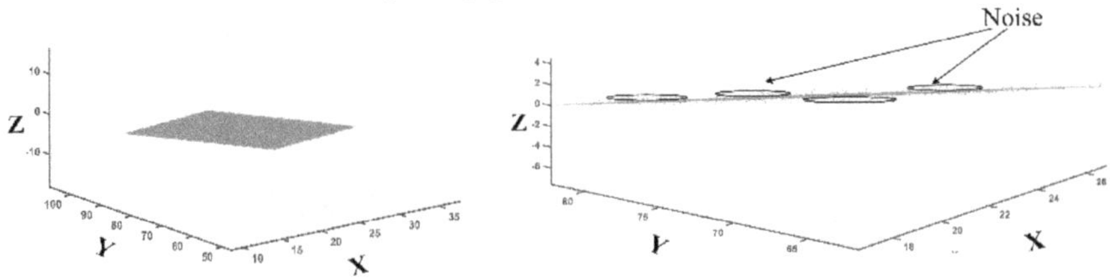

Fig. 3. The PC of a planar surface with general noise.

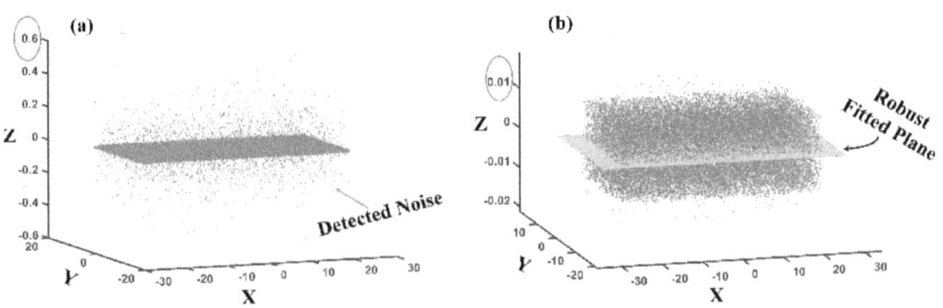

Fig. 4. Noise removal and robust plane fitting by MCD method of the PC with general noise.

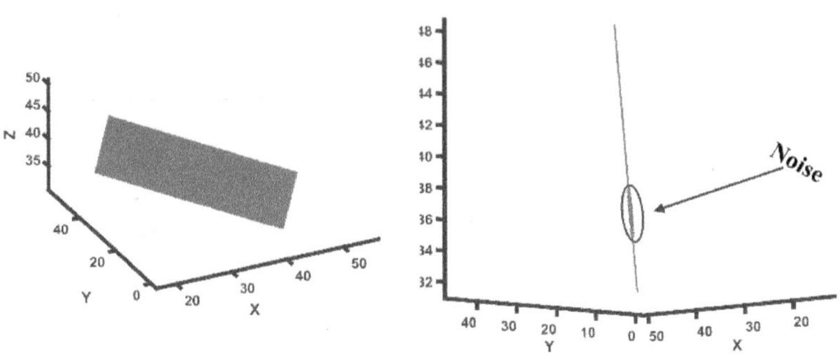

Fig. 5. The PC of a planar surface with local noise.

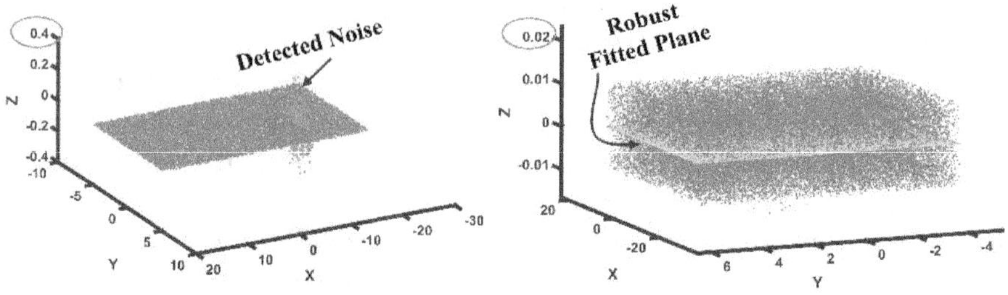

Fig. 6. Noise removal and robust plane fitting of the PC with local noise.

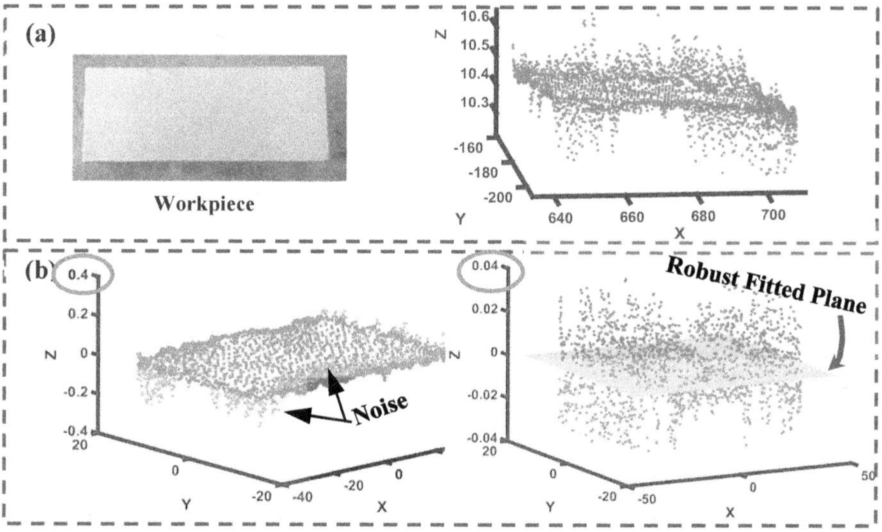

Fig. 7. (a) The PC of a workpiece with ground surface, (b) noise removal and robust plane fitting.

Table 1. The results of the case studies.

Surface Number	Number of Points	Number of Noise	STD of deviations (mm)	TP	FP	FN	Mean of deviations of FN (mm)
1	33000	3000	Before 0.034 After 0.003	2545	0	455	0.0077
2	44660	660	Before 0.0151 After 0.003	601	0	59	0.008

Figure 7(a) shows an aluminum workpiece whose surface has been machined and ground to achieve a good surface finish. This surface was scanned, and the acquired PC was filtered as real measurement data. Figure 7(b) Shows the result of noise removal. The standard deviation of the PC decreased from 0.04 mm to 0.008 mm after using the proposed procedure, demonstrating the capability of the method for real data processing.

5. Conclusion

Minimum covariance determinant as a robust estimator is used for concurrent robust surface fitting and noise removal of unorganized point cloud acquired from the planar surfaces. Robust SGE is highly essential in order to provide accurate information for precise evaluation of the actual surface in DZE and feedbacking correct information to PMP cyber-component of IIS. The results of robust fitting and also noise removal of three case studies have been discussed; applying the proposed method on a PC with general noise and on a PC with local noise has shown 85% and 91% noise reduction, respectively. Moreover, it was demonstrated that the remained noisy points in the filtered PC have trivial effects. The two latter case studies have also shown considerable reduction in STD of the deviations for the PC due to robust surface fitting. Besides, the noise removal of a PC from a ground aluminium surface, demonstrated the method's efficiency in filtering real data.

References

[1] E. Ghanbary Kalajahi, M. Mahboubkhah, A. Barari, On detailed deviation zone evaluation of scanned surfaces for automatic detection of defected regions, *Measurement*, 221 (2023) 113462.

[2] H. Gohari, A. Barari, Finding optimal correspondence sets for large digital metrology point clouds using anisotropic diffusion analogy, *International Journal of Computer Integrated Manufacturing*, 35 (2022) 462–482.

[3] E.G. Kalajahi, M. Mahboubkhah, A. Barari, Reducing Noises in Digital Surface Inspection Using a Data Clustering Approach, *IFAC-PapersOnLine*, 54 (2021) 193–198.

[4] C. Berry, M.S.G. Tsuzuki, A. Barari, Data Analytics for Noise Reduction in Optical Metrology of Reflective Planar Surfaces, *Machines*, 10 (2021) 25.

[5] X.-F. Han, J.S. Jin, M.-J. Wang, W. Jiang, L. Gao, L. Xiao, A review of algorithms for filtering the 3D point cloud, *Signal Processing: Image Communication*, 57 (2017) 103–112.

[6] J. Digne, C. de Franchis, The Bilateral Filter for Point Clouds, *Image Processing On Line*, 7 (2017) 278–287.

[7] F. Zhang, C. Zhang, H. Yang, L. Zhao, Point Cloud Denoising with Principal Component Analysis and a Novel Bilateral Filter, *Traitement du Signal*, 36 (2019) 393–398.

[8] X.-F. Han, J.S. Jin, M. Wang, W. Jiang, Guided 3D point cloud filtering, *Multimedia Tools and Applications*, 77 (2018) 17397–17411.

[9] X.-F. Han, J.S. Jin, M.-J. Wang, W. Jiang, Iterative guidance normal filter for point cloud, *Multimedia Tools and Applications*, 77 (2017) 16887–16902.

[10] M. Alexa, J. Behr, D. Cohen-Or, S. Fleishman, D. Levin, C.T. Silva, Computing and rendering point set surfaces, *IEEE transactions on visualization and computer graphics*, 9 (2003) 3–15.

[11] Z. Xu, A. Foi, Anisotropic Denoising of 3D Point Clouds by Aggregation of Multiple Surface-Adaptive Estimates, *IEEE transactions on visualization and computer graphics*, 27 (2021) 2851–2868.

[12] O. Schall, A. Belyaev, H.P. Seidel, Robust filtering of noisy scattered point data, Proceedings Eurographics/IEEE VGTC Symposium Point-Based Graphics, 2005, 2005, pp. 71–144.

[13] P.J. Rousseeuw, Least Median of Squares Regression, *Journal of the American Statistical Association*, 79 (1984) 871–880.

[14] P.J. Rousseeuw, K. van Driessen, A Fast Algorithm for the Minimum Covariance Determinant Estimator, *Technometrics*, 41 (1999) 212–223.

[15] B. De Ketelaere, M. Hubert, J. Raymaekers, P.J. Rousseeuw, I. Vranckx, Real-time outlier detection for large datasets by RT-DetMCD, *Chemometrics and Intelligent Laboratory Systems*, 199 (2020) 103957.

[16] A. Nurunnabi, D. Belton, G. West, Robust statistical approaches for local planar surface fitting in 3D laser scanning data, *ISPRS Journal of Photogrammetry and Remote Sensing*, 96 (2014) 106–122.

[17] M. Hubert, M. Debruyne, Minimum covariance determinant, *WIREs Computational Statistics*, 2 (2009) 36–43.

[18] K. Pearson, LIII. On lines and planes of closest fit to systems of points in space, *The London, Edinburgh, and Dublin Philosophical Magazine and Journal of Science*, 2 (2010) 559–572
DOI: https://doi.org/10.1080/14786440109462720.

[19] G.H.a.V.L. Golub, Charles F., Matrix Computations, Fourth ed., The Johns Hopkins University Press, 2013.

F Pavese, A Bošnjaković, S Eichstädt, A B Forbes, J Alves e Sousa (eds.):
Advanced Mathematical and Computational Tools in Metrology and Testing XIII
Series on Advances in Mathematics for Applied Sciences, Vol. 94

https://doi.org/10.1142/9789819800674_0015

Machine learning data augmentation by electrical contact resistance technique for edge effect correction in nanoindentation*

Jasurkhuja Kholkhujaev†, Giacomo Maculotti, Gianfranco Genta and Maurizio Galetto

Department of Management and Production Engineering, Politecnico di Torino, Corso Duca degli Abruzzi 24, 10129 Torino, Italy
†E-mail: jasurkhuja.kholkhujaev@polito.it
http://www.qualityengineering.polito.it/

This research describes a novel method to correct for the biased measurement of contact area due to edge-effect phenomena in Instrumented Indentation Test (IIT) process. IIT is a widely used depth-sensing technique to evaluate the mechanical characteristics of materials e.g., Indentation modulus, Indentation hardness, residuals stresses, damping, creep. The proposed procedure utilizes the Electrical Contact Resistance approach to augment the data on contact area. Monte Carlo Markov Chain model was used to randomly generate the relatively small dataset to robustly predict projected area. The advantage of the method is demonstrated through the application for different conductive metals.

Keywords: Instrumented Indentation Test; uncertainty; electric contact resistance; quality inspection.

1. Introduction

Ongoing industrial transition towards green initiatives is forcing industries to develop novel technological processes, materials, and quality controls [1]. Particularly, the surface technologies have been largely studied due to miniaturization of manufacturing, semiconductor industry and advancements in multi-layered coating technologies. For example, physical vapor deposition (PVD) coatings have been studied to increase the surface mechanical and chemical properties of surgical instruments increasing the service live and corrosion resistance [2]. In addition to innovative surface technologies, the application of machine learning models to understand, control and optimize the mechanical properties of the materials and surface processes are increasingly exploited [3]. Furthermore, sensor fusion, data augmentation and enrichment are

*This work is supported by the European Union Next-Generation EU (PIANO NAZIONALE DI RIPRESA E RESILIENZA (PNRR) – MISSIONE 4 COMPONENTE 2, INVESTIMENTO 1.3 – D.D. 1551.11-10-2022, PE00000004).

often resorted to improve accuracy and precision of characterization and quality controls. Monte Carlo Markov Chain (MCMC) approach in the context of machine learning and computational statistics have been recently proposed to enrich small datasets [4]. Indeed, robust quality inspection and quality assurance methods are required to deal with advanced surface technologies. Instrumented Indentation Test (IIT) emerged as a non-conventional mechanical surface characterization method for scales ranging from nano to macro. IIT is a depth-sensing technique that enables the extraction of mechanical properties of the materials [5].

The IIT procedure continuously measures the applied force, *F*, and the indenter's displacement h, allowing construction of load-depth curve. Indentation hardness H_{IT} of the materials can be evaluated exploiting the relation $A_p(h)$ between the projected contact area of the indenter with the sample and the indentation depth h, $H_{IT} = F/A_p(h_c)$ using hc: $h_c = h - h_0 - \varepsilon F/S - C_f F$.

Such relationship allows resolving mechanical characterization at the nano and micro scales. The correction of the contact depth h_c is required to cater for indenter machine compliance C_f, elastic recovery of the sample material proportional to the sample contact stiffness S, and "non-zero" contact point h_0 [6]. The parameter ε is related to the indenter geometry and equal to 0.75 for Berkovich indenter, while the calibrated machine compliance C_f, to correct for elastic displacement of instrument during the indentation process, is crucial to ensure the traceability of characterization [7].

The area shape function $A_p(h)$ is a polynomial function with several terms accounting for different irregularities of the indenter geometry, e.g., tip rounding and geometrical imperfections, and is calibrated either by direct or indirect methods. However, despite the calibration, IIT application for some elastic-plastic materials is troublesome, due to edge localized plasticity at relatively high loads and indentation size effect (ISE) at low loads [8]. Edge-effect tends to induce the calibrated $A_p(h)$ to systematically underestimate or overestimate the actual contact area as in the Fig. 1(b). This phenomenon is related to the material mechanical response, well modeled by Young's modulus, yield stress and work-hardening exponent [9]. Traditional approach to correct the area for the edge-effect is post-indentation analysis with surface-topographical measurement of the indentation areas by Atomic Force Microscopy (AFM), which is highly impractical [8].

A novel approach based on electrical contact resistance (ECR) enables direct traceable correction of the edge effects, through continuous measurement of resistance *R* during IIT as in Fig. 1(a). The resistance measurement augments the informativeness of mechanical quantities conventionally measured introducing a

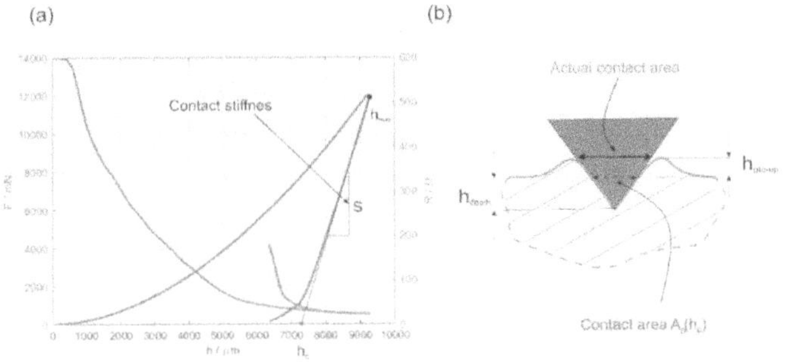

Fig. 1. (a) Force and electrical resistance is used for electromechanical characterization of materials. (b) Biased evaluation of material effected by pile-up by standard Oliver and Pharr approach.

redundancy that enables the bias correction. The method calibrates the $A_p(R)$ relationship of a reference material that is not influenced by edge effects, e.g., Aluminum. Then, it exploits normalization with respect to the reference material to predict actual projected area, thus correcting the bias, while removing the effect of the resistivity. This work proposes a methodology to enrich the dataset numerosity and improve robustness of the data augmentation procedure based on MCMC data simulation method which will also enable the evaluation of the measurement uncertainty of the corrected quantity.

2. Methodology

The ECR edge effect correction is based on resistance-based data augmentation obtained thanks to a current passing through a conductive indenter into the material, as can be seen in Fig. 2. The electrical contact between the conductive indenter and the metal can be modeled as a series of resistances:

$$R_{tot} = R_c + R_{spread} + R_{tip} + R_{elc}, \tag{1}$$

where R_{tip} and R_{elc} are bulk resistance components of the indenter and the electronics. R_{spread} is used to describe the current spreading phenomena due to narrow conducting spot and it is proportional to the reciprocal of contact radius. R_c is a contact resistance that is inversely proportional area of contact. R_{spread} and R_c contributions can prevail total resistance measurement in semiconductors and metals [10].

The current flows through so-called *a-spots* when two conductors are pressed one onto the other to form an electric contact. The current flowing between the indenter and conductive material has a non-uniform density that causes the spreading phenomenon [12]. In theory, the spreading resistance is a function that

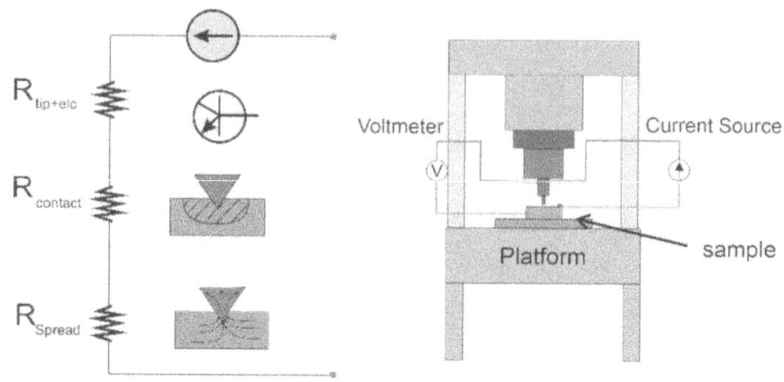

Fig. 2. Electric resistance model of IIT on the right and ECR set-up on the left.

depends on the resistivities of the indenter and the sample, ρ_{ind} and ρ_s, on the ratio of the contact radius to the thickness of the members t_{ind} and t_s, the radius of contact a_c, and a contact geometry constant c_g as in the Eq. (2):

$$R_{spred} = c_g \frac{\rho_{ind}}{a_c} \arctan \frac{2t_{ind}}{a_c} + c_g \frac{\rho_s}{a_c} \arctan \frac{2t_s}{a_c}. \quad (2)$$

Further approximation could be applied for thick samples such that $t_{ind} \gg a_c$ [10], resulting in Eq. (3). As for R_c, the relationship is given according to the Eq. (4), where σ_c is the specific contact resistance.

$$R_{spred} = c_g \frac{\rho_{ind}}{a_c} \frac{\pi}{2} + c_g \frac{\rho_s}{a_c} \frac{\pi}{2} = c_g \pi \frac{\rho_{aver}}{a_c} \quad (3)$$

$$R_c = c_g \frac{\sigma_c}{a_c^2}. \quad (4)$$

The serial resistance model of IIT process is widely adopted by recent works on ECR [11] and can be seen in the Eq. (5).

$$R_t = c_g \pi \frac{\rho_{aver}}{a_c} + c_g \frac{\sigma_c}{a_c^2} + R_{tip} + R_{elec}. \quad (5)$$

The proportionality according to Eq. (6) can be derived based on the projected area A_p and radius of contact a_c relationship for IIT procedure. Mathematical inversion of the Eq. (6) allows analyzing the contact area as a function of resistance, as in Eq. (7):

$$R_t = \frac{c_0}{A_p} + \frac{c_1}{\sqrt{A_p}} + R_0 \quad (6)$$

$$A_p(R) = \frac{c_2}{\sqrt{R}} + \frac{c_3}{R^2} + c_4. \quad (7)$$

The parameters c_2, c_3 and c_4 could be evaluated from a regression based on experimental data and it is crucial to estimate the projected area without an a-posteriori measurement with optical instruments to allow resolving nanoscales.

The regression can be performed by collecting the indentation data on the different force levels for F_{max}, $R(h_{c,max})$ and $h_{c,max}$ from a sample within a force range not significantly affected by the edge effect, i.e., unbiased. $A_p(R)$ must be evaluated based on the previously calibrated $A_p(h)$ relationship. The proposed methodology relies on a reference material that does not suffer from edge effect for the force levels at which Eq. (7) is applied. As the approach depends on the reference material, a "normalization" is required to cope with difference of resistivities of the unknown material needing bias correction and the reference material. It has also been observed in the Eq. (6) that c_0 is an order of magnitude smaller than c_1 for semiconductor and metals for indentation depths $h_c > 20$ nm [11] so that the simplifications can follow as:

$$R_t - R_0 \approx \frac{c_1}{\sqrt{A_p}} \tag{8}$$

$$A_p(\Delta R) \approx \frac{c_1^2}{\Delta R^2}. \tag{9}$$

Orthogonal Distance Regression (ODR) is applied to evaluate the coefficients of Eq. (7), since it allows robust evaluation compared to conventional least-squares-regression (LSQ) when the regressor's uncertainty cannot be neglected. The general proposed methodology is illustrated in Fig. 3.

Subscripts "*m*" and "*ref*" indicate the unknown conductive material that requires the correction and the reference material, respectively. After estimating reference material parameters which calibrates the relationship $A_{p,ref}(R)$, the dataset of the resistance R_m and projected area $A_{p,m}(h_c)$ are collected for an unbiased range. Indeed, this requires a wide range of forces to be investigated, and to exclude data affected by ISE at the lower end and by edge effect at the higher end. Typically, this investigation may be time consuming, and few data per load and sparse load levels can be investigated. Thus, an enrichment of dataset numerosity is performed by applying Monte Carlo Markov Chain (MCMC) approach to a gaussian process with deterministic part as per step 3 in Fig. 3.

Specifically, the model assumes a gaussian process for contact area $A_p \sim GP(A_{p,m}(R), N(0, \Sigma_A))$ and a stochastic contribution with variance estimated from reproducibility of projected area $A_{p,m}$ and resistance R_m. The enriched dataset is used to fit the ODR model to estimate accurately the projected area $A_{p,m}$ of the unknown material. The expanded uncertainty of the projected area $A_{p,m}$ is evaluated according to GUM [8] considering contribution of the RMSE of the residuals, resolution of measuring instruments and standard deviation of parameters. The sensibility to the sampling and the force level selection will be investigated by randomly subsampling the force range and applying the methodology, and relying on the measurement uncertainty of the corrected area hypothesis test to identify systematic effects introduced by the force levels will be performed. Last, the resistivity normalization can be carried out relying on the proportionality simplification of Eq. (9), considering that the indentation depth in the case at hand that is $h_c > 9$ µm. The normalization is performed for the neighborhoods of $R_m \pm \delta R$ in an unbiased range of area measurements $A_{p,m}(h_c)$. Indeed, the reference material regression model as for Eq. (7) is used to generate the dataset for the unknown material that is based on the physical behavior of the unbiased material, thus needing a correction for the resistivity that is achieved through the normalization [12]. Last, the prediction of the unbiased area in the edge-effect affected range, i.e. the corrected area, is relying on the ODR fitting on normalized and MCM-increased dataset in the unbiased force range.

1. Estimation by **ODR** the parameters c_2, c_3 **and** c_4 of the reference material

↓

$$A_{p,ref}(R) = \frac{c_{2,ref}}{\sqrt{R}} + \frac{c_{3,ref}}{R^2} + c_{4,ref}$$

2. Measurement of electrical resistance and contact area of unknown material based on conventional **Oliver and Phar** approach

↓

$$A_{p,m}(R_m)$$

3. Random increase of dataset numerosity according to **MCMC** for the contact area with deterministic part based on normalization of contact area of unknown material in a neighborhood of each ΔR in an unbiased range.

↓

$$A_{p,MCMC}(\Delta R_m \pm \delta R) = A_{p,ref}(\Delta R_m \pm \delta R)\frac{A_{p,m}(h_{c,m})}{A_{p,ref}(\Delta R_m)}$$

4. Estimation by **ODR** the parameters c_2, c_3 **and** c_4 for unknown material for unbiased force range

↓

$$A_{p,un}(R) = \frac{c_{2,m}}{\sqrt{R}} + \frac{c_{3,m}}{R^2} + c_{4,m}$$

Fig. 3. ECR based edge-effect correction procedure.

3. Experimental setup

Anton Paar MCT[3] STeP6 indentation platform hosted in Mind4Lab laboratory at Politecnico di Torino has been used to apply and validate the method proposed in Sec. 2. The platform has built-in voltage data acquisition set-up with conductive Berkovich indenter. The force transducer of the indentation machine is calibrated with an uncertainty of 0.5 mN and equipped with LVDT displacement sensor with uncertainty of 0.6%. Area shape function and frame compliance has been calibrated according to conventional approach [7].

Three conductive bulk materials have been examined: Stainless Steel (SS) that suffers from pile-up for application range in hand, Aluminum (Al) and Brass (Br) that are not affected by edge-effect [13]. Calibration of the materials has been performed with AXHU09 by AXIOTEK, at macro-scale ranges at which edge-effect has almost no impact on measurements [13].

The forces ranging from 1 N to 30 N have been analyzed with 10 replications per force level. Al has been used as a reference material, Br to validate the approach and Ss to correct for edge-effect.

Analysis of the raw data of H_{IT} measured by standard approach for three materials in Fig. 4 shows the systematic effect in SS data for $F > 10$ N, evidence of the edge-effect. At low loads ($F < 3$ N), ISE can be appreciated for the three materials. The data collected and shown in Fig. 4, in terms of $R(h_{c,max})$, $h_{c,max}$ and F_{max}, are used to apply the methodology to correct the edge effect bias according to Sec. 2. Specifically, according to Fig. 4, Al is selected as a reference material and Br to validate the methodology.

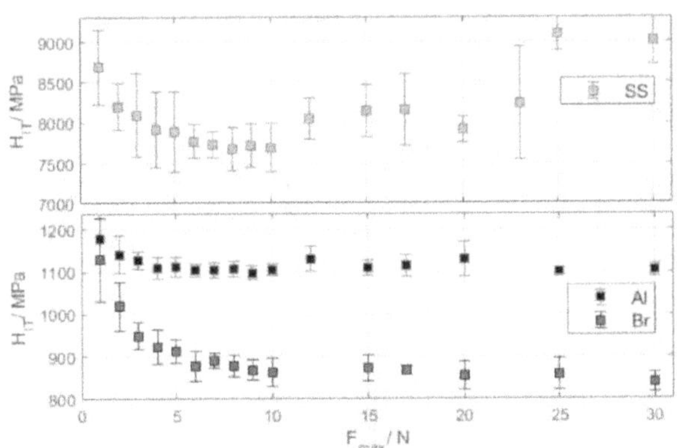

Fig. 4. Indentation hardness H_{IT} measurement by standard Oliver and Phar approach against the F_{max}. Notice that it is possible to divide the figures into three phases e.g. ISE at low loads, unbiased measurement of hardness and systematic effect of pile-up.

4. Results

Figure 5 illustrates the electromechanical characterization results obtained by interpolation of resistance and force measurements with indentation depth on Al.

ODR parameters of reference material Al have been evaluated based on the Eq. (7) yielding the RMSE of 4.9e-10 m^2 and R^2 equal 0.9946. Validation of the methodology is performed on the Br, because the material is also not biased by edge-effect as Al. MCMC based dataset points were generated according to the step 3 in Fig. 5 on the neighborhood of $\delta R = 0.5\ \Omega$.

ODR fitting of Br dataset yields RMSE equal 10.2e-12 m^2 and R^2 of 0.99. Figure 5(a) shows a random choice of force cluster to generate the MCMC dataset.

The ODR fitting illustrates the expected results on raw data of Br falling within the 95% confidence intervals showing no systematic difference between the fitting and raw data. Random force level extraction is performed between 1 N–30 N varying the subsamples, and no systematic difference is detected as illustrated in Fig. 5(b).

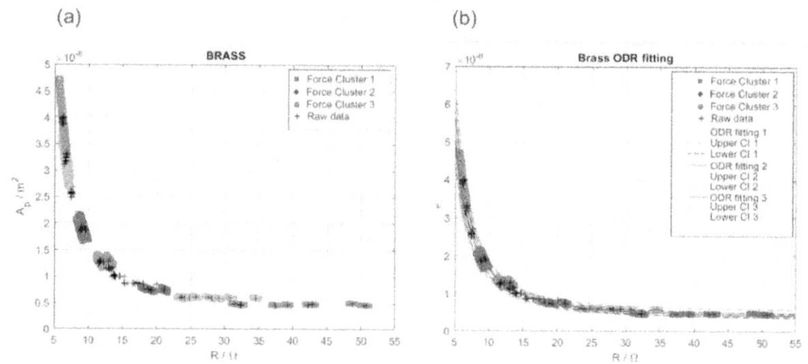

Fig. 5. Validation of ECR-based approach on Br. (a) Force clusters of data is chosen randomly. (b) Notice that 3 different force clusters are chosen but no systematic difference can be seen for the confidence interval (CI) overlap.

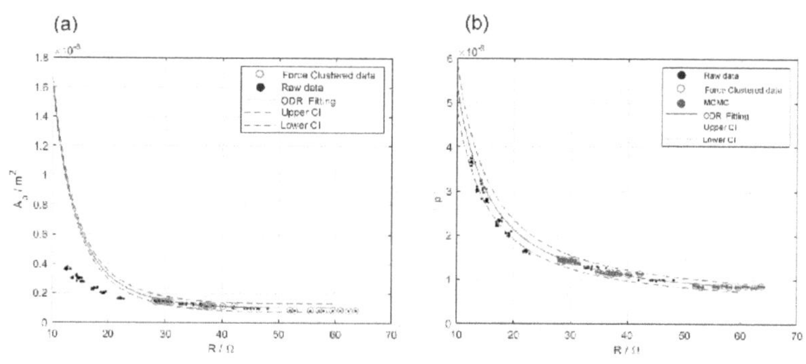

Fig. 6. Prediction of corrected area on edge-effect affected SS. (a) ODR applied on the raw data. (b) MCMC data generation.

Figure 6 shows the necessity of application of MCMC method and compares the results of standard ODR application. The prediction based on ODR method overestimates the projected area function while MCMC data generation and further fitting yields compatible prediction. MCMC dataset generation on the randomly selected force clusters in an unbiased range that is defined as “No Pile-up zone” in Fig. 7(a) is performed. ODR fitting on obtained MCMC-enriched dataset successfully predicts the actual area for the zone effected by pile-up, with a RMSE of 3.77e-11 and R^2 of 0.99 and identifies biased results in Fig. 7(a) as the experimental points falling outside the prediction interval. Such beneficial effect of the data enrichment by MCMC is induced by the introduction of local trends, calibrated on the unbiased material and normalized, which thanks to the ODR allows to robustly cater for that [14].

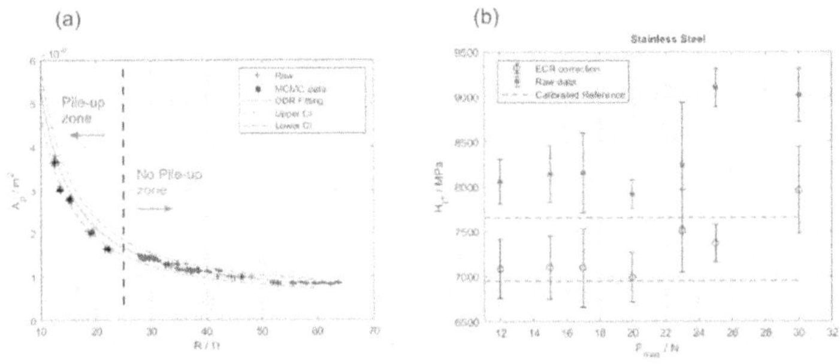

Fig. 7. Application of ECR method to correct the pile-up on Ss. Notice that the pile-up zone has biased by "flatness" of curve.

The bias correction based on ECR data augmentation and machine learning enrichment method in the Fig. 7(b) shows that there is no statistically significant difference between the corrected value and the reference at 95% confidence level, thus demonstrating the approach. The proposed correction method has a relative expanded uncertainty of 4.63% and accuracy of 4.06%. In comparison, AFM based topography correction yields expanded uncertainty of 4.77%.

5. Conclusion

This work highlights metrological performances of a bias correction method for IIT based on in-situ resistance measurement data augmentation and machine learning modelling. The method allows to improve accuracy and precision of the bias correction. In fact, thanks to the sensor fusion providing the data augmentation and the redundancy, the accuracy of IIT can be greatly improved. Similarly, the machine learning approach based on the gaussian process regression modelling and MCMC data enrichment allows to increase the robustness of the model predicting the correct contact area. The method, which requires calibration on a reference material, and the identification of ranges of forces affected by the edge-effect bias allows to set up the correction independently from the resistivity thanks to a normalization approach and relying on few data thanks to the MCMC data enrichment. The obtained results show that ECR method detects and successfully corrects the systematic errors, resulting a powerful practical tool for in-line accurate measurement of the contact area, for it does not require expensive optical post-indentation analysis performed by microscopes. Future work will apply the methodology for semiconductor materials and the nano-range.

References

[1] S. Wolf, J. Teitge, J. Mielke, F. Schütze and C. Jaeger, "The European Green Deal — More Than Climate Neutrality," *Intereconomics*, Vol. 56, no. 2, pp. 99–107 (2021).

[2] L. Geyao, D. Yang, C. Wanglin and W. Chengyong, "Development and application of physical vapor deposited coatings for medical devices: A review," in *Procedia CIRP*, Elsevier B.V. (2020), pp. 250–262.

[3] G. Maculotti, G. Genta and M. Galetto, "Optimisation of laser welding of deep drawing steel for automotive applications by Machine Learning: A comparison of different techniques," *Qual. Reliab. Eng. Int.* (2023).

[4] R. Bardenet, A. Doucet and C. Holmes, "On Markov chain Monte Carlo methods for tall data," 2017. Available: http://jmlr.org/papers/v18/15-205.html.

[5] D. A. Lucca, K. Herrmann and M. J. Klopfstein, "Nanoindentation: Measuring methods and applications," *CIRP Annals*, Vol. 59, no. 2, pp. 803–819 (2010).

[6] G. Barbato *et al.*, "Uncertainty evaluation of indentation modulus in the nano-range: Contact stiffness contribution," *CIRP Annals*, Vol. 66, no. 1, pp. 495–498 (2017).

[7] M. Galetto, G. Genta and G. Maculotti, "Single-step calibration method for nano indentation testing machines," *CIRP Annals*, Vol. 69, no. 1, pp. 429–432, Jan. (2020).

[8] J. Kholkhujaev, G. Maculotti, G. Genta and M. Galetto, "Metrological Comparison of Available Methods to Correct Edge-Effect Local Plasticity in Instrumented Indentation Test," *Materials*, Vol. 16, no. 12, Jun. (2023).

[9] Y. T. Cheng and C. M. Cheng, "Effects of 'sinking in' and 'piling up' on estimating the contact area under load in indentation," *Philos. Mag. Lett.* Vol. 78, no. 2, pp. 115–120 (1998).

[10] R. Holm, "The contact resistance. General theory," in *Electric Contacts: Theory and Application*, R. Holm, Ed., Berlin, Heidelberg: Springer Berlin Heidelberg, (1967), pp. 9–11.

[11] D. J. Sprouster *et al.*, "Quantitative electromechanical characterization of materials using conductive ceramic tips," *Acta Mater*, Vol. 71, pp. 153–163 (2014).

[12] M. Galetto, J. Kholkhujaev and G. Maculotti, "Improvement of instrumented indentation test accuracy by data augmentation with electrical contact resistance," *CIRP Annals*, (2023).

[13] J. Kholkhujaev, G. Maculotti, G. Genta and M. Galetto, "Calibration of machine platform nonlinearity in Instrumented Indentation Test in the macro range," *Precis. Eng.* Vol. 81, pp. 145–157, May 2023.

[14] Jared S. Murray and Jerome P. Reiter, "Multiple Imputation of Missing Categorical and Continuous Values via Bayesian Mixture Models with Local Dependence", *Journ. of Amer. Stat Assoc.*, 111:516, 1466–1479 (2016).

F Pavese, A Bošnjaković, S Eichstädt, A B Forbes, J Alves e Sousa (eds.):
Advanced Mathematical and Computational Tools in Metrology and Testing XIII
Series on Advances in Mathematics for Applied Sciences, Vol. 94
 (pp. 180–193)
https://doi.org/10.1142/9789819800674_0016

How to efficiently propagate p-box uncertainty

O. M. Kosheleva

Department of Teacher Education, University of Texas at El Paso
El Paso, Texas 79968, USA
E-mail: olgak@utep.edu

V. Kreinovich

Department of Computer Science, University of Texas at El Paso
El Paso, Texas 79968, USA
E-mail: vladik@utep.edu, www.cs.utep.edu/vladik

In many practical situations, to get the desired estimate or prediction, we need to process existing data. This data usually comes from measurements, and measurements are never 100% accurate. Because we only know the input values with uncertainty, the results of processing this data also comes with uncertainty. To make an appropriate decision, we need to know how accurate is the resulting estimate, i.e., how the input uncertainty "propagates" through the data processing algorithm. In the ideal case, when we know the probability distribution of each measurement error, we can, in principle, use Monte-Carlo simulations to describe the uncertainty of the data processing result. In practice, however, we often only have partial information about the measurement uncertainty: for example, instead of the exact values of the cumulative distribution function $F(X)$, we only know bounds on $F(X)$. Such information is known as the probability box (p-box, for short). In this paper, we provide feasible algorithms for propagating p-box uncertainty.

1. A very brief introduction: what is the main problem and what are the intended audiences

The main purpose of this paper is to provide new efficient uncertainty propagation algorithms for situations when we only have partial information about probabilities — namely, for the case when this partial information is described by probability boxes (also known as p-boxes).

This paper is intended both for specialists in metrology and for researchers in uncertainty who may not very familiar with metrological practices but who are interested in metrological applications. The following two section describe techniques and ideas that are mostly well-known in metrology, so specialists in metrology can skim these sections and start reading attentively with Sec. 4.

2. Uncertainty propagation: formulation of the general problem

The main objectives of science and engineering can be roughly summarized as follows: to understand the current state of the world, to predict the future state of the world, and to come up with a proper strategy to make this future state as good for us as possible.

How can we describe this in precise terms? To describe the current or future state of the world, we need to list the values of all the numerical characteristics of this state — i.e., in other words, the values of the corresponding physical quantities. For example, to describe the current state of the weather in a given area, we need to list the temperature, wind speed, wind direction, atmospheric pressure, humidity, etc. The state of a mechanical system can be characterized by the locations and velocities of all the particles that form this system, etc.

Similarly, to describe the strategy for changing the state, we need to list the numerical characteristics of the corresponding control. For example, if we want to drive from Point A to Point B, we need to know the exact path, i.e., distances, angles, speeds, etc.

Our information about the values of different physical quantities comes from measurements — this is, in effect, a definition of measurement: a process to supply us with the values of different physical quantities.

Some quantities we can measure directly. For example, we can directly measure the width of an office desk, we can directly measure the weight of a person or even of a truck. However, for some physical quantities, it is very difficult — or even impossible — to measure them directly. For example, we cannot directly measure the distance to a faraway star or the overall amount of oil in an oil field. Since we cannot measure the corresponding quantity y directly, the only way to measure them is to measure them *indirectly*, i.e.:

- to measure some easier-to-measure quantities $x_1, \ldots, x_n$ that are related to the desired quantity y by a known dependence $y = f(x_1, \ldots, x_n)$, and then
- apply the algorithm f to the results $\widetilde{x}_1, \ldots, \widetilde{x}_n$ of these measurements and thus get an estimate $\widetilde{y} = f(\widetilde{x}_1, \ldots, \widetilde{x}_n)$ for the desired quantity y.

This is an important particular case of what is called *data processing.*

Data processing is also needed to predict the future value y of each quantity of interest based on the current values $x_1, \ldots, x_n$ of this and related quantities. Data processing is needed to determine the appropriate control values based on the current state. In all these cases, we apply an appropriate algorithm $y = f(x_1, \ldots, x_n)$ to the results $\widetilde{x}_i$ of measuring some quantities x_i.

Measurement are never 100% accurate; see, e.g.,[15]. Each measurement result $\widetilde{x}_i$ is, in general, different from the actual (unknown) value x_i of the corresponding quantity. The difference $\Delta x_i \stackrel{\text{def}}{=} \widetilde{x}_i - x_i$ — known as the *measurement error* — is, in general, different from 0.

Because $\widetilde{x}_i \neq x_i$, the result $\widetilde{y}$ of applying the algorithm f to measurement results $\widetilde{x}_i$ is, in general, different from the value $y = f(x_1, \ldots, x_n)$ that we would have gotten if we knew the exact values of the measured quantities $x_1, \ldots, x_n$.

In practice, it is important to know how big the difference $\Delta y \stackrel{\text{def}}{=} \widetilde{y} - y$ can be. For example, if we estimated the amount of oil in a given field as 150 million tons, then, if it is 150 ± 20, this is good news, we can start exploring the field. However, if it is 150 ± 200, then maybe there is no oil in this field at all, so we need to perform additional measurements before investing millions of dollars into this exploitation.

In general, we need to know how uncertainty in x_i affects the uncertainty of the result of applying the algorithm f to the inputs x_i, i.e., how uncertainty *propagates* from the inputs x_i to the output y. In other words, we need to solve the following problem.

We *know*:

- the data processing algorithm $y = f(x_1, \ldots, x_n)$,
- the measurement results $\widetilde{x}_1, \ldots, \widetilde{x}_n$, and
- some information about the measurement errors $\Delta x_i = \widetilde{x}_i - x_i$.

Based on this information, we *need to get* the information about the estimation error $\Delta y = \widetilde{y} - y$, where $\widetilde{y} = f(\widetilde{x}_1, \ldots, \widetilde{x}_n)$, and $y = f(x_1, \ldots, x_n)$.

3. How uncertainty is usually described and how it is usually propagated

To analyze how uncertainty propagates, we need to analyze how this uncertainty is described. To fully describe the uncertainty of each measurement, we need to know what values of measurement error are possible, and how frequent are different possible values. In other words, we need to know the probability distribution of each measurement error Δx_i.

In metrological practice, we usually assume that these distributions are known. The most widely used GUM methodologies[1–3] assume, for the most part, that the distributions involved are known, standard distributions such as Gaussian, rectangular, gamma, etc. It is indeed true that in many practical cases, the distributions are close to Gaussian; see, e.g.,[12,13]. This empirical fact has a natural explanation: in many cases, the measurement error comes from the joint effect of many small independent factors, and in such situations, according to the Central Limit Theorem (see, e.g.,[16]), the resulting probability distribution is indeed close to Gaussian.

Bayesian methods (see, e.g.,[4,8]) also assume that the distributions for the likelihood and for the priors are known.

There are many ways to represent a probability distribution: by the probability density function (pdf) $f(X)$, by the cumulative distribution function $F(X)$, by moments, etc. It is important to mention that most of these representations are not universal:

- some distributions do not have finite moments — e.g., Cauchy distribution;
- some distributions do not have the probability density function — e.g., distribution located at a single value with probability 1.

The only universal representation is by using a cumulative distribution function (cdf) $F_i(X) \stackrel{\text{def}}{=} \text{Prob}(\Delta x_i \le X)$.

When using cdf, it is also important to take into account that there are infinitely many real numbers X, and thus, infinitely many values $F_i(X)$. Of course, we cannot store infinitely many values. The usual way to describe a general function $F_i(X)$ is to store its values $F_i(X_{i,1}), \ldots, F_i(X_{i,N})$ for some inputs $X_{i,1} < X_{i,2} < \ldots < X_{i,N}$ that form a dense grid on the real line.

Measurement errors corresponding to different measurements are usually caused by different factors and are, thus, independent.

The objective of uncertainty propagation is to use all this information about Δx_i (and their independence) to provide the probability distribution for the desired estimation error Δy.

How can we solve this problem? In the general case, we can, e.g., use Monte-Carlo simulations to produce the desired distribution for Δy. Namely, several times $k = 1, \ldots, K$, we:

- simulate the measurement errors $\Delta x_i^{(k)}$ according to the known distribution for each such error,
- compute the simulated values $x_i^{(k)} = \widetilde{x}_i - \Delta x_i^{(k)}$, and
- compute $y^{(k)} = f\left(x_1^{(k)}, \ldots, x_n^{(k)}\right)$ and $\Delta y^{(k)} = y^{(k)} - \widetilde{y}$.

This technique is described in Supplement 1 to the GUM.[2] One can check that the resulting values $y^{(k)}$ are distributed according to the desired distribution of Δy.

In many cases, faster methods are possible. These methods utilize the fact that usually, measurements are reasonably accurate. In general, this means that terms which are quadratic in terms of Δx_i are much smaller than linear terms and can, thus, be safely ignored. For example, if $\Delta x_i \approx 10\%$, then $(\Delta x_i)^2 \approx 1\% \ll 10\%$. Thus, we can use the usual linearization idea (frequent in physics[7,17] and in processing measurement results in general[2,15]): expand the expression

$$\Delta y = \widetilde{y} - y = f(\widetilde{x}_1, \ldots, \widetilde{x}_n) - f(\widetilde{x}_1 - \Delta x_1, \ldots, \widetilde{x}_n - \Delta x_n)$$

in Taylor series in terms of Δx_i, ignore terms which are quadratic (and higher order) in Δx_i and keep only linear terms.

Thus, we arrive at the linear dependence:

$$\Delta y = \sum_i c_i \cdot \Delta x_i$$

for some coefficients c_i. This linearized formula leads to the known GUM formulas (see, e.g.,[2]) describing the mean m and standard deviation σ of the uncertainty Δy in terms of means m_i and standard deviations σ_i of each measurement error Δx_i:

$$m = \sum_i c_i \cdot m_i \text{ and } \sigma = \sqrt{\sum_i c_i^2 \cdot \sigma_i^2}.$$

4. Remaining problem

In many real-life cases, we often only have partial information about the probabilities. This means that for each possible value X_i of the measurement error Δx_i:

- instead of knowing the exact value $F_i(X_{i,j})$,
- we only have partial information about $F_i(X_{i,j})$ — i.e., we have *bounds* on this value.

Usually, possible values of $F_i(X_{i,j})$ form an interval $[\underline{F}_i(X_{i,j}), \overline{F}_i(X_{i,j})]$. So, a natural way to describe such cases is to have a function that assigns such interval to each value $X_{i,j}$. This function is known as a *probability box*, or *p-box*, for short; see, e.g.,[6].

Usually, when we estimate a characteristic c of a probability distribution — be it mean, standard deviation, or cdf — from a sample:

- we have an estimate $\widetilde{c}$ based on the this sample — e.g., the sample mean, the sample standard deviation, etc. — and
- we also have bounds Δ_c (with some sufficient certainty) on the absolute value of the difference $\Delta c = \widetilde{c} - c$ between the sample-based estimate $\widetilde{c}$ and the actual (unknown) value c of this characteristic.

In this case, we can conclude that the actual value c lies somewhere in the interval $[\widetilde{c} - \Delta_c, \widetilde{c} + \Delta_c]$.

Some methods for estimating the value of a statistical characteristic directly produce the interval of possible values $[\underline{c}, \overline{c}]$. This interval can also be represented in the form $[\widetilde{c} - \Delta_c, \widetilde{c} + \Delta_c]$ if we take

$$\widetilde{c} = \frac{\underline{c} + \overline{c}}{2} \text{ and } \Delta_c = \frac{\overline{c} - \underline{c}}{2}.$$

So, in the following text, we will inter-changingly use both ways of representing the interval of possible values of each statistical characteristic.

So, we arrive at the following problem: We *know*:

- the data processing algorithm $y = f(x_1, \ldots, x_n)$,
- the measurement results $\widetilde{x}_1, \ldots, \widetilde{x}_n$, and
- n p-boxes $[\underline{F}_i(X_{i,j}), \overline{F}_i(X_{i,j})] = [\widetilde{F}_i(X_{i,j}) - \Delta_{i,j}, \widetilde{F}_i(X_{i,j}) + \Delta_{i,j}]$ that describe possible probability distributions for the measurement errors $\Delta x_i = \widetilde{x}_i - x_i$.

We are interested in the information about a statistical characteristic c describing the difference $\Delta y = \widetilde{y} - y$, where $\widetilde{y} = f(\widetilde{x}_1, \ldots, \widetilde{x}_n)$ and $y = f(x_1, \ldots, x_n)$. This characteristic can be the mean, the standard deviation, or the value $F(Y_j)$ of the cdf of Δy.

Based on the known information, we *need to estimate* the range

$$[\widetilde{c} - \Delta_c, \widetilde{c} + \Delta_c]$$

of possible values of c for all possible distributions $F_i(X)$ of Δx_i for Δx_i which $F_i(X_{i,j}) \in [\underline{F}_i(X_{i,j}), \overline{F}(X_{i,j})]$ for all i and j.

It is important to mention that by using the Monte-Carlo techniques, we can estimate the value $\widetilde{c}$ corresponding to the sample-based distributions $\widetilde{F}_i(X)$. So, the main task to estimate the value Δ_c — that described possible deviations from this value.

There exist feasible algorithms for propagating p-box uncertainty for many important cases; see, e.g., [6]. However, there is no general efficient algorithm for such propagation. In this paper, we provide a feasible algorithm for solving this problem.

5. Possible straightforward approaches to this problem, and why they do not always work

Situations when we only partial information about the probability distribution are ubiquitous. In such situations, we do not know the exact probability distribution, we only know the *class* of possible probability distributions. A usual statistical approach to such situations is to select one of these distributions. A natural criterion for selecting this distribution is that we should maximally preserve uncertainty: we do not want to add unjustified certainty when there was none. For example, if all we know is that the measurement error is located on some interval $[-\Delta, \Delta]$, we do not want to select a distribution which is located only on a subinterval of this interval, because this will add unjustified certainty.

The uncertainty of a probability distribution can be characterized by the average number of "yes"-"no" questions that one need to ask to determine the value with a given accuracy ε. This number is known to be proportional to the entropy $S = -\int f(X) \cdot \ln(f(X))\, dX$ of the corresponding distribution (see, e.g.,[5,11]). So, the above natural criterion means that, out of all possible distributions, we select the one with the largest entropy. This *Maximum Entropy* approach often leads to reasonable results (see, e.g.,[9]). For example, when all we know about the measurement error is that it is located on an interval, then this approach leads to the uniform distribution on this interval. This conclusion makes perfect sense: if we have no reason to assume that some points on this interval are more probable than others, then it makes sense to assume that all these points are equally probable.

However, from the metrological viewpoint, limiting ourselves to a single distribution may drastically underestimate uncertainty. We can show this on a very simple example when $f(x_1, \ldots, x_n) = x_1 + \ldots + x_n$, the measured values $\widetilde{x}_i$ of all n quantities are 0s, and about each of the measurement errors all we know is that the error is located somewhere on the interval $[-\Delta, \Delta]$. In this case, the only thing we can guarantee about y is that y is located on the interval $[-n \cdot \Delta, n \cdot \Delta]$. For example, it is possible that $\Delta x_1 = \ldots = \Delta x_n = \Delta$, in which case we have $y = n \cdot \Delta$.

On the other, if we assume that all measurements errors are uniformly distributed on the interval $[-\Delta, \Delta]$, then, since we assumed that these errors and independent, we can apply the Central Limit Theorem (see, e.g.,[16]) and conclude that the distribution of y is close to Gaussian. Based on the mean (0) and variance ($\Delta^2/3$) of each measurement error, we can conclude that y has mean 0 and variance $n \cdot \Delta^2/3$ and thus, standard deviation $\sqrt{n} \cdot \Delta/\sqrt{3}$.

In practical applications of statistics, we usually assume that deviations larger than 3 sigma are practically impossible — OK, we can make it 6 sigma to make them really impossible. Thus, the Maximum Likelihood approach would make us conclude that, with probability close to 1, the absolute value of y cannot exceed $\sqrt{n} \cdot \Delta$. For large n, this bound is indeed much smaller than the actual bound $n \cdot \Delta$.

OK, we cannot just limit ourselves to a single distribution, we have to take into account several possible distributions. But why cannot we just use Monte-Carlo simulations? The main difference between the p-box case and the ideal case — when we know all the probability distributions — is that for each input x_i, we may have several different probability distributions. So, a natural idea is to try many of them.

A simple analysis shows that this is not feasible. Even if we consider 2 values for each of N points $X_1, \ldots, X_N$, this means 2^N options. Trying 2^N options is not feasible: already for reasonable $N \approx 300$, this requires more computational steps than the lifetime of the Universe; see, e.g., [10,14]. It is usually assumed that only algorithm that require no more than polynomial number of $O(n^k)$ computational steps are feasible.

So, we cannot use this straightforward approach, we need to come up with new ideas.

6. Analysis of the problem

For different possible distributions of Δx_i, we have different values of the cdf $F_i(X)$, and thus, different values of $\Delta F_{i,j} \stackrel{\text{def}}{=} \widetilde{F}_i(X_{i,j}) - F_i(X_{i,j})$. As a result, in general, we get different values of the desired characteristic c, and thus, different values of the difference $\Delta c = \widetilde{c} - c$.

The value Δc depends on all the differences $\Delta F_{i,j}$:

$$\Delta c = D(\Delta F_{1,1}, \ldots, \Delta F_{1,N}, \ldots, \Delta F_{i,j}, \ldots, \Delta F_{n,N}),$$

for some function D. Our objective is to find this function D.

When the sample is reasonable large, sample-based estimates of all the characteristics — including cdf — are reasonably accurate. This means that the differences $\widetilde{F}_i(X_{i,j}) - F_i(X_{i,j})$ are reasonably small. In general, this means that terms which are quadratic in terms of $\Delta F_{i,j}$ are much smaller than linear terms and can, thus, be safely ignored. For example if $\Delta F_{i,j} \approx 10\%$, then $(\Delta F_{i,j})^2 \approx 1\% \ll 10\%$.

Thus, we can use the same linearization idea as above:

- we expand the dependence D of Δc on $\Delta F_{i,j}$ in Taylor series, and

- we ignore terms which are quadratic (and higher order) in $\Delta F_{i,j}$ and keep only linear terms.

Thus, we arrive at the linear dependence:

$$\Delta c = \sum_{i,j} a_{i,j} \cdot \Delta F_{i,j}, \tag{1}$$

for some coefficients $a_{i,j}$.

So, the question of funding the function D is reduced to the question of finding the appropriate coefficients $a_{i,j}$.

How can we find the coefficients $a_{i,j}$? For each $i = 0, 1, \ldots, n$, and $j = 1, \ldots, N$, we form a cdf $F_i^{(j)}(X)$ for which we have $F_{i'}^{(j)}(X_{i',j}) = \widetilde{F}_{i'}(X_{i,j})$ for all $i' \neq i$, while for $i' = i$:

- we have $F_i^{(j)}(X_{i,k}) = \underline{F}_i(X_{i,k})$ for $k \leq j$, and
- we have $F_i^{(j)}(X_{i,k}) = \overline{F}_i(X_{i,k})$ for $k > i$.

We can use Monte-Carlo (or any other) method to find the value $\Delta c_{i,j}$ corresponding to the cdf's $F_i^{(j)}(X)$.

Because of linearity (1), we have

$$\Delta c_{i,j} - \Delta c_{i,j-1} = a_{i,j} \cdot (\overline{F}_i(X_j) - \underline{F}_i(X_{i,j})) = 2a_{i,j} \cdot \Delta_{i,j}.$$

So, we can estimate $a_{i,j}$ as

$$a_{i,j} = \frac{\Delta c_{i,j} - \Delta c_{i,j-1}}{2\Delta_{i,j}}.$$

Once we know $a_{i,j}$, how can we estimate Δ_c? We know that each value $\Delta F_{i,j}$ lies somewhere in the interval $[-\Delta_{i,j}, \Delta_{i,j}]$. We also an additional constraint on these values – that the resulting values $F_i(X_{i,j}) = \widetilde{F}_i(X_{i,j}) + \Delta F_{i,j}$ must be non-strictly increasing in j;

$$\widetilde{F}_i(X_{i,j}) + \Delta F_{i,j} \leq \widetilde{F}_i(X_{i,j+1}) + \Delta F_{i,j+1}.$$

Thus, we arrive at the following algorithm for computing Δ_c.

7. Algorithm for propagating p-box uncertainty: description

- First, we use the Monte-Carlo method — or any other method — to find the value $\widetilde{c}$ of the characteristic c when each Δx_i is distributed according to the cdf $\widetilde{F}_i(X)$.

- Then, for each $i = 0, 1, \ldots, n$, and $j = 1, \ldots, N$, we form a cdf $F_i^{(j)}(X)$ for which we have $F_{i'}^{(j)}(X_{i',j}) = \widetilde{F}_{i'}(X_{i,j})$ for all $i' \neq i$, while for $i' = i$:
  - we have $F_i^{(j)}(X_{i,k}) = \underline{F}_i(X_{i,k})$ for $k \leq j$, and
  - we have $F_i^{(j)}(X_{i,k}) = \overline{F}_i(X_{i,k})$ for $k > i$.
- We use Monte-Carlo (or any other) method to find the value $\Delta c_{i,j}$ corresponding to the cdf's $F_i^{(j)}(X)$.
- Then, we estimate

$$a_{i,j} = \frac{\Delta c_{i,j} - \Delta c_{i,j-1}}{2\Delta_{i,j}}.$$

- Finally, we find the desired value Δ_c by solving the following optimization problem:

$$\Delta c = \sum_{i,j} a_{i,j} \cdot \Delta F_{i,j} \to \max$$

under the following constraints:

$$-\Delta_{i,j} \leq \Delta F_{i,j} \leq \Delta_{i,j} \text{ for all } i \text{ and } j, \text{ and}$$

$$\widetilde{F}_i(X_{i,j}) + \Delta F_{i,j} \leq \widetilde{F}_i(X_{i,j+1}) + \Delta F_{i,j+1} \text{ for all } i \text{ and } j.$$

Comments.

- In the last optimization problem, we maximize a linear expression under linear constraints — which means that this problem is a particular case of linear programming. There are many feasible algorithms for solving linear programming problems; see, e.g.,[18]. Specifically, solving a linear programming problem with u unknowns requires $O(u^{2+1/18})$ computational steps. In our problem, we have $u = N \cdot n$ unknowns, so this is indeed feasible.
- Our algorithm requires $N \cdot n + 1$ calls to Monte-Carlo simulation procedure, which is also feasible.

8. Discussion: how accurate are the results of the proposed algorithm?

Our algorithms uses Monte-Carlo simulations, and Monte-Carlo simulations only provide an estimate for a statistical characteristic. It is therefore desirable to know how accurate are the resulting estimates for Δ_c — and what can we do to get the estimate of Δ_c with given relative accuracy ε.

It is known that if we run the simulations T times, the resulting uncertainty decreases as $1/\sqrt{T}$; see, e.g.,[16].

Let Δ denote the size of the differences $\Delta F_{i,j} = \overline{F}_i(X_{i,j}) - \underline{F}_i(X_{i,j})$. So, in linear approximation, the value Δ_c is proportional to Δ.

Let ε be the relative accuracy with which we want to estimate the value Δ_c. For example, we can take $\varepsilon = 20\%$:

- remember, this is accuracy with which we determine accuracy;
- measuring instrument can have accuracy 10%, but 11.6% accuracy does not make too much practical sense.

This means that we need absolute accuracy $\varepsilon \cdot \Delta$. In general, if we use values at N points, a monotonic function is represented with accuracy $\sim 1/N$. Thus, we need to have $N \sim 1/(\varepsilon \cdot \Delta)$.

Let δ be the accuracy with which we determine each value $\Delta c_{i,j}$. Each term in the linear dependence (1) is close to the difference $\Delta c_{i,j} - \Delta c_{i,j-1}$. Thus, the value of each term is of order δ.

The standard deviation of the sum of $N \cdot n$ independent terms grows as $\sqrt{N \cdot n}$. So, the accuracy with which we determine Δc is $\delta \cdot \sqrt{N \cdot n}$. Thus, to reach accuracy $\varepsilon \cdot \Delta$, we need to select $\delta = \varepsilon \cdot \Delta/\sqrt{N \cdot n}$.

Let M denote the number of calls to f that we use to estimate each $\Delta c_{i,j}$. In general, M iterations provide relative accuracy $\sim 1/\sqrt{M}$. To get $1/\sqrt{M} \sim \delta = \varepsilon \cdot \Delta/\sqrt{N \cdot n}$, we thus need:

$$M \sim \varepsilon^{-2} \cdot \Delta^{-2} \cdot N \cdot n \sim \varepsilon^{-3} \cdot \Delta^{-2} \text{ calls to } f.$$

We need to compute $N \cdot n \sim n \cdot \varepsilon^{-1} \cdot \Delta^{-1}$ values $\Delta c_{i,j}$. Thus, overall, we need

$$N \cdot n \cdot M \sim (\varepsilon^{-1} \cdot \Delta^{-1}) \cdot (\varepsilon^{-3} \cdot \Delta^{-2}) = \varepsilon^{-4} \cdot \Delta^{-3} \text{ calls to } f.$$

This is polynomial in ε^{-1} and Δ^{-1} and thus, still feasible.

9. Conclusions

This paper addresses the issue of uncertainty evaluation in circumstances in which we only have partial information about the probability distribution that govern the random effects influencing the measurement data. The usual GUM methodologies assume, for the most part, that the distributions involved are known. Bayesian methods also assume that the distributions for the likelihood and the priors are known. The assignment of these distributions amount to model assumptions and the resulting uncertainty estimates are conditional on these assumptions being valid.

There has been much interest in the statistical community and — lately — in the metrology community to see if we can still provide uncertainty estimates with much less definitive information about the underlying probability distributions. P-box uncertainty is one of these imprecise probability technologies that offer a solution. This paper considers the question of p-box uncertainty propagation. The main result of this paper is a new algorithm for propagating p-box uncertainty that uses linear programming and is, as a result, computationally feasible — in the sense that it finishes computations in polynomial time.

Acknowledgements

This work was supported in part by the National Science Foundation grants 1623190 (A Model of Change for Preparing a New Generation for Professional Practice in Computer Science), HRD-1834620 and HRD-2034030 (CAHSI Includes), EAR-2225395 (Center for Collective Impact in Earthquake Science C-CIES), and by the AT&T Fellowship in Information Technology.

It was also supported by the program of the development of the Scientific-Educational Mathematical Center of Volga Federal District No. 075-02-2020-1478, and by a grant from the Hungarian National Research, Development and Innovation Office (NRDI).

The authors are greatly thankful to Franco Pavese for his encouragement, to all the participants of the International Conference on Advanced Mathematical Tools in Metrology and Testing AMCTM 2023 (September 26–28, 2023) for valuable discussions, and to the anonymous referees for very helpful suggestions.

References

1. BIPM, IEC, IFCC, ILAC, ISO, IUPAC, IUPAP and OIML, *Evaluation of Measurement Data. Guide to the Expression of Uncertainty in Measurement*, Joint Committee for Guides in Metrology, JCGM 100:2008, 2008.
2. BIPM, IEC, IFCC, ILAC, ISO, IUPAC, IUPAP and OIML, *Evaluation of Measurement Data. Supplement 1 to the "Guide to the Expression of Uncertainty pn Measurement": Propagation of Distributions Using a*

Monte Carlo Method, Joint Committee for Guides in Metrology, JCGM 101:2008, 2008.

3. BIPM, IEC, IFCC, ILAC, ISO, IUPAC, IUPAP and OIML, *Evaluation of Measurement Data. Supplement 2 to the "Guide to the Expression of Uncertainty in Measurement": Extension to Any Number of Output Quantities*, Joint Committee for Guides in Metrology, JCGM 102:2011, 2011.
4. G. E. P. Box and G. C. Tiao, *Bayesian Inference in Statistical Analysis*, Wiley, New York, 2014.
5. B. Chokr and V. Kreinovich. "How far are we from the complete knowledge: complexity of knowledge acquisition in Dempster-Shafer approach." In R. R. Yager, J. Kacprzyk and M. Pedrizzi (Eds.), *Advances in the Dempster-Shafer Theory of Evidence*, Wiley, N.Y., 1994, pp. 555–576.
6. S. Ferson, V. Kreinovich, L. Ginzburg, D. S. Myers and K. Sentz, *Constructing Probability Boxes and Dempster-Shafer Structures*, Sandia National Laboratories, Report SAND2002-4015, January 2003.
7. R. Feynman, R. Leighton and M. Sands, *The Feynman Lectures on Physics*, Addison Wesley, Boston, Massachusetts, 2005.
8. A. Gelman, J. B. Carlin, H. S. Stern, D. B. Dunson, A. Vehtari and D. B. Rubin, *Bayesian Data Analysis*, Chapman & Hall/CRC, Boca Raton, Florida, 2014.
9. E. T. Jaynes and G. L. Bretthorst, *Probability Theory: The Logic of Science*, Cambridge University Press, Cambridge, UK, 2003.
10. V. Kreinovich, A. Lakeyev, J. Rohn and P. Kahl, *Computational Complexity and Feasibility of Data Processing and Interval Computations*, Kluwer, Dordrecht, 1998.
11. H. T. Nguyen, V. Kreinovich, B. Wu and G. Xiang, *Computing Statistics under Interval and Fuzzy Uncertainty*, Springer Verlag, Berlin, Heidelberg, 2012.
12. P. V. Novitskii and I. A. Zograph, *Estimating the Measurement Errors*, Energoatomizdat, Leningrad, 1991 (in Russian).
13. A. I. Orlov, "How often are the observations normal?", *Industrial Laboratory*, 1991, Vol. 57, No. 7, pp. 770–772.
14. C. Papadimitriou, *Computational Complexity*, Addison-Wesley, Reading, Massachusetts, 1994.
15. S. G. Rabinovich, *Measurement Errors and Uncertainty: Theory and Practice*, Springer Verlag, New York, 2005.

16. D. J. Sheskin, *Handbook of Parametric and Nonparametric Statistical Procedures*, Chapman and Hall/CRC, Boca Raton, Florida, 2011.
17. K. S. Thorne and R. D. Blandford, *Modern Classical Physics: Optics, Fluids, Plasmas, Elasticity, Relativity, and Statistical Physics*, Princeton University Press, Princeton, New Jersey, 2021.
18. R. J. Vanderbei, *Linear Programming: Foundations and Extensions*, Springer, New York, 2014.

F Pavese, A Bošnjaković, S Eichstädt, A B Forbes, J Alves e Sousa (eds.):
Advanced Mathematical and Computational Tools in Metrology and Testing XIII
Series on Advances in Mathematics for Applied Sciences, Vol. 94

https://doi.org/10.1142/9789819800674_0017

Relative sensitivities and correlation of factors introducing uncertainty in radiotherapy dosimetry audits

P. Krishnadas[†], S. A. Thomas, J. Goldring, N. A. S. Smith and M. Hussein

National Physical Laboratory,
London, TW11 0LW, United Kingdom
[†]*E-mail: padmini.krishnadas@npl.co.uk*
www.npl.co.uk

Dosimetry audits are carried out to determine how well radiotherapy is delivered to the patient. It is also used to understand the uncertainty introduced into the measurement result when using different computational models. As measurement procedures are becoming increasingly complex with technological advancements, it is harder to establish sources of variability in measurements and understand if they stem from true differences in measurands or in the measurement pipelines themselves. The gamma index calculation is a widely accepted metric used for the comparison of measured and predicted doses in radiotherapy. However, various steps in the measurement pipeline can introduce variation in the measurement result. In this paper, we perform a sensitivity and correlation analysis to investigate the influence of various input factors (i.e. setting) in gamma index calculations on the uncertainty introduced in dosimetry audits. We identify a number of factors where standardization will improve measurements by reducing variability in outputs. Furthermore, we also compare gamma index metrics and similarities across audit sites.

Keywords: sensitivity analysis; dosimetry audits; gamma index; uncertainty.

1. Introduction

Measurement pipelines are becoming increasingly complex with technological advancements making it difficult to establish where exactly uncertainty in measurements stems from. Radiotherapy is a major modality for the treatment of cancer. The implementation of audits in radiotherapy pipelines is important to assure quality and safety in radiotherapy to patients.[1] By using tissue-mimicking material (or phantoms) in combination with passive or active detectors, it is possible to compare the measured radiation dose with the dose predicted by the hospital, providing a validation of how well the radiotherapy equipment and software have been implemented. Conducting these audits using the same measurement equipment in multiple centers allows for benchmarking and improving standards. One detector that is commonly used is called radiochromic film.[2] In an audit setting, measurements from the detector can be compared to the predicted dose generated by a radiotherapy Treatment Planning System (TPS). TPS is software used to simulate the radiation dose deposition in patients and to design the optimal radiotherapy approach. Gamma index[3,4] analysis is typically used to compare film and TPS dose, as it is a dimensionless metric which gives a quality indicator on the agreement between these distributions. However, there are many factors that affect the uncertainty of the measured film dose.

Sensitivity analysis[5] examines the extent of impact of individual factors, i.e. measurement setting, and the effect of interaction between the factors. It does so by identifying the factors introducing most variability to the measurement result or output. The results of sensitivity analyses are then used as a proxy for uncertainty introduced by the respective factors, i.e. measurement settings. Understanding the impact of different factors on the measurement result can help ensure the audit methodology's accuracy and improve radiotherapy quality. In this paper, we conduct a sensitivity analysis of Gamma index calculations made using the in-house Versatile Independent Gamma Analysis sOftware (VIGO) on dosimetry audit data[3] collected from Stereotactic Ablative Radiotherapy treatment (SABR) performed on phantom objects.

2. Methods

2.1. *Gamma index calculations and VIGO*

VIGO is NPL's in-house software developed in MATLAB[6] for the analysis of radiotherapy dosimetry audit data, producing 12 outputs. There are four Gamma Index Criteria (GIC) metrics, which are reported as both a median Gamma value and a Gamma Passing Rate (%), corresponding to eight of the 12 outputs. The other four are measurements for the mean dose difference (%); median dose difference (%); distance to agreement (DTA) (mm) and center of mass (CoM) distance (mm). Each of the GICs are computed at a specified dose difference (%) and distance difference (mm) criterion, these are; $GIC1$: 5 % / 2 mm, $GIC2$: 3 % / 2 mm, $GIC3$: 2 % / 2 mm and $GIC4$: 5 % / 1 mm. Figure 1 represents the steps involved in NPL's dosimetry audit highlighting the complexity in the workflow. The steps in blue represent where uncertainty can be introduced and propagated, which can have a significant impact on the output. The factors, levels and their combinations required for the sensitivity analyses were decided based on these blue steps, illustrated in Fig. 1. It is interesting to note that the main factors found to introduce variability to measurement results were those associated with the Region of Interest (ROI) where the radiotherapy was delivered.

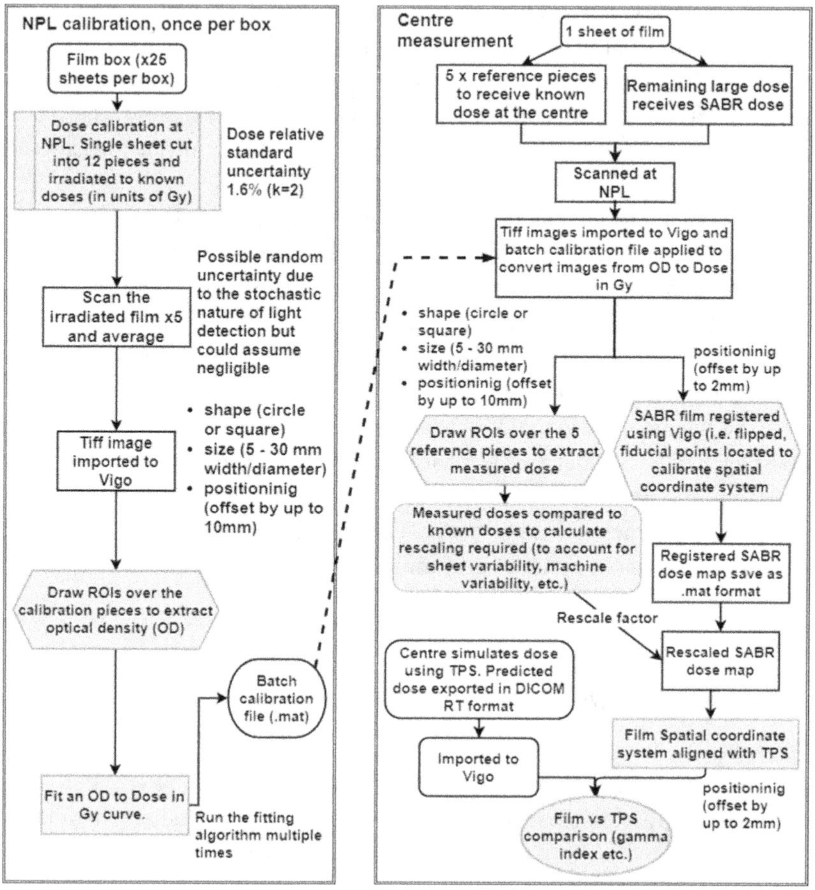

Fig. 1. Steps involved in an NPL dosimetry audit; blue steps indicate the focus of our sensitivity analysis in this work. Green steps indicate steps where measurement results are collected. VIGO compares the delivered dosage to the reference dosage (available in the Treatment Planning System or TPS) to assess the quality of delivered radiation. The input is in the form of DICOM RT images (Digital and Imaging Communications in Medicine – Radiotherapy).

2.2. *Sensitivity analysis*

Sensitivity analysis is a method used to identify and quantify the influential factors in an experimental design. Relative sensitivity[5,7] refers to the sensitivity of a measurement result to a factor in a measurement pipeline relative to its sensitivity to all other factors. We extend the concept of relative sensitivities by calculating the Type III sum of squares[7–9] to obtain a relative sensitivity value. The Type III sum of squares (also known as partial sum of squares) tests each term in the model against every other term in the model irrespective of order to calculate relative sensitivities for factors and their interactions over various levels.

For further information about calculating relative sensitivities, we refer the readers to the following key texts.[5,7–11]

2.3. *Experimental design*

We analyze the output from VIGO based on various settings for the steps in blue boxes shown in Fig. 1. This provides 12 outputs for each of the configurations of VIGO investigated for each center that has been audited. We calculate the relative sensitivities using a balanced design to provide insights on which factors are most influential in introducing uncertainties to measurement results, i.e. the VIGO metrics. We then compute the Pearson correlations between each of the metrics to understand the impact of different factors when using different GIC. This helped provide insights on which GICs produce measurements with similar characteristics. We also compute the correlation between factors themselves when using different GICs. This provided insights to which factors behave similarly in terms of introducing uncertainty.

3. Results

We performed a correlation analysis of the sensitivity analysis results as a proxy for uncertainties associated with factors. The sensitivity analysis results quantify the amount of variability introduced by each factor. Performing a correlation analysis on these factors after the sensitivity analysis allows us to understand which factors behave similarly in terms of introducing uncertainty. This in turn helps us understand if we can reduce uncertainty but controlling or standardizing specific factors. We can also ascertain if uncertainty can be reduced in several factors by standardizing a single factor; thus, reducing the variability introduced to the overall measurement result.

3.1. *Correlation between all centers for all metrics and measurements*

Across all centers and metrics, the results for the Gamma Passing Rate for each of the four GICs are strongly correlated with each other. Similarly, the mean Gamma metrics for different GICs are also closely correlated. The mean and median dose difference measurements are perfectly correlated with each other, however DTA and CoM are not correlated (Pearson's Coefficient = 0.65). Moreover, DTA and CoM difference measurement results are not correlated with any other measurement results, with the exception of DTA and Gamma Passing Rates for GIC1 and GIC4. Overall, this implies that measurement results from DTA and CoM are less likely to be influenced by the choice of GIC. These results are illustrated by the correlation matrix in Fig. 2.

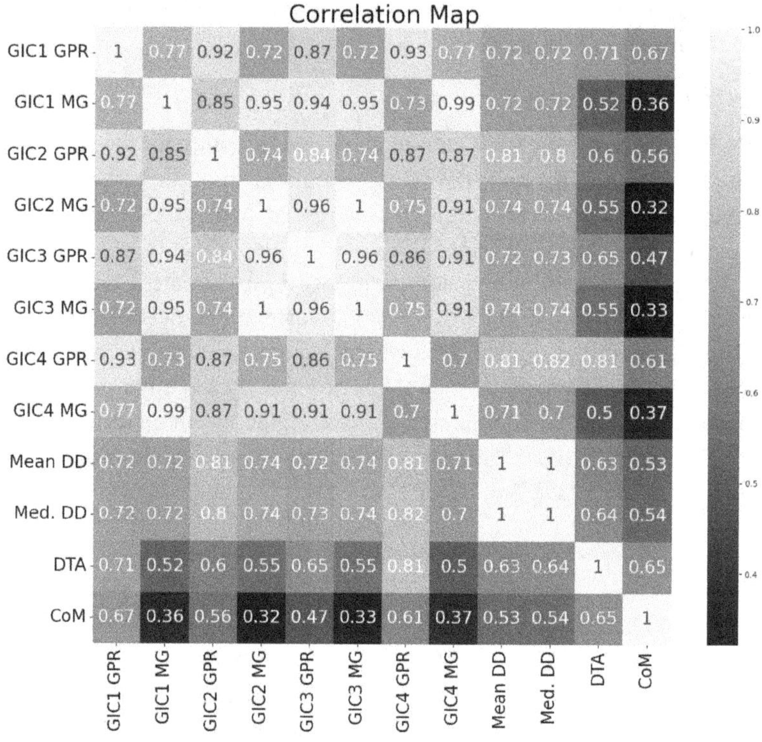

Fig. 2. Correlation matrix between different measurands for all centers. GIC = Gamma Index Criteria, GPR = Gamma Passing Rate, MG = median Gamma, DD = dose difference, DTA = distance to agreement, CoM = centre of mass distance.

3.2. *Correlation between centers for Gamma Passing Rate when using different Gamma Index Criteria*

Figure 3 is a matrix representing the correlation between centers for Gamma Passing Rates measurements made using GIC1. We observe that certain centers (such as c6 and c8) are closely related with a correlation score of 0.9 or above. We also observe some centers are poorly correlated with any of the other centers with correlation scores lower than 0.6. Specifically, we observe that c2 is poorly correlated with c1, c3, c4 and c7, while c5 is poorly correlated with c6, c8 and c9.

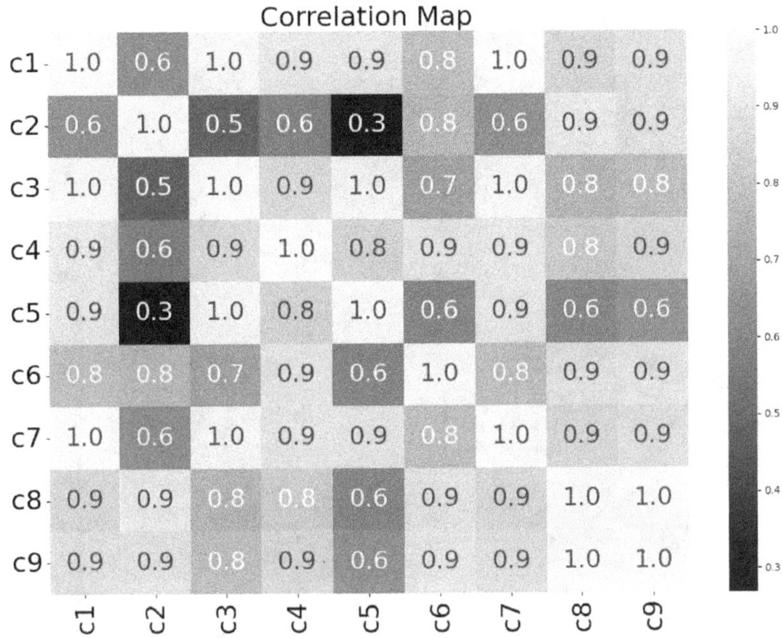

Fig. 3. Correlation matrix between centers for Gamma Passing Rates (GPR) measurements when using GIC1. Centers are marked as c1, c2, etc.

We performed these correlation analyses for Gamma Passing Rates when using GICs 1, 2, 3 and 4. We found that c6 and c8; and c1 and c7 were strongly correlated and we can expect measurement results from these centers to have similar characteristics, whereas c2 and c5 were poorly correlated with most other centers when using all for GICs. This indicates that there are uncertainties within the measurement pipelines used at these centers, and they may benefit from further inspection in the interest of standardization of measurement results produced.

3.3. *Correlation between factors for Gamma Passing Rates using different Gamma Index Criteria*

Figure 4 is a correlation matrix representing the correlation between factors for Gamma Passing Rates made using GIC1. From Fig. 4, we observe that the calibrated shape has a strong negative correlation with higher order interactions. This implies that if the uncertainty introduced to measurement results by the calibrated shape is small, uncertainty introduced by higher order interactions will be large. The calibrated size of region of interest was also observed to be negatively correlated with higher order interactions.

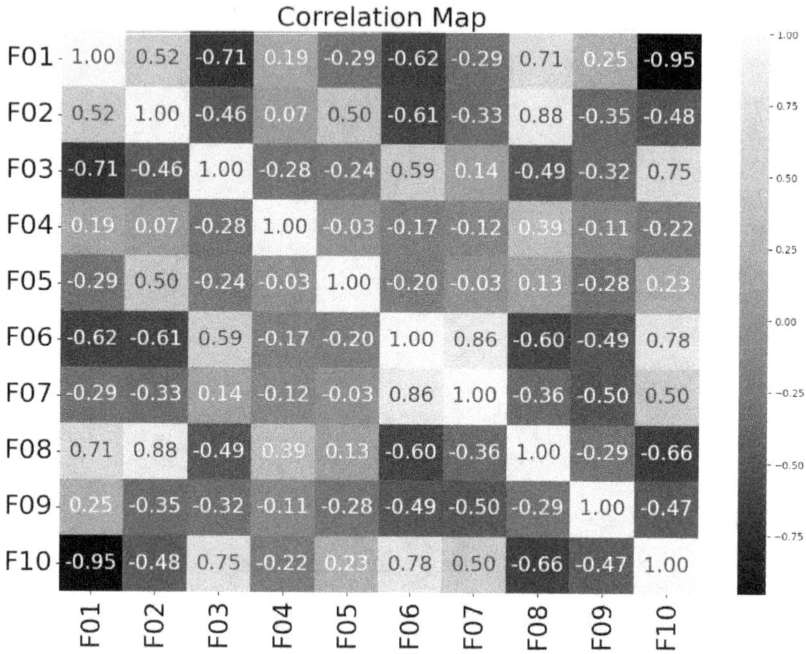

Fig. 4. Correlation between factors for Gamma Passing Rates (GPR) made using GIC 1. Factors are indicated as: Calibrated shape = F01, Reference shape = F02, Calibrated offset dosage = F03, Calibrated offset on x axis = F04, calibrated offset on y axis = F05, reference offset on x axis = F06, reference offset on y axis = F07, Reference Size = F08, calibrated size = F09, higher order interactions = F10.

We performed these correlation analyses for Gamma Passing Rates using GICs 1, 2, 3 and 4. We found that the calibrated shape and the calibrated size share a positive correlation while both the calibrated shape and size are negatively correlated with higher order interactions (nonlinear combinations of individual factors). Higher order interactions are positively correlated with the calibrated offset dosage, reference offset on the x- and y-axis.

4. Inference and discussions

On interpreting the above analyses, we can better understand how the factors (measurement settings) in audits, centers conducting the measurements, and outputs investigated, can contribute to uncertainty introduced in radiotherapy audit measurements.

From the center correlation analysis, considering the trends displayed across the four GICs, we conclude that center 1 and center 7 display a strong positive correlation, as well as center 6 and center 8. This indicates that results from these centers may be produced from similar pipelines and may be similar in terms of introduced uncertainty in measurement results. However, center 2 and center 5 were poorly correlated with many other centers suggesting their results will diverge from regular trends. Measurement pipelines at these centers may benefit from further investigation to ensure standardization.

From the factor correlation analyses conducted and the trends displayed by the factors across the four GICs, we conclude that the calibrated shape and size of the regions of interest share a strong positive correlation. These two factors will have similar behavioral characteristics in terms of introducing uncertainty to measurement output. We also observe that the calibrated shape has a strong negative correlation with higher order interactions. This indicates that they do not behave similarly in terms of introducing uncertainty. Higher order interactions, however, were observed to be positively correlated with the calibrated offset dosage, reference offset on the x-axis and the reference offset on the y-axis. Therefore, standardizing these factors can help reduce uncertainty introduced not only through the factors directly, but also through the higher order interactions, and thus reduce over uncertainty associated with the measurement.

Determining the influence of each of the factors through sensitivity analysis and their correlation can help reduce the variability in measurement results. Understanding correlations between the factors highlights which factors are responsible for introducing variability to measurement results and influence on other factors. The correlation results help better ascertain which factors should be standardized (such as the shape and size of region of interest) as well as which factors should be controlled on priority to reduce the overall variability introduced to measurement results. Coupled with domain knowledge, we can use the correlation results to reduce the variability of factors that are difficult to control, e.g. higher order interactions, by standardizing factors that are correlated with that may be easier to control in measurements, e.g. calibration shape.

Acknowledgements

This work was funded by the UK Government's Department for Science, Innovation and Technology through the UK's National Measurement System programmes. The authors would also like to acknowledge Dr Xavier Loizeau and Dr Peter Harris from the National Physical Laboratory for their contributions to this paper.

References

1. Clark C.H., Aird E.G., Bolton S. *et al.*, Radiotherapy dosimetry audit: three decades of improving standards and accuracy in UK clinical practice and trials. *Br. J. Radiol.* 2015;88(1055):20150251. doi:10.1259/bjr.20150251.
2. Devic S., Radiochromic film dosimetry: Past, present, and future. *Phys. Medica. Eur. J. Med. Phys.* 2011;27(3):122–134. doi:10.1016/j.ejmp.2010.10.001.
3. Hussein M., Clark C.H., Nisbet A., Challenges in calculation of the gamma index in radiotherapy – Towards good practice. *Phys. Med.* 2017;36:1–11. doi:10.1016/j.ejmp.2017.03.001.
4. Das, S., Kharade, V., Pandey, V., Anju, K.V., Pasricha, R.K. Gupta, M., 2002. Gamma index analysis as a patient-specific quality assurance tool for high-precision radiotherapy: A clinical perspective of single institute experience. *Cureus*, 14(10).
5. Esward T.J., Matthews C.E., Wright L., Yang X.S., Sensitivity analysis, optimisation, and sampling methods applied to continuous models. Published November 2010. Accessed March 20, 2023. https://eprintspublications.npl.co.uk/4783/.
6. Symbolic Math Toolbox User's Guide.pdf. Accessed October 16, 2023. https://citeseerx.ist.psu.edu/document?repid=rep1&type=pdf&doi=eb378f78e809267ea393c3c98b93d432933f38a6.
7. Partarrieu I.X., Smith N., Harris P., What Am I Measuring? Using Experimental Design to Understand and Improve Measurement Pipelines, with an Example Application to Radiomics Measurements. Published January 2022. Accessed March 20, 2023. https://doi.org/10.47120/npl.MS35.
8. Weisstein E.W., ANOVA. Accessed March 20, 2023. https://mathworld.wolfram.com/.
9. pingouin.anova — pingouin 0.5.3 documentation. Accessed March 20, 2023. https://pingouin-stats.org/build/html/generated/pingouin.anova.html.

10. Yu L., Tang T.L.S., Cassim N. *et al.*, Analysis of dose comparison techniques for patient-specific quality assurance in radiation therapy. *J. Appl. Clin. Med. Phys*. 2019;20(11):189–198. doi:10.1002/acm2.12726.
11. Eta-Squared and Partial Eta-Squared in Fixed Factor Anova Designs - Jacob Cohen, 1973. Accessed March 20, 2023. https://journals.sagepub.com/doi/abs/10.1177/001316447303300111?journalCode=epma.
12. Benesty J., Chen J., Huang Y., Cohen I., Pearson Correlation Coefficient. In: Cohen I., Huang Y., Chen J., Benesty J., eds. *Noise Reduction in Speech Processing*. Springer Topics in Signal Processing. Springer; 2009:1–4. doi:10.1007/978-3-642-00296-0_5.

F Pavese, A Bošnjaković, S Eichstädt, A B Forbes, J Alves e Sousa (eds.):
Advanced Mathematical and Computational Tools in Metrology and Testing XIII
Series on Advances in Mathematics for Applied Sciences, Vol. 94

https://doi.org/10.1142/9789819800674_0018

GUM-compliant uncertainty propagation for deep neural networks

B. Ludwig, B. Jung and S. Eichstädt*

Physikalisch-Technische Bundesanstalt,
Abbestr. 2–12, 10585 Berlin, Germany
**E-mail: sascha.eichstädt@ptb.de*
www.ptb.de

Advances in algorithms and computing capacities continue to increase the popularity of neural networks. In particular, their use in safety-critical applications (e.g. aviation, transport, telecommunications) poses new challenges with regard to the consideration of uncertainties in their input data. If these consist of sensor measurements, metrology provides a standard reference for dealing with such uncertainties in the internationally harmonised and recognised Guide to the expression of uncertainty in measurements (GUM). In this work, a solid mathematical foundation is established for applying the principles of the GUM to simple neural networks, so-called multilayer perceptrons.

Keywords: measurement uncertainty; uncertainty propagation; neural networks; GUM.

1. Introduction

Advances in algorithms and computing capacities are continuing to increase the popularity of neural networks. However, their use in safety-critical applications (aviation, traffic, telecommunication) also poses ever new challenges for current research. Especially in the areas where sensor readings are used as input data for networks, there is a growing need to incorporate uncertainties into the determination of results. Treatment of uncertainty in deep learning methods nowadays already is a wide field.

The Guide to the expression of uncertainty in measurement (GUM) [1] and its two supplements, together with the International vocabulary of metrology (VIM) [2] are the fundamental documents to provide guidelines on the metrological treatment of measurement uncertainty. The formulation of the requirements placed on neural networks by the GUM and especially its Supplement 2 [3] for multivariate, multistage models for the propagation of the uncertainty of the input data — i.e. the measurement uncertainty — is a field of active research. Consequently, a basic toolbox to propagate uncertainties in accordance with the demands of the GUM from the input data to the output data of neural networks is also missing at present. The main tool to this end is the law of propagation of uncertainty (LPU) in its form for explicit, multivariate measurement models.

The GUM's guidelines result in additional, mathematical constraints on all parts of the involved data transforming model. In this work, the requirements for the networks are first formulated in a mathematically stringent manner, which are only implicitly given in the GUM. Based on this, the LPU for multilayer perceptrons is formulated and applied by way of example. Formally, the solved task reads:

Task 1 (input uncertainty propagation for deep neural network). *Let* $\mathbf{X} = (X_j)_{j=1,\ldots,N}$ *be a real vector quantity representing the input quantities of a measurement function* $f\colon \mathbb{R}^N \to \mathbb{R}^M : \mathbf{x} \mapsto f(\mathbf{x}) = \mathbf{y}$*, which takes the form of an already trained, fully connected, feedforward, deep neural network (DNN) as per Definition 2.1. Let* $(x_j)_{j=1,\ldots,N} \in \mathbb{R}^N$ *be the best estimates representing the input quantities and* $\mathbf{U}_x \in \mathbb{R}^{N\times N}_{+,0}$ *the associated covariance matrix. In parallel with the usual forward propagation of inputs* x_j *through the neural network* f *to form an estimate* $\mathbf{y}$ *of the output quantity* $\mathbf{Y}$*, the associated uncertainties in the form of the covariance matrix* $\mathbf{U}_x$ *are to be processed in a suitable way according to the model to form the covariance matrix* $\mathbf{U}_\mathbf{y}$ *of the estimate* $\mathbf{y}$ *as stated in the GUM's Supplement 2*[3] *paragraph 6.2.*

2. Law of propagation of uncertainty for deep neural networks

In this section, the foundation for the uncertainty propagation in compliance with the GUM and its supplements for deep neural networks is presented, which are defined for the rest of this work as follows:

Definition 2.1. [Fully connected, feedforward deep neural network (DNN)] Let $\ell \in \mathbb{N}$ be the number of layers not including the input layer, $n^{(0)}, \ldots, n^{(\ell)} \in \mathbb{N}$ the number of neurons per layer, $x^{(0)} \in \mathbb{R}^{n^{(0)}}$ the input vector, $h^{(i)}\colon \mathbb{R}^{n^{(i)}} \to \mathbb{R}^{n^{(i)}}, i = 1, \ldots, \ell$ the vector of the activation functions of the ith layer and $x^{(i)}\colon \mathbb{R}^{n^{(i-1)}} \to \mathbb{R}^{n^{(i)}}, i = 1, \ldots, \ell$ the hidden and output layers' activations defined as

$$x^{(i)}\colon x^{(i-1)} \mapsto x^{(i)}(x^{(i-1)}) := (h^{(i)} \circ z^{(i)})(x^{(i-1)}), \quad i = 1, \ldots, \ell,$$

(ith layer's activation)

where the ith layer's affine transformation $z^{(i)}\colon \mathbb{R}^{n^{(i-1)}} \to \mathbb{R}^{n^{(i)}}$ is defined as

$$\begin{aligned} z^{(i)}\colon x^{(i-1)} \mapsto z^{(i)}(x^{(i-1)}) &:= W^{(i)}x^{(i-1)} + b^{(i)} \\ &= \left(b_k^{(i)} + \sum_{j=1}^{n^{(i-1)}} w_{k,j}^{(i)} x_j^{(i-1)}\right)_{k=1,\dots,n^{(i)}}, i = 1,\dots,\ell, \end{aligned} \tag{ith layers affine transformation}$$

with weight matrices $W^{(i)} = (w_{k,j}^{(i)})_{\substack{k=1,\dots,n^{(i)} \\ j=1,\dots,n^{(i-1)}}} \in \mathbb{R}^{n^{(i)} \times n^{(i-1)}}$ and bias vectors $b^{(i)} \in \mathbb{R}^{n^{(i)}}, i = 1,\dots,\ell$. For the sake of readability, we will omit the indication of the argument $x^{(i-1)}$ in the following if it does not serve the purpose of comprehensibility. Then $f\colon \mathbb{R}^{n^{(0)}} \to \mathbb{R}^{n^{(\ell)}} : x^{(0)} \mapsto f(x^{(0)})$ defined as

$$\begin{aligned} f(x^{(0)}) &:= x^{(\ell)} \\ &= \underbrace{h^{(\ell)}\left(W^{(\ell)}\cdots \underbrace{h^{(1)}\left(W^{(1)}x^{(0)} + b^{(1)}\right)}_{x^{(1)}} \cdots + b^{(\ell)}\right)}_{\substack{\vdots \\ x^{(\ell)}}} \end{aligned} \tag{DNN}$$

is called a *fully connected, feedforward deep neural network (DNN).*

A concrete instance of a DNN with $\ell = 3$ layers and $n^{(0)} = n^{(3)} = 2, n^{(1)} = n^{(2)} = 5$ neurons for the respective layers is exemplified in Fig. 1.

For the central definition of a law of propagation of uncertainty suited for neural network applications, first the notion of a *sensitivity matrix* needs to be defined.

Definition 2.2 (Sensitivity matrix). *Let $N, M \in \mathbb{N}, \Omega \subseteq \mathbb{R}^N$ be an open set, $\mathbf{x} = (x_1,\dots,x_N) \in \Omega$ the best estimate of the vector input quantity $\mathbf{X} = (X_i)_{i=1,\dots,N}$ for a multivariate measurement function $f\colon \mathbb{R}^N \to \mathbb{R}^M : \mathbf{x} \mapsto f(\mathbf{x}) := \mathbf{y}$. Let $f \in C^1(\Omega)$, i.e. continuously differentiable on Ω. Then the sensitivity matrix $\mathbf{C_x} \in \mathbb{R}^{M\times N}$ with respect to the input of this measurement is the Jacobian matrix of the measurement function*

$$\mathbf{C_x} := \begin{pmatrix} \frac{\partial f_1}{\partial x_1}(\mathbf{x}) & \dots & \frac{\partial f_1}{\partial x_N}(\mathbf{x}) \\ \vdots & \ddots & \vdots \\ \frac{\partial f_M}{\partial x_1}(\mathbf{x}) & \dots & \frac{\partial f_M}{\partial x_N}(\mathbf{x}) \end{pmatrix} = J_f(\mathbf{x}). \tag{1}$$

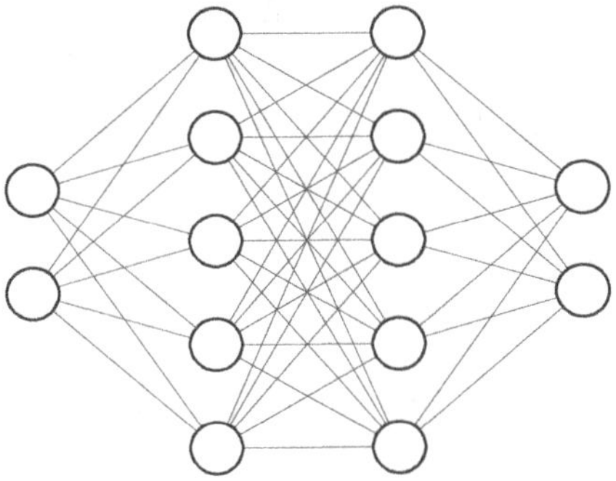

Fig. 1. A schematic view on a DNN with each two input and output neurons and two hidden layers, where each connecting line corresponds to one $x_k^{(i-1)}$ and each neuron to one mapping $h_k^{(i)} \circ z_k^{(i)}, i = 1, 2, 3$.

Now a formal definition with explicit requirements for the law of propagation of uncertainties for real, multivariate measurement models can be presented.

Definition 2.3 (Multivariate law of propagation of uncertainty). *Let $N, M \in \mathbb{N}, \Omega \subseteq \mathbb{R}^N$ be an open set, $\mathbf{x} \in \Omega$ the best estimate of the vector input quantity $\mathbf{X} = (X_i)_{i=1,\dots,N}$ for a multivariate measurement function $f \colon \mathbb{R}^N \to \mathbb{R}^M \colon \mathbf{x} \mapsto f(\mathbf{x}) := \mathbf{y}$ and $\mathbf{U_x} \in \mathbb{R}^{N \times N}_{+,0}$ the associated covariance matrix. Let $f \in C^1(\Omega)$ and $\mathbf{C_x} \in \mathbb{R}^{M \times N}$ the sensitivity matrix with respect to the input as per Definition 2.2. Then the covariance matrix $\mathbf{U_y} \in \mathbb{R}^{M \times M}_{+,0}$ associated with $\mathbf{y}$ is*

$$\mathbf{U_y} = \mathbf{C_x} \mathbf{U_x} \mathbf{C_x}^T. \tag{2}$$

Remark 2.1. A neural network as defined by Definition 2.1 from the perspective of the GUM is a real, multivariate, multistage measurement model with the measurement function f as in Definition 2.1, best estimate $x^{(0)}$ for the real, vector valued input quantity $\mathbf{X}^{(0)}$, intermediate vector quantities $x^{(i)}, i = 1, \dots, \ell - 1$ and the best estimate $x^{(\ell)}$ for the vector measurand $\mathbf{X}^{(\ell)}$. As such, the LPU 2.3 requires continuous differentiability of each layer's mapping $x^{(i)} \colon \mathbb{R}^{n^{(i-1)}} \to \mathbb{R}^{n^{(i)}} \colon x^{(i-1)} \mapsto x^{(i)}(x^{(i-1)}) = (h^{(i)} \circ z^{(i)})(x^{(i-1)}), i = 1, \dots, \ell$ in a neighbourhood of the previous layer's activations $x^{(i-1)}$. As these in general are not known in advance, the continuous differentiability prerequisite of f implies the prerequisite of continuous differentiability of each of the $x^{(i)}$ on its entire domain $\mathbb{R}^{n^{(i-1)}}, i = 1, \dots, \ell$ to ensure the applicability of (2).

Theorem 2.1 (LPU for deep neural networks). *Let f be a DNN according to Definition 2.1, $\ell \in \mathbb{N}$ the number of layers in addition to the input layer, $n^{(0)}, \ldots, n^{(\ell)} \in \mathbb{N}$ the number of neurons per layer, $x^{(0)} = \left(x_j^{(0)}\right)_{j=1}^{n^{(0)}} \in \mathbb{R}^{n^{(0)}}$ the input vector comprised of the best estimates of the input quantities $X_j, \mathbf{U}_{x^{(0)}}$ the associated covariance matrix and $W^{(i)}, i = 1, \ldots, \ell$ the ith layer's weight matrix. Further let each activation function $h_k^{(i)} : \mathbb{R} \to \mathbb{R}, i = 1, \ldots, \ell, k = 1, \ldots, n^{(i)}$ be continuously differentiable on $\mathbb{R}$. Then the LPU according to Definition 2.3 can be applied to propagate the covariance matrix $\mathbf{U}_{x^{(0)}}$ associated with the inputs to compute the covariance matrix $\mathbf{U}_{x^{(\ell)}}$ associated with the outputs $x^{(\ell)}$ as*

$$\mathbf{U}_{x^{(\ell)}} = \Big(\prod_{i=0}^{\ell-1} H^{(\ell-i)} W^{(\ell-i)}\Big) \mathbf{U}_{x^{(0)}} \Big(\prod_{i=1}^{\ell} {W^{(i)}}^T H^{(i)}\Big), \qquad \text{(LPU for DNN)}$$

with the diagonal matrices $H^{(i)}$

$$H^{(i)} := \operatorname{diag}\left(\frac{\mathrm{d}h_k^{(i)}}{\mathrm{d}z}(z_k^{(i)})\right)_{k=1,\ldots,n^{(i)}}, \quad i = 1, \ldots, \ell, \tag{3}$$

where diag *denotes the* diag *operator for vectors.*

Remark 2.2. In the notation of Eq. (LPU for DNN) the matrix products must be expanded in the specified order as matrix multiplication does not commute.

Proof of Theorem 2.1. It is to show that

- indeed this leads to the sensitivity matrices as stated by Eqs. (LPU for DNN) and (3), and
- Eq. (LPU for DNN) then provides the outputs' covariance matrix.

Let f be a DNN as per Definition 2.1 with input layer and additional $\ell \in \mathbb{N}$ layers. The applicability of the LPU 2.3 and Eq. (LPU for DNN) are shown by means of complete induction on the number of layers, denoted by n.

For $n = 1$, it is

$$\begin{aligned} f(x^{(0)}) &= x^{(1)} \\ &= (h^{(1)} \circ z^{(1)})(x^{(0)}) \\ &= \begin{pmatrix} h_1^{(1)} \circ z_1^{(1)} \\ \vdots \\ h_{n^{(1)}}^{(1)} \circ z_{n^{(1)}}^{(1)} \end{pmatrix}. \end{aligned} \tag{4}$$

As affine map $z^{(1)}$ is differentiable with constant derivative, thus continuously differentiable on $\mathbb{R}^{n^{(0)}}$ with JACOBIAN

$$J_{z^{(1)}} = \begin{pmatrix} \frac{\partial z_1^{(1)}}{\partial x_1} & \cdots & \frac{\partial z_1^{(1)}}{\partial x_{n^{(0)}}} \\ \vdots & \ddots & \vdots \\ \frac{\partial z_{n^{(1)}}^{(1)}}{\partial x_1} & \cdots & \frac{\partial z_{n^{(1)}}^{(1)}}{\partial x_{n^{(0)}}} \end{pmatrix}. \tag{5}$$

From Eq. (ith layers affine transformation) it follows immediately

$$\frac{\partial z_k^{(1)}}{\partial x_j} = w_{k,j}^{(1)}, \quad , k = 1, \ldots, n^{(1)}, j = 1, \ldots, n^{(0)}. \tag{6}$$

Since $h_k^{(1)}, k = 1, \ldots, n^{(1)}$ is continuously differentiable on $\mathbb{R}$ the composition $h_k^{(1)} \circ z_k^{(1)}$ is continuously differentiable on $\mathbb{R}^{n^{(0)}}$ as well with its JACOBIAN being $J_{h_k^{(1)} \circ z_k^{(1)}} = J_{h_k^{(1)}} J_{z_k^{(1)}}$ with

$$J_{h_k^{(1)}} = \frac{\mathrm{d}h_k^{(1)}}{\mathrm{d}z}(z_k^{(1)}), \quad k = 1, \ldots, n^{(1)}. \tag{7}$$

As all the coordinate functions of f are continuously differentiable on $\mathbb{R}^{n^{(0)}}$ so is f and its derivative f', i.e. sensitivity matrix according to Definition 2.2, is

$$\mathbf{C}_{x^{(0)}} = f' = \begin{pmatrix} h_1^{(1)} \circ z_1^{(1)} \\ \vdots \\ h_{n^{(1)}}^{(1)} \circ z_{n^{(1)}}^{(1)} \end{pmatrix}' = H^{(1)} W^{(1)}, \tag{8}$$

where $H^{(1)} := \operatorname{diag}\left(\frac{\mathrm{d}h_k^{(i)}}{\mathrm{d}z}(z_k^{(i)})\right)_{k=1,\ldots,n^{(1)}}$ with the diag operator for vectors as in Definition 0.9.1.[4] Now the LPU for multivariate real measurement functions in Definition 2.3 can be applied which provides for the covariance matrix $\mathbf{U}_{x^{(1)}}$ associated with the neural network's outputs $x^{(1)}$

$$\begin{aligned}\mathbf{U}_{x^{(1)}} &= \mathbf{C}_{x^{(0)}}\mathbf{U}_{x^{(0)}}\mathbf{C}_{x^{(0)}}^T \\ &= H^{(1)}W^{(1)}\mathbf{U}_{x^{(0)}}\left(H^{(1)}W^{(1)}\right)^T \\ &= H^{(1)}W^{(1)}\mathbf{U}_{x^{(0)}}{W^{(1)}}^T H^{(1)},\end{aligned}$$

where $H^{(1)} = {H^{(1)}}^T$ was used. Assume now, that the LPU is applicable for each of the n first layers and $\mathbf{U}_{x^{(i)}}$ is the covariance matrix associated with the ith layer's activations $x^{(i)}, i = 1, \ldots, n$ given by

$$\mathbf{U}_{x^{(i)}} = \mathbf{C}_{x^{(i-1)}}\mathbf{U}_{x^{(i-1)}}\mathbf{C}_{x^{(i-1)}}^T = H^{(i)}W^{(i)}\mathbf{U}_{x^{(i-1)}}{W^{(i)}}^T H^{(i)}.$$

For the $(n+1)$th layer, it again is

$$x^{(n+1)} = (h^{(n+1)} \circ z^{(n+1)})(x^{(n)}) \tag{9}$$

$$= \begin{pmatrix} h_1^{(n+1)} \circ z_1^{(n+1)} \\ \vdots \\ h_{n^{(n+1)}}^{(n+1)} \circ z_{n^{(n+1)}}^{(n+1)} \end{pmatrix}. \tag{10}$$

Now by the same reasoning as above, the LPU can be applied to $x^{(n+1)}$ and the resulting covariance matrix $\mathbf{U}_{x^{(n+1)}}$ associated with the output layer's activations $x^{(n+1)}$ is

$$\mathbf{U}_{x^{(n+1)}} = \mathbf{C}_{x^{(n)}}\mathbf{U}_{x^{(n)}}\mathbf{C}_{x^{(n)}}^T = H^{(n+1)}W^{(n+1)}\mathbf{U}_{x^{(n)}}{W^{(n+1)}}^T H^{(n+1)}.$$

Applying this result to f and replacing the covariance matrices recursively leads to

$$\mathbf{U}_{x^{(\ell)}} = \prod_{i=1}^{\ell}\mathbf{C}_{x^{(\ell-i)}}\mathbf{U}_{x^{(0)}}\prod_{i=0}^{\ell-1}\mathbf{C}_{x^{(i)}}^T \tag{11}$$

$$= \left(\prod_{i=0}^{\ell-1} H^{(\ell-i)}W^{(\ell-i)}\right)\mathbf{U}_{x^{(0)}}\left(\prod_{i=1}^{\ell}{W^{(i)}}^T H^{(i)}\right). \tag{12}$$

□

3. Uncertainty propagation for sigmoid activation

In the following, networks with exclusively one distinct activation function are considered, although the same approach can be readily adopted for networks with varying activation functions per neuron. Due to the smoothness prerequisite on the activation functions employed (see Remark 2.1), in the following we consider a continuously differentiable activation function to exemplify the application of the LPU for DNN.

The sigmoid activation function is a common choice for DNNs when continuous differentiability is desired and the output of the neurons should lie in the open interval $(0, 1)$, as in the case of Bernoulli output probability distributions Chapter 6.2.2.2.[5]

Definition 3.1 (Sigmoid activation function). *The mapping $\sigma\colon \mathbb{R} \to (0,1)$ defined as*

$$\sigma(x) := \frac{e^x}{1+e^x} = \frac{1}{1+e^{-x}}, \qquad \text{(sigmoid)}$$

is called the sigmoid activation function (or just sigmoid).

Figure 2 shows the graph of the sigmoid activation function.

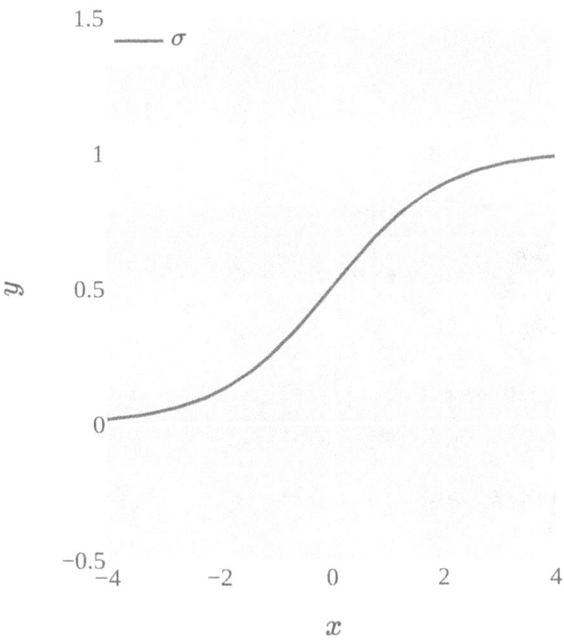

Fig. 2. Graph of sigmoid activation function for $x \in [-4, 4]$.

Proposition 3.1 (LPU for sigmoid activation function). *Let f be a DNN in the same situation as in Theorem 2.1 exclusively equipped with the* *sigmoid activation function σ. Then the LPU according to Definition 2.3 can be applied with $H^{(i)}$ equal to*

$$H^{(i)} = \operatorname{diag}\sigma(z^{(i)})\operatorname{diag}(\mathbf{1}-\sigma(z^{(i)})), \quad i=1,\dots,\ell,$$

with $\mathbf{1} := (1,\dots,1) \in \mathbb{R}^{n^{(i)}}$, such that for the covariance matrix $\mathbf{U}_{x^{(\ell)}}$ associated with the output layer's activations $x^{(\ell)}$ it is

$$\mathbf{U}_{x^{(\ell)}} = \Big(\prod_{i=0}^{\ell-1}\operatorname{diag}\sigma(z^{(\ell-i)})\operatorname{diag}(\mathbf{1}-\sigma(z^{(\ell-i)}))W^{(\ell-i)}\Big)\mathbf{U}_{x^{(0)}}$$
$$\Big(\prod_{i=1}^{\ell}W^{(i)^T}\operatorname{diag}\sigma(z^{(i)})\operatorname{diag}(\mathbf{1}-\sigma(z^{(i)}))\Big).$$

Proof. σ is a composition of the exponential $\exp\colon \mathbb{R}\to\mathbb{R}^+ : x\mapsto \exp(x) := e^x$ and the affine function $1+x$, which are both continuously differentiable on $\mathbb{R}$, and $\frac{1}{x}$, which is continuously differentiable on $\mathbb{R}^+$. As such, σ is continuously differentiable on $\mathbb{R}$ with derivative

$$\frac{\mathrm{d}\sigma}{\mathrm{d}x}(x) = \frac{e^{-x}}{(1+e^{-x})^2} = \frac{1}{1+e^{-x}}\left(1-\frac{1}{1+e^{-x}}\right) = \sigma(x)(1-\sigma(x)). \quad (13)$$

With $z_k^{(i)}, i = 1,\dots,\ell, k = 1,\dots,n^{(i)}$ as per (ith layers affline transformation) and Eqs. (3) and (13), it is

$$H^{(i)} = \operatorname{diag}\frac{\mathrm{d}\sigma}{\mathrm{d}z}(z^{(i)}) = \operatorname{diag}\sigma(z^{(i)})\operatorname{diag}(\mathbf{1}-\sigma(z^{(i)})), \quad i=1,\dots,n^{(i)}.$$

Theorem 2.1 then yields

$$\begin{aligned}\mathbf{U}_{x^{(\ell)}} &= \Big(\prod_{i=0}^{\ell-1}H^{(\ell-i)}W^{(\ell-i)}\Big)\mathbf{U}_{x^{(0)}}\Big(\prod_{i=1}^{\ell}W^{(i)^T}H^{(i)}\Big)\\ &= \Big(\prod_{i=0}^{\ell-1}\operatorname{diag}\sigma(z^{(\ell-i)})\operatorname{diag}(\mathbf{1}-\sigma(z^{(\ell-i)}))W^{(\ell-i)}\Big)\mathbf{U}_{x^{(0)}}\\ &\qquad\Big(\prod_{i=1}^{\ell}W^{(i)^T}\operatorname{diag}\sigma(z^{(i)})\operatorname{diag}(\mathbf{1}-\sigma(z^{(i)}))\Big).\end{aligned}$$

□

4. Conclusions and outlook

The successes of this work are primarily the formulation of the mathematical definitions and statements from Secs. 2 and 3, first and foremost

Theorem 2.1, with the accompanying application in Proposition 3.1. This provides a starting point for experts or those interested in the domain of metrological uncertainty propagation in continuously differentiable neural networks.

One aspect that was not considered at all in the present work is the quality of the uncertainty propagation with regard to the comparison with the Monte Carlo method according to the GUM's Supplement 1.[6] However, due to the highly non-linear nature of DNNs, it must be assumed that, especially for an increasing number of layers and neurons, the deviations between the approach presented here and GUM's Supplement 1[6] will steadily increase.

References

1. JCGM, Evaluation of measurement data — Guide to the expression of uncertainty in measurement (2008).
2. JCGM, International Vocabulary of Metrology — Basic and general concepts and associated terms (2012).
3. JCGM, Evaluation of measurement data — Supplement 2 to the "Guide to the expression of uncertainty in measurement" — Extension to any number of output quantities (2011).
4. R. A. Horn and C. Johnson, Matrix analysis, second edition (Cambridge University Press, Erscheinungsort nicht ermittelbar, 2012), OCLC: 1151459555.
5. I. Goodfellow, Y. Bengio and A. Courville, Deep Learning (2016).
6. JCGM, Evaluation of measurement data — Supplement 1 to the "Guide to the expression of uncertainty in measurement" — Propagation of distributions using a Monte Carlo method (2008).

F Pavese, A Bošnjaković, S Eichstädt, A B Forbes, J Alves e Sousa (eds.):
Advanced Mathematical and Computational Tools in Metrology and Testing XIII
Series on Advances in Mathematics for Applied Sciences, Vol. 94

https://doi.org/10.1142/9789819800674_0019

A SQC proposal for monitoring the shipyard panel line

D. A. Maisano[†], M. Trombini and A. Pagani
Politecnico di Torino,
Corso Duca degli Abruzzi 24, Turin, 10129, Italy
[†]*E-mail: domenico.maisano@polito.it*

Constructing cruise-ship hull is complex and requires precise metalwork, welding and assembly. Early anomaly detection and correction are vital for cost and time management. This paper presents a novel *statistical quality control* (SQC) methodology for monitoring the *panel line* in cruise-ship shipyards, i.e., one of the initial manufacturing workshops, with a relatively high level of automation. The proposed methodology adopts a standardized p control chart with samples of variable size, incorporating two elements: (i) it accounts for the high level of customization of panels, and (ii) it takes into consideration the measurement uncertainty associated with the *large-volume metrology* instrument employed for conformity verification (such as a state-of-the-art total station), following the ISO 14253-1:2017 standard. A real-world case study demonstrates its practical application.

Keywords: shipbuilding; cruise-ship hull; panel line; conformity verification; ISO 14253-1:2017; large-volume metrology; total station; standardized p-chart.

1. Introduction

Cruise-ship hull construction involves multiple operations on large metal parts, divided into four phases: (i) preparation of *panels* by cutting, shaping, bending, and joining plates to rigid elements, (ii) assembly of panels to make *units* with dimensions around (20–40 m) × (20–40 m) × (3–5 m), (iii) stacking of two or three units to make *modules*, and (iv) final assembly of modules to erect the complete hull [1]. This process is challenging due to several factors, such as [2]:

(i) The large size and the relatively high level of customization of the parts manufactured complicate the organization of the production process and supply chain.

(ii) The unpredictable sequence of operations, due to the availability of a large number of operators, equipment and the *rework* that becomes systematically necessary.

(iii) The inevitable presence of deviations from the nominal design dimensions, which can lead to nonconformities.

Nonconformities, if not detected and corrected in time, can lead to significant additional costs, especially in the final stages of the process. In some cases, they can even alter the structural aspects of the ship or generate aesthetic anomalies [1]. Dimensional specifications related to the geometry of parts manufactured in cruise-ship shipyards are usually around a few millimetres from nominal (design) values [3, 4].

To prevent the "propagation" of nonconformities, it is important to monitor the production process in real time using *statistical quality control* (SQC) tools. Timely feedback on the conformity of the work performed is highly beneficial, not only to prevent the propagation of errors but also to provide guidance to operators [5]. Traditional SQC methods are not always suitable for the construction of cruise ships, as (i) they are typically designed for mass production, with a limited level of customisation, and (ii) they do not consider the *measurement uncertainty* of the quality characteristics of interest.

This paper proposes a novel SQC methodology for monitoring the panel line of cruise-ship shipyards. The proposed methodology is based on a standardized *p*-chart that takes into account the measurement uncertainty of the instrument used for conformity verification, in line with the ISO 14253-1:2017 standard, and accommodates the fact that panels can be highly customized [6]. Dimensional verifications are performed through a state-of-the-art instrument for *large-volume metrology* (LVM), namely a Leica Nova TS60 scanning total station equipped with a contact probe [7, 8].

This paper is organized into three sections. The first one contains preliminary information, including a case study from a Fincantieri S.p.A. shipyard, technical instrument details, and ISO 14253-1:2017 standard overview. The second section details the SQC methodology, incorporating ISO 14253-1:2017-based conformity verifications and a standardized *p*-chart, illustrated through the case study. The third section illustrates the practical implications of this research, highlighting the methodology's potential, limitations, and future research insights.

2. Preliminary information

Case study

Let's delve into the panel line, probably the shipyard's only automated workshop, in which steel plates (i.e., *panels*) are joined, marked with lines/curves for alignment, cut, and reinforced with longitudinal stiffeners, transverse beams, and longitudinal girders. Panels move rapidly on roller conveyors between various stations, each with brief stationary times. The total throughput time of panels (i.e., from entering to leaving the panel-line workshop) is of the order of a few hours.

The panel line's automation offers flexibility for producing uniquely shaped components efficiently. In addition, ensuring dimensional accuracy is crucial for quality and timely manufacturing. Let's consider the panel line at a Fincantieri S.p.A. shipyard, producing panels for various ships with a dynamic sequence. In this context, quality engineers define several (geometric) quality characteristics to monitor, which may vary across panels. Table 1 provides an example of (geometric) quality characteristics for a specific panel, primarily distances between reference positions, with specifications of ±2 mm around their nominal value. Operators use advanced total stations for verifications.

Table 1. Dimensional verifications for a specific panel. Thirty-eight quality characteristics (i.e., distances between reference positions on the panel surface) are verified using a Leica Nova TS60 scanning total station (with standard uncertainty $u \approx 0.45$ mm). Conformity verification follows ISO 14253-1:2017, with a guard band of $g \approx 0.74$ mm around each specification limit. "Δ" indicates deviation from the nominal value. Symbols "✓", "✗", and "?(✓)" represent undoubted conformity, undoubted nonconformity and dubious conformity, respectively.

Qual. characteristic no.	Specifications [mm]			y [mm]	$\Delta = y - NV$ [mm]	Conforming?
	NV	*LSL*	*USL*			
1	2500	2498	2502	2500.7	0.7	✓
2	2740	2738	2742	2738.7	-1.3	?(✓)
3	13460	13458	13462	13459.9	-0.1	✓
4	9410	9408	9412	9408.9	-1.1	✓
5	11510	11508	11512	11509.6	-0.4	✓
6	5120	5118	5122	5116.8	-3.2	✗
7	16240	16238	16242	16238.0	-2.0	?(✓)
⋮	⋮	⋮	⋮	⋮	⋮	⋮
36	3180	3178	3182	3180.1	0.1	✓
37	10940	10938	10942	10940.0	0.0	✓
38	14990	14988	14992	14993.0	3.0	✗

LVM instrument

In recent decades, shipyard measuring tools have evolved from traditional equipment (e.g., plumb bobs, steel tapes and transits) to advanced LVM instruments, like laser scanners, trackers, and total stations equipped with contact/non-contact probes, which enable the rapid acquisition of 3D data for surface reconstruction [7].

For panel-line inspections, modern total stations with contact probe accessories offer a practical balance of convenience and precision [8]. In our case study, we used the Leica Nova TS60 scanning total station with a "mini-vector" probe [9]. The instrument's performance was assessed, yielding a conservative $u \approx 0.45$ mm standard uncertainty in single distance measurements.

ISO 14253-1:2017 standard

ISO 14253-1:2017, part of the GPS standards family, can be used to verify conformity with quality-characteristic specifications while considering measurement uncertainty [6]. Verification is sensitive near lower and upper specification limits (i.e., LSL and USL), with the risk of resulting in *false nonconformings* (i.e., real conforming items that are misclassified as nonconforming ones) or *false conformings* (i.e., real nonconforming items that are misclassified as conforming ones). Figure 1 shows doubt in conformity or nonconformity verification outside or inside the specification range; this doubt arises in the "?" zones around specification limits, requiring the application of a suitable decision rule.

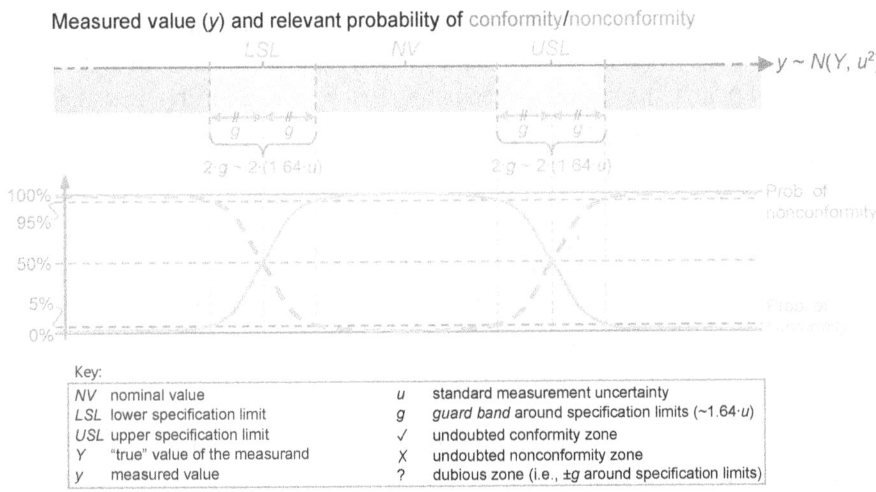

Fig. 1. Conformity verification according to ISO 14253-1:2017, based on (i) the measurement result (y) of a quality characteristic of interest and (ii) the decision rule adopted.

ISO 14253-1:2017 offers two decision rules, depending on whether the need to limit the risk of false nonconformings (e.g., to limit unnecessary repair work) or the risk of false conformings prevails (e.g., to prevent potential failures). Rule #1 bilaterally extends the conformity range of the specification limits by a guard band of semi-width g (i.e., conformity range equal to zone "✓" plus zone "?", in Fig. 1), while Rule #2 bilaterally narrows it by the same amount (i.e., conformity range equal to zone "✓" only). When the standard uncertainty (u) is ¼ of the specification interval (USL – LSL) or less, a $g \approx 1.64 \cdot u$ guard band maintains risks (of false nonconfirmings and false conformings) below 5% for both rules (see Fig. 1) [6].

3. SQC methodology

Control chart selection

Selecting a control chart for monitoring the shipyard panel line should consider its unique characteristics: fast panel transition, high customization, and diverse conformity verifications (in type and number). The standardized p control chart for attributes is chosen, for which each panel represents an ith sample with variable quality characteristics, assessed against specifications [6]. For example, Table 1 shows an ith sample (panel) with $n_i = 38$ quality characteristics, verified using the Leica Nova TS60 scanning total station. Verifications are done collectively at the panel line exit, and each quality characteristic is marked as conforming ("✓") or nonconforming ("✗"). The total number of defectives ($d_i \in [0, n_i]$) and the panel defectiveness ($p_i = d_i/n_i \in [0, 1]$) can then be calculated for the ith specific panel.

The verification of different quality characteristics and the determination of an overall panel-by-panel defectiveness (d_i) are justified when the *natural variability* among quality characteristics is relatively homogeneous, specification ranges are similar (e.g., ±2 mm around nominal values), and measurement uncertainty for the different measurements is comparable. In this case study, these conditions are met.

Construction of the standardized p control chart

The construction of the standardized p control chart requires a dataset of at least 15–20 samples (i.e., panels) of no less than 20–25 elements (i.e., quality characteristics) each. In the case study, 21 panels were used, each with different quality characteristics (in number and type). The fact that the measurement uncertainty ($u \approx 0.45$ mm, cf. Sec. 2) is not negligible compared to the specification interval (i.e., to be negligible it should be at least one order of magnitude lower) leads to the adoption of ISO 14253-1:2017, with rule #1 chosen for the specific case of interest. Quality characteristics classified as conforming are further categorized as undoubtedly conforming (in case the measurement result falls in the "✓" zone) or doubtfully conforming (in case the measurement result falls in the "?(✓)" zone) (cf. Fig. 1).

Table 1 (in the last column) exemplifies the outcome of some conformity verifications for the specific panel no. 1, revealing $d_1 = 6$ nonconforming quality characteristics out of n_1 = 38, resulting in a sample defectiveness of p_1 = 6/38 ≈ 15.8%. Among the 32 conforming quality characteristics, half are undoubtedly conforming ("✓"), and half are doubtfully conforming ("?(✓)"). Extending these verifications to the remaining 20 panels generates the data in Table 2. The number of defectives (d_i) in an ith panel with n_i quality characteristics follows a binomial distribution, with mean μ_d and variance σ_d^2. Based on available data, the best estimate of the defectiveness (p) of the whole panel-line process is [5]:

$$\bar{p} = \frac{\sum_{i=1}^{21} d_i}{\sum_{i=1}^{21} n_i} \approx 14.2\%. \tag{1}$$

Table 2. Data for constructing the standardized p control chart, including the calculation of $\bar{p}$, i.e., the best estimate of p based on available data.

Sample (panel) no.	n_i	d_i	p_i	z_i
1	38	6	15.8%	0.280
2	39	6	15.4%	0.212
3	35	8	22.9%	1.467
4	35	7	20.0%	0.983
5	46	6	13.0%	-0.225
6	37	4	10.8%	-0.591
7	45	7	15.6%	0.260
8	38	5	13.2%	-0.184
9	44	6	13.6%	-0.107
10	44	3	6.8%	-1.403
11	35	7	20.0%	0.983
12	43	8	18.6%	0.827
13	40	3	7.5%	-1.214
14	44	9	20.5%	1.188
15	36	9	25.0%	1.856
16	47	9	19.1%	0.972
17	41	4	9.8%	-0.816
18	37	2	5.4%	-1.533
19	42	3	7.1%	-1.311
20	40	6	15.0%	0.145
21	46	3	6.5%	-1.492
	$\sum_{i=1}^{m} n_i = 852$	$\sum_{i=1}^{m} d_i = 121$	$\bar{p} = 14.2\%$	

If $n_i \cdot p_i \geq 5$, the binomial distribution of d_i can be approximated by a normal distribution with the same parameters [5]. Since p_i is linearly related to d_i (with n_i considered constant for an individual ith panel), it can also be approximated by a normal distribution with identical parameters (i.e., $\mu_p = p, \sigma_p = \sqrt{\frac{p \cdot (1-p)}{n_i}}$).

In addition, a standardisation of p_i values can be introduced by means of the transformation:

$$z_i = \frac{p_i - p}{\sigma_p} = \frac{p_i - p}{\sqrt{\frac{p \cdot (1-p)}{n_i}}}, \quad (2)$$

z being the standard normal variable with zero mean and unit variance.

Table 2 displays d_i, p_i, and z_i values for each ith sample. Except for isolated cases (samples 3, 4, and 11), the condition $n_i \cdot p_i \geq 5$, which is necessary for approximating d_i and p_i as normally distributed variables, is consistently met [5, 10].

Figure 2 presents the standardized p-chart with z_i values from Table 2, along with three-sigma control limits ($UCL = 3$, $CL = 0$, and $LCL = -3$); three-sigma limits are commonly used in SQC because they offer a good compromise between responsiveness in detecting anomalies and reducing the risk of false alarms [5]. All data points fall within control limits, displaying a seemingly random pattern. This randomness is statistically confirmed through the traditional Western Electric rules and the Anderson-Darling normality test [5, 10]. Therefore, this control chart effectively monitors the process under stable conditions, free from "assignable" sources of variability, and can be used to monitor future process evolution.

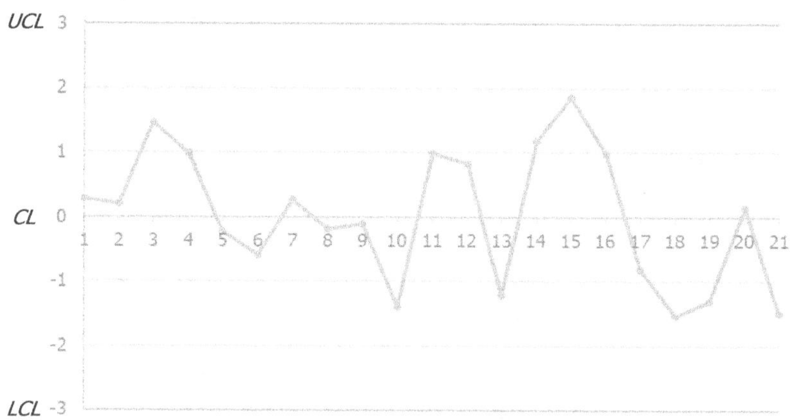

Fig. 2. Standardized p-chart related to the z_i values in Table 2.

4. Discussion

The proposed methodology is valuable for overseeing manufacturing operations in the shipyard's panel line, operating at two intertwined levels:

Product Conformity Verification This activity aims at promptly identifying anomalies in manufactured products, enabling corrective actions to limit error propagation and excessive repair work in final assembly. An innovative aspect is handling the measurement uncertainty through ISO 14253-1:2017, allowing the distinction between undoubted cases of conformity/nonconformity and doubtful cases, in which measured values fall in the guard band around specification limits. Undoubted nonconformities demand immediate action, while doubtful cases can be documented for cautious handling in subsequent stages; for instance, measurements could be reinforced around the area of concern and/or more precise instruments might be employed. The approach also provides flexibility, allowing quality managers to choose the most suitable decision rule. Although in the case study at a Fincantieri S.p.A. shipyard's panel line, a Leica Nova TS60 total station was used for conformity verifications, this method can adapt to other LVM instruments.

Process Stability Monitoring The standardized *p*-chart makes it possible to continuously monitor the manufacturing process, identifying steady progress or potential disturbances from abnormal factors requiring investigation. Any out-of-control situation detected by the control chart prompts investigations into root causes. For instance, increased panel defectiveness may stem from processing errors (human-induced or machinery/material-related), while reduced defectiveness could result from errors or deliberate manipulation of conformity verification by operators [5]. The standardized *p*-chart is straightforward to create and manage, with variables d_i, n_i, and p_i having practical meanings for non-statisticians. However, interpreting z_i requires basic statistical knowledge [5, 10].

The proposed methodology has some limitations. It assumes that conformity verifications pertain to quality characteristics with reasonably similar defectiveness rates; otherwise, the control chart model becomes more complex [5, 10]. Defining the specific quality characteristics for each panel is delicate, as they should constitute an adequate number of features that are representative of the process quality, without redundancy. Expertise from quality engineers is crucial. Additionally, all conformity verifications for a panel condense into a single data point, making the control chart not very responsive to gradual process shifts, which typically require multiple data points for detection.

References

1. Mandal N.R. (1989) *Ship construction and welding*, Vol. 329, Springer Series on Naval Architecture, Marine Engineering, Shipbuilding and Shipping 2. Singapore: Springer.
2. Bay M., Crama Y., Langer Y. *et al.* (2010) Space and time allocation in a shipyard assembly hall. *Annals of Operations Research*, 179(1): 57–76.

3. Storch R.L. and Giesy P.J. (1998) The use of computer simulation of merged variation to predict rework levels on ship's hull blocks. *Journal of ship production*, 4(03): 155–168.
4. ISO 15583:2005. Ships and marine technology — Maritime standards list, International Organization for Standardization, 2005, https://www.iso.org/standard/36696.html.
5. Montgomery D.C. (2019) *Introduction to statistical quality control* (8th edition), John Wiley & Sons.
6. ISO 14253-1:2017 (2017) *Geometrical product specifications (GPS) – Inspection by measurement of workpieces and measuring equipment – Part 1: Decision rules for verifying conformity or nonconformity with specifications.*
7. Franceschini F., Galetto M., Maisano D. *et al.* (2011) *Distributed large-scale dimensional metrology: new insights*. Springer Science & Business Media, London.
8. Maisano D.A., Mastrogiacomo L., Franceschini F. *et al.* (2023) Dimensional measurements in the shipbuilding industry: on-site comparison of a state-of-the-art laser tracker, total station and laser scanner. *Production Engineering*, 17(3–4): 625–642.
9. Hexagon Geosystems (2023) Leica Nova TS60 Totla Station. Available online: https://leica-geosystems.com/it-it/products/total-stations/robotic-total-stations/leica-nova-ts60.
10. Ross S.M. (2020) *Introduction to probability and statistics for engineers and scientists* (6th edition). Academic press (Elsevier).

F Pavese, A Bošnjaković, S Eichstädt, A B Forbes, J Alves e Sousa (eds.):
Advanced Mathematical and Computational Tools in Metrology and Testing XIII
Series on Advances in Mathematics for Applied Sciences, Vol. 94

https://doi.org/10.1142/9789819800674_0020

LNE Uncertainty: A user-friendly software for a validated implementation of the Guide to the Expression of Uncertainty in Measurement

S. Marmin*, L. Coquelin and N. Fischer

Data Science and Uncertainty Department,
Laboratoire National de Métrologie et d'Essais,
29 av. Hennequin Trappes (78190) France
**E-mail: sebastien.marmin@lne.fr*
www.lne.fr

LNE Uncertainty allows for the evaluation of measurement uncertainties by propagating variances or distributions in accordance with the GUM and GUM Supplement 1. LNE Uncertainty is capable of handling multiple output quantities, perform sensitivity analysis and goodness of fit tests. The software has been developed following the recommendations of European Metrology Network for Mathematics and Statistics (Mathmet) to ensure reliability. The software, a standalone application, is freely available. Its user-friendly interface, guiding through the evaluation process, democratises access to statistical methods, requiring no programming skills, yet demanding experience in the measurement uncertainty evaluation process.

Keywords: measurement uncertainty; software; Monte-Carlo method; sensitivity analysis.

1. Introduction

The Guide to the Expression of Uncertainty in Measurement (GUM)[1] has long been the cornerstone in measurement uncertainty evaluation in scientific and industrial processes. The reference method provided is centred around the Law of Propagation of Uncertainty (LPU), which involves calculating a variance term of a Taylor polynomial. Despite its wide adoption due to its straightforward implementation, the GUM's LPU method shows limitations, particularly in the context of strongly non-linear models where the linear approximation does not yield satisfactory estimates of the output quantity's variance. To address these limitations, GUM Supplement 1 (GUM-S1)[2] advocates the Monte-Carlo Method, a stochastic approach involving randomly selected values from assigned probability laws.

Historically, the GUM's evolution and its supplements reflect a growing refinement in measurement uncertainty evaluation. Since its first publication in 1993, subsequent modifications and publications, notably the GUM-S1 in 2008 and new perspectives introduced in 2015, have expanded the scope and applicability of these guidelines. LNE-Uncertainty, released in 2022, offers a comprehensive tool for uncertainty evaluation through both LPU and Monte-Carlo methods. It is released after a previous software[3] from 2015, that is focused only on the Monte-Carlo method and is no longer maintained.

LNE-Uncertainty, is developed using MATLAB programming language, but it is freely available without requiring a MATLAB license for its usage. It combines the integration of robust statistical methodologies, featuring a modern pseudo-random number generator aligned with standards set in GUM-S1. It supports fourteen probability distribution laws for input quantities and accommodates correlations between inputs modelled as multivariate Gaussian distributions or using Student distributions.

An essential aspect of LNE-Uncertainty is its adherence to quality standards, being developed and validated according to Mathmet's Quality Assessment Tool (QAT) procedure.[4] This not only ensures reliability but also aligns with contemporary needs for rigorous data analysis in metrology.

The software also includes multiple measurands scenarios as addressed in GUM Supplement 2.[5] Its capabilities extend to sensitivity analysis, employing Spearman rank correlation coefficients and Sobol's indices for diverse model complexities. Moreover, it incorporates the Kolmogorov–Smirnov test[6] for fitting output quantities to various theoretical laws.

This paper on the LNE Uncertainty software provides an overview of its functionality and applications in measurement uncertainty evaluation. Section 2 presents the specific methodologies chosen for the framework of LNE-Uncertainty. Section 3 presents more in details some supplementary features: multiple measurands handling, sensitivity analysis and test of goodness-of-fit. In Sec. 4, a practical case study on mass calibration demonstrates the software with a reference case. This paper concludes by confirming the software's utility and accessibility for calculating measurement uncertainties.

2. Methods of uncertainty calculation

2.1. *LPU and Monte-Carlo*

The Monte-Carlo Method serves as an alternative to the Law of Propagation of Uncertainty method from Guide to the Expression of Uncertainty in Measurement (GUM-LPU), particularly when the latter's applicability conditions are not met. In GUM-LPU method, measurement uncertainty is typically evaluated using a first-order Taylor expansion. This approach, however, is problematic for highly non-linear models. Moreover, when the output quantity's distribution significantly deviates from a Gaussian distribution, using an enlargement factor $k = 2$ fails to yield an expanded interval encompassing approximately 95% of the values attributable to the measurand. The Monte-Carlo Method, on the other hand, provides two types of expanded intervals independent of the output quantity's distribution: the symmetric probability interval and the shortest interval. LNE-Uncertainty allows the user to compute these intervals and compare them visually with the LPU interval. Although the LPU method is widely used due to its simplicity and immediate results, its validity is narrower compared to the Monte-Carlo Method. GUM-S1 defines a validation criterion based on comparing the expanded intervals obtained by both methods. This reference validation tool is also implemented.

2.2. *Four steps for calculating the output uncertainty*

Similar to the GUM-LPU method, the Monte-Carlo Method can be summarised in four major steps. The first step involves analysing the measurement process to define the measurand, enumerate input quantities, and determine the measurement model. Subsequently, based on available information (e.g. calibration certificates, repeated measurements, expert opinions, etc.), probability densities are assigned to the input quantities. In the third step, these distributions are propagated through the mathematical model (via a variance calculation or Monte-Carlo simulations). Finally, the results are summarised using statistical parameters such as mean, standard deviation, and expanded intervals. The software interface follows clearly these four steps to facilitate the user navigation during the process. It is important to note that the software is a calculation tool, not a modelling tool; it does not allow users to establish the mathematical model of the measurement process nor determine the input quantities and their potential correlations. GUM-S1 addresses correlations between input quantities modelled by normal laws. However, in practice, correlations may exist between different sets of measurements. If these are modelled using a Student's distribution, the software can also handle such correlations.

3. Advanced features for uncertainty calculation

3.1. *Multiple measurands*

While GUM-S1 focuses on the Monte-Carlo Method for a single measurand, GUM Supplement 2[5] published in 2011 by the JCGM (Joint Committee for Guides in Metrology) extends both the variance propagation formula and distribution propagation to cases of multiple measurands or any number of output quantities. In reality, many measurement processes involve multiple measurands. When uncertainty sources are common to several measurands, they lead to covariances between the measurand estimates (e.g., in the field of chemistry, measurands may be the concentration of different elements in a same solution). If these are then used in a new measurement model, it is crucial to consider the existing correlations between their estimates. Indeed, determining an expanded region rather than juxtaposing individual expanded intervals provides a more faithful representation of the measurands' knowledge. According to GUM Supplement 2, an expanded region $\mathcal{R}_{\text{exp}}$ assigned to the vector measurand $\mathbf{y}$ is an m-dimensional space such that the probability of $\mathbf{y}$ residing in $\mathcal{R}_{\text{exp}}$ equals the stipulated coverage probability p. Three particular cases of expanded regions are proposed in GUM Supplement 2: the multivariate ellipse, rectangle, and the smallest confidence region. In the LNE-Uncertainty software, only the ellipsoidal set are calculated. The quantities determining the position, shape and the orientation of the ellipsoid (i.e. the affine transformation of the unit hypersphere) are evaluated from the Monte-Carlo simulated measurand values.

3.2. *Sensitivity analysis*

For metrologists, the final uncertainty budget is an important tool for analysing their measurement process. It helps in understanding the contributions of different uncertainty sources to the overall uncertainty. It also aids in recognising which sources have minimal influence and could potentially be treated as constants in the mathematical model of the measurement process.

Three methods are available in the software. The first method involves calculating the uncertainty budget through variance propagation, requiring partial derivatives of the output quantity with respect to each input quantity. This approach, while straightforward, does not compute sensitivity coefficients in cases of distribution propagation via Monte-Carlo simulations. GUM-S1 recommends an alternative method of varying each input quantity individually and comparing the resulting standard uncertainties, but this method is limited as it necessitates additional simulations and fails to account for interactions in non-linear models.

To address these limitations, the software incorporates two global sensitivity methods: the Spearman correlation coefficient and Sobol's indices. The Spearman correlation coefficient measures the relationship's intensity and direction between two variables, relying on their sorting ranks from Monte-Carlo simulations. This method is advantageous due to its simplicity and lack of need for extra simulations. On the other hand, Sobol's indices are generally preferred for more complex models where input interactions are significant. While this method is computationally demanding due to additional simulations, its indices provide a comprehensive decomposition of the output quantity's variance, reflecting individual and interactive contributions of input variables in complex measurement models.

3.3. *Goodness-of-fit test*

The Kolmogorov–Smirnov test is facilitated by the software. This test is commonly used for determining the most suitable probability distribution law for an output quantity, particularly when this output is used as an input in another measurement model. The Kolmogorov–Smirnov test works by comparing the empirical cumulative distribution function (CDF) of the output quantity, derived from Monte-Carlo simulations, with the CDF of a theoretical distribution. The test statistic, measures the maximum deviation between these two distribution functions. A significant difference indicated by this statistic suggests that the output quantity does not follow the specified theoretical distribution at a given risk level. The software computes and presents both the test statistic and the p-values for various theoretical distributions.

However, the sensitivity of the Kolmogorov–Smirnov test, especially in scenarios with a large number of simulations, can lead to predominantly low p-values due to minor deviations from the theoretical law. Therefore, alongside the p-value, the test statistic itself is crucial in determining the closeness of the fit between the empirical and theoretical distributions. In cases where the empirical distribution does not precisely match any of the proposed theoretical distributions, the software assists in selecting the theoretical distribution that most closely aligns with the observed data. This approach ensures a more accurate and representative choice of probability distribution for the output quantity, which is crucial for its subsequent use in measurement models.

4. Mass calibration example

The functionalities of the LNE-Uncertainty software are illustrated using the example treated in the GUM-S1, which deals with mass calibration. In this example, the measurand δm [kg], representing the deviation of the conventional mass from its nominal value, is expressed as a function of five input quantities

$$\delta m = (m_{\mathrm{R,c}} + \delta m_{\mathrm{R,c}}) \left[1 + (\rho_{\mathrm{a}} - \rho_{\mathrm{a_0}}) \left(\frac{1}{\rho_{\mathrm{W}}} - \frac{1}{\rho_{\mathrm{R}}}\right)\right] - m_{\mathrm{nom}}, \qquad (1)$$

with

- δm, the deviation between conventional and nominal masses,
- $m_{\mathrm{R,c}}$, the conventional mass of the reference weight,
- $\delta m_{\mathrm{R,c}}$, the conventional mass of the adjustment,
- ρ_{a}, the density of the air,
- ρ_{W}, the density of the weight to be calibrated and
- ρ_{R}, the density of the reference weight.

The quantities m_{nom}, the nominal mass and ρ_{a_o}, the conventional density are treated as constants (100 g and 1.2 kg/m^3). The inputs distributions for $m_{\mathrm{R,c}}$ and $\delta m_{\mathrm{R,c}}$ are normal, with means of 100 g and 1.234 mg and standard deviations of 50 and 20 µg. The mass densities follow uniform distributions with lower and upper bounds of 1.1 and 1.3 for ρ_{a}, 7000 and 9000 for ρ_{W}, 7950 and 8050 kg/m^3 for ρ_{R}.

Figure 1 shows a screenshot of the software showing the results from this case study. It informs about the mean and standard deviation of the simulated measurand values, with a total of $M = 10^6$ simulations. Additionally, users can customise the expanded uncertainty interval by selecting the coverage probability and choosing between the shortest or symmetric probability interval. The graph displaying the empirical probability density along with the chosen interval boundaries allows users to visualise the distribution of the measurand and to position it in relation to an evaluation made by the GUM method.

5. Conclusion

The LNE-Uncertainty software facilitates the implementation of distribution propagation via LPU and Monte-Carlo methods in accordance with the GUM and GUM-S1. This freely available software on the LNE website has been designed for an intuitive and straightforward evaluation of measurement uncertainty.

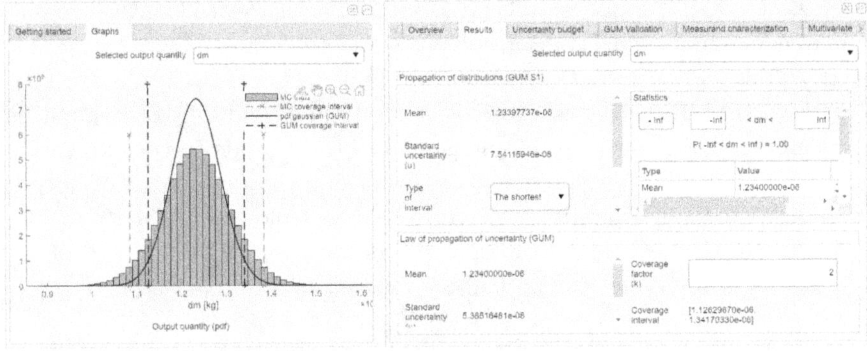

Fig. 1. Screen capture of the LNE Uncertainty software showing the results of the case study of Sec. 4.

In this software, additional features are available to further exploit the obtained results, enhancing the understanding of the studied measurement process. Notably, sensitivity analysis identifies the main uncertainty sources associated with the output quantity. The goodness-of-fit test compares the simulated sample to a known theoretical law, providing important information for using the output quantity as an input in a new measurement process. The software can also be used to validate assumptions of the GUM method (linearity of the model, Gaussian law for the output quantity) and to consider cases of multiple measurands, treating them as not necessarily independent.

References

1. BIPM, IEC, IFCC, ILAC, ISO, IUPAC, IUPAP and OIML, Evaluation of measurement data — Guide to the expression of uncertainty in measurement Joint Committee for Guides in Metrology, JCGM 100:2008.
2. BIPM, IEC, IFCC, ILAC, ISO, IUPAC, IUPAP and OIML, Evaluation of measurement data — Supplement 1 to the "Guide to the expression of uncertainty in measurement" — Propagation of distributions using a Monte Carlo method Joint Committee for Guides in Metrology, JCGM 101:2008.
3. G. Ebrard, A. Allard and N. Fischer, Un logiciel simple d'utilisation pour évaluer l'incertitude de mesure par la méthode de Monte-Carlo, *Revue française de métrologie* 27–36 (2017).

4. K. Lines, J.-L. Hippolyte, I. George and P. Harris, A MATHMET quality management system for data, software, and guidelines, *Acta IMEKO* **11**, 1 (2022).
5. BIPM, IEC, IFCC, ILAC, ISO, IUPAC, IUPAP and OIML, Evaluation of measurement data — Supplement 2 to the "Guide to the expression of uncertainty in measurement" — Extension to any number of output quantities Joint Committee for Guides in Metrology, JCGM 102:2011.
6. F. J. Massey Jr., The Kolmogorov–Smirnov test for goodness of fit, *Journal of the American statistical Association* **46**, 68 (1951).

F Pavese, A Bošnjaković, S Eichstädt, A B Forbes, J Alves e Sousa (eds.):
Advanced Mathematical and Computational Tools in Metrology and Testing XIII
Series on Advances in Mathematics for Applied Sciences, Vol. 94

https://doi.org/10.1142/9789819800674_0021

Check standard uncertainty impact on capability indices of higher generation in measurement process

J. Palenčár† and R. Palenčár
Faculty of Mechanical Engineering, Slovak University of Technology in Bratislava, Bratislava, Slovakia
†E-mail: jakub.palencar@stuba.sk

M. Dovica
Faculty of Mechanical Engineering, Technical University of Košice, Košice, Slovakia

The expression of measurement quality is generally solved by measurement uncertainties. In addition to uncertainties, capability indexes are increasingly used to measure measurement quality, especially in industrial measurements. Although the first generation capability indices are predominantly used for measurement processes, second and third generation indices can also be used. The problem of measuring processes might be the uncertainty of so-called check standard on which the measurements in the capability assessment are being executed. The present paper addresses the issue of the influence of the uncertainty of the check standard on the capability indices.

Keywords: measurement process; capability index; check standard error; measurement uncertainty.

1. Introduction

Capability indices compare the required (prescribed) process accuracy (uncertainty) with true process variability.

In practice, the so-called first generation indexes C_{g} and C_{gk} are being used. The calculated values of these indices from the measured data should have a value greater than 1.33 in order to declare the measurement process suitable for the measurement for which it was determined and should always be calculated in pair, because of lack of the sensitivity to process centralization of index C_{g}.

In addition, the 2nd and 3rd generation indices C_{gm} and C_{gmk} are mainly used in manufacturing processes, which also should be calculated in pair and are more sensible to process centralization than previous generation [2–8]. These can also be taken into account for assessing the capability of measurement processes. The evaluation capability approach does not require expressing the measurement result in line with the GUM method [1], however it requires the check standard (CS) on which control measurements we implemented [9–12].

CS is a certain artefact reproducing the value of the measurand that we know with sufficient precision. There are some specificities concerning the probability of validity of the capability indices due to the uncertainty of the used CS of measurement, which characterizes its accuracy. The problems are caused by the fact that the uncertainty consists of two components, the source of which is the random variation of the value of the measurand and the systematic effects on the determination of the value of the CS, which are manifested in a bias of the value of the CS against the declared value. This CS uncertainty U_{CS} below 10% of the overall (required) process uncertainty U is usually neglected and it is not considered in actual index calculation, similar to the practice in manufacturing processes [2,6].

For the first generation capability indexes C_{g} and C_{gk}, this is shown in [11,12].

2. Theoretical background

We assume that the measurement process (MP) has the distribution $X \sim N(\mu, \sigma^2)$, error of the CS $X_{\mathrm{CS}} \sim N(\delta_{\mathrm{CS}}, \sigma_{\mathrm{CS}}^2)$, where δ_{CS} represents the systematic error of the CS. The result of the measurement Y on the CS will have the distribution $Y \sim N(\mu_Y = \mu + \mu_{\mathrm{CS}}, \sigma_Y^2 = \sigma^2 + \sigma_{\mathrm{CS}}^2)$.

Assumption of normality is not stringent, practically speaking, all that is required is that the distribution of measurements be bell-shaped and symmetric.

What we know about the CS is an estimate of its mean (nominal, indicated value) $\hat{\mu}_{\mathrm{CS}}$, where $\mu_{\mathrm{CS}} = \hat{\mu}_{\mathrm{CS}} + \delta_{\mathrm{CS}}$, and sometimes an estimate of the standard deviation of the variation in the data, which we will take to be identical to the type A uncertainty of the CS $\hat{\sigma}_{\mathrm{CS}} = u_{\mathrm{A(CS)}}$.

We define parameter λ_{A} as $\lambda_{\mathrm{A}} = 2\sigma_{\mathrm{CS}} / U$ and parameter λ_{B} as $\lambda_{\mathrm{B}} = |\delta_{\mathrm{CS}}| / U$, where U is the overall required MP uncertainty, λ_{A} represents a relative variation of the CS value in the control measurements on the CS due to the required process uncertainty and parameter λ_{B} represents a relative deviation of the CS (systematic error of CS) due to the required process uncertainty.

The true value the CS is around the declared value $\hat{\mu}_{\mathrm{CS}}$ within $\pm U_{\mathrm{CS}}$, where U_{CS} is extended CS uncertainty. Thus, the CS error can be in the range of U_{CS}. When we assess the measurement process capability, we need the most unfavorable situation to consider. This is when the CS error is equal to the value of the U_{CS} and at the same time will be aiming in the opposite direction as the true deflection of the evaluated MP.

3. Capability index C_{gm}

The C_{gm} index belongs to the so-called second generation of capability indices and is based on the Taguchi approach to quality assessment and its loss function. The index quantifies process capability in terms of variability of quality indicator X around target value T (nominal value of check standard, $T = \hat{\mu}_{\mathrm{CS}}$).

The C_{gm} index for measurement process capability with 95% coverage interval is calculated using the formula:

$$C_{gm} = \frac{U}{2\sqrt{\sigma^2+(\mu-T)^2}} = \frac{U}{2\sqrt{\sigma^2+(\mu-\hat{\mu}_{CS})^2}} = C_g \frac{1}{\sqrt{1+\xi^2}}, \tag{1}$$

where $C_{\mathrm{g}} = U / 2\sigma$, $\xi = |\mu - T| / \sigma$, $T = \hat{\mu}_{\mathrm{CS}}$.

The empirical capability index C_{gm}^{Y} represents the theoretical value of the index based on calculation from the given data traditional way and is given by substituting μ for μ_Y and σ for σ_Y.

If we denoted $\vartheta = \delta_{\mathrm{CS}} / \sigma$, will be for the CS systematic error (CS deviation) we can calculate "real" corrected value of the index with CS systematic error in mind.

We assume that $\delta_{\mathrm{CS}} > 0$, if it increases the difference μ from desired value $\hat{\mu}_{\mathrm{CS}}$ and $\delta_{\mathrm{CS}} < 0$ otherwise.

$$C_{\mathrm{gm}}^{Y} = \frac{U}{2\sqrt{\sigma^2+\sigma_{\mathrm{CS}}^2+\left((\mu+\delta_{\mathrm{CS}})-\hat{\mu}_{\mathrm{CS}}\right)^2}} = C_{\mathrm{g}} \frac{1}{\sqrt{1+\lambda_{\mathrm{A}}^2 C_{\mathrm{g}}^2+(\xi+\vartheta)^2}}. \tag{2}$$

Let us equal $C_{\mathrm{g}}^{Y} = U / 2\sqrt{\sigma^2 + \sigma_{\mathrm{CS}}^2} = C_{\mathrm{g}} / \sqrt{1 + \lambda_{\mathrm{A}}^2 C_{\mathrm{g}}^2}$, which is the basic capability index of the first generation, respectively $C_{\mathrm{g}} = C_{\mathrm{g}}^{Y} / \sqrt{1 + \lambda_{\mathrm{A}}^2 C_{\mathrm{g}}^2}$ for $\lambda_{\mathrm{A}} C_{\mathrm{g}} < 1$.

Then

$$C_{\mathrm{gm}} = C_{\mathrm{gm}}^{Y} \sqrt{\frac{1+\xi^2}{1-\lambda_{\mathrm{A}}^2 C_{\mathrm{g}}^{Y2}} + \frac{(\xi+\vartheta)^2}{1+\xi^2}}. \tag{3}$$

The ratio of empirical to the actual capability index is described by the following formula:

$$\frac{C_{\mathrm{gm}}^{Y}}{C_{\mathrm{gm}}} = \sqrt{\frac{1+\xi^2}{1+\lambda_{\mathrm{A}}^2 C_{\mathrm{g}}^2+(\xi+\vartheta)^2}}. \tag{4}$$

Figures 1 and 2 show the ratios of the empirical capability indices C_{gm}^{Y} to the actual ones for $\lambda_{\mathrm{A}} = 0.0$, respectively $\lambda_{\mathrm{A}} = 0.2$ and different ξ. The figures allow to interpret the obtained values of the empirical fitness indices.

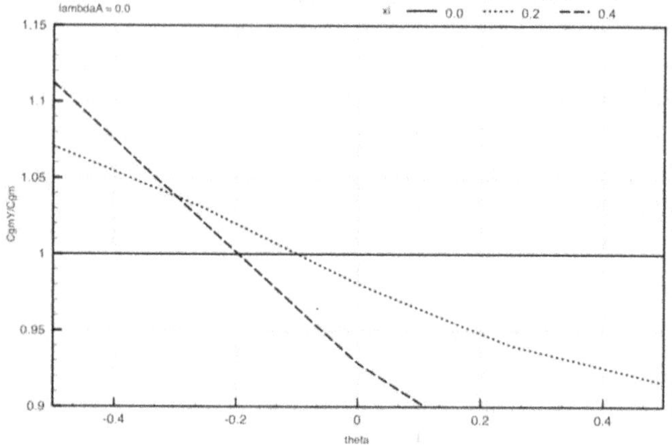

Fig. 1. Ratio of empirical a corrected C_{gm} index for $\lambda_{\mathrm{A}} = 0.0$.

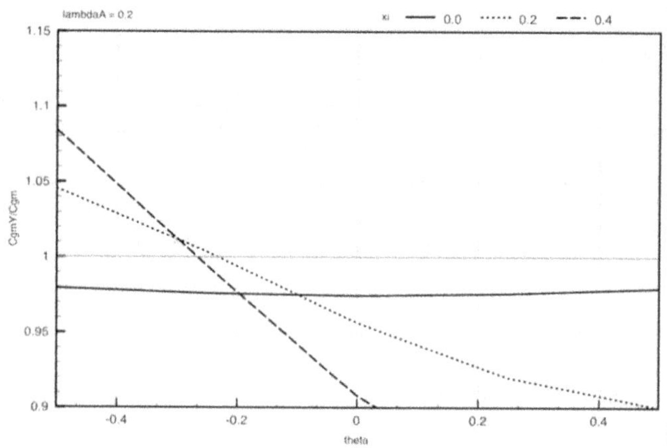

Fig. 2. Ratio of empirical a corrected C_{gm} index for $\lambda_{\mathrm{A}} = 0.2$.

We can see that if we take as the target value the actual value of the control standard for symmetric requirement on the measurement process (symmetric uncertainty) at parameter value $\vartheta < 0$ and if the systematic error of the control standard is in the opposite direction to the bias of the measurement process, the empirical index $C_{\mathrm{gm}}^{Y} > C_{\mathrm{gm}}$, will be overestimated, and thus it is overestimated due to the uncertainty of the check standard. As λ_{A} increases with increasing ξ and C_{g} increases, this trend decreases for $\vartheta < 0$. we can consider the empirical index practically accurate enough.

For $C_{\mathrm{g}} \geq 1.33$ for $\xi \leq 0.3$, the capability index C_{gm} will differ from the empirical one by up to 2%.

For the measured values Y_i the estimated empirical index $\hat{C}_{\text{gm}}^{Y}$ can by calculated with the help of Eq. (5) from the empirical data

$$\hat{C}_{\text{gm}}^{Y} = \frac{U}{2\sqrt{\tilde{s}_n^2 + (\bar{Y} - \hat{\mu}_{\text{CS}})^2}}, \tag{5}$$

where $\bar{Y} = 1 / n \sum_{i=1}^{n} Y_i$, $\tilde{s}_n^2 = 1 / n \sum_{i=1}^{n} (Y_i - \bar{Y})^2$, $\hat{\mu}_{\text{CS}}$ is the estimated value of CS gained from its corresponding certificate.

4. Capability index C_{gmk}

The C_{gmk} index is more sensitive to variability around the target value T as the indexes C_{gk} and C_{gm} [3,4] and actually is a combination of these two indices.

The index C_{gmk} with 95% coverage interval is defined as follows:

$$C_{\text{gmk}} = \min\left(\frac{(\hat{\mu}_{\text{CS}} + U) - \mu}{2\sqrt{\sigma^2 + (\mu - T)^2}}, \frac{\mu - (\hat{\mu}_{\text{CS}} - U)}{2\sqrt{\sigma^2 + (\mu - T)^2}}\right), \tag{6}$$

or

$$C_{\text{gmk}} = \frac{U - |\delta|}{2\sqrt{\sigma^2 + (\mu - T)^2}} = C_{\text{g}} \frac{1 - v}{\sqrt{1 + \xi^2}}. \tag{7}$$

If we consider $\delta = \mu - \hat{\mu}_{CS}$, $\lambda_{\text{B}} = |\delta_{\text{CS}}| / U$, $v = |\delta| / U$ the parameter h is defined by:

(i) δ_{CS} and δ acting in the same direction:

$$h = -\frac{\lambda_{\text{B}}}{1 - v}, \tag{8}$$

(ii) δ_{CS} and δ acting in the opposite direction:

(a) If $\delta_{\text{CS}} \leq \delta$, then

$$h = \frac{\lambda_{\text{B}}}{1 - v}, \tag{9}$$

(b) If $\delta_{\text{CS}} > \delta$, then

$$h = \frac{2v - \lambda_{\text{B}}}{1 - v}. \tag{10}$$

The empirical capability index C_{gmk}^{Y} represents the theoretical value of the index based on calculation from the given data traditional way and is given by substituting μ for μ_Y and σ for σ_Y.

If we denoted $\vartheta = \delta_{\text{CS}} / \sigma$, will be for the CS systematic error (CS deviation) and using the parameter h we can calculate "real" corrected value of the index with CS systematic error in mind.

$$C_{\mathrm{gmk}} = C^{Y}_{\mathrm{gmk}} \frac{1}{1+h} \sqrt{\frac{\frac{1}{1-\lambda_{\mathrm{A}}^{2} C_{\mathrm{g}}^{Y^{2}}} + (|\xi| - |\vartheta|)^{2}}{1+\xi^{2}}}. \tag{11}$$

In standard conditions, we have no knowledge of the direction of the CS deflection and the MP deflection, and hence we have to consider the worst case, i.e., the case where the deflections act against each other, when determining the parameter h.

The correction often depends practically only on the ratio h under normal conditions.

The parameter h for values of $v < 0.8$ and $\lambda_{\mathrm{B}} < 0.15$ takes values $h \in (0; 0.75)$.

Table 1 shows the values of the empirical capability index for the corrected index $C_{\mathrm{gmk}} = 1.33$ for different ratios of h.

Table 1. Values of the empirical capability index C^{Y}_{gmk} with a value of corrected capability index of 1.33 as a function of h (conservative approach).

(a)

h	C^{Y}_{gmk}
0.00	1.13
0.10	1.24
0.20	1.36
0.30	1.47
0.40	1.58
0.50	1.69
0.60	1.81
0.75	1.98

with condition
$\xi = 0.10$
$\vartheta = 0.10$

(b)

h	C^{Y}_{gmk}
0.00	0.84
0.10	0.93
0.20	1.01
0.30	1.09
0.40	1.18
0.50	1.26
0.60	1.35
0.75	1.47

with condition
$\xi = 1.00$
$\vartheta = 1.00$

(c)

h	C^{Y}_{gmk}
0.00	1.33
0.10	1.46
0.20	1.60
0.30	1.73
0.40	1.86
0.50	2.00
0.60	2.13
0.75	2.33

with condition
$\xi = 0.00$
$\vartheta = 0.00$

In Table 1(a) are the minimum values of the calculated index for different values of h and up to the fractions of the parameter ξ and the parameter ϑ up to 10%.

If we keep the parameter ξ and the parameter ϑ close to 1, which means that the biases of the control standard and the measurement process are close to matching, illustrated in Table 1(b), where we can see that in this case, there may often be situations where the measurement process would be falsely judged to be non-compliant.

If we approximate the parameter ξ and the parameter ϑ to 0 we can see the dependence of the index purely on the parameter h. We can observe in Table 1(c) fact that in this case the minimum value of the empirical index C^{Y}_{gmk} is 2.33, which in practice should ensure the certainty of the capability of the observed measurement process.

For measured values Y_i the empirical capability index estimate $\hat{C}^{Y}_{\mathrm{gmk}}$ will be calculated from empirical data by

$$\hat{C}^{Y}_{\mathrm{gmk}} = \frac{U-|\Delta|}{2\sqrt{s^2+(\bar{Y}-\hat{\mu}_{\mathrm{CS}})^2}} = \hat{C}_{\mathrm{g}}\left(1-\frac{|\bar{Y}-\hat{\mu}_{\mathrm{CS}}|}{U}\sqrt{1+\left(\frac{|\bar{Y}-\hat{\mu}_{\mathrm{CS}}|}{s}\right)^2}\right), \qquad (12)$$

with conditions $\bar{Y} = (1/n)\sum_{i=1}^{n} Y_i$, $\Delta = \bar{Y} - \hat{\mu}_{\mathrm{CS}}$, $s^2 = (1/(\mathrm{n}-1))\sum_{i=1}^{n}(Y_i - \bar{Y})^2$.

5. Conclusion

We discussed indices of capability of the 2nd and 3rd generation C_{gm} and C_{gmk} for needs of measuring process capability. Due to their advantages, we assume that they will be extended to measurement processes. For this purpose, we examined the influence of the uncertainty of the control standard on the values of empirical indexes and we provided correction factors for each situation.

In practice, we are faced with the task of assessing the impact of the control standard uncertainty on the value of the capability index.

If we know that it is control standard with significantly greater uncertainty due to fluctuations in control standard values, then we will use statistical approach for indexes C_{gm} and C_{gmk} with initial conditions $\lambda_{\mathrm{A}} = \lambda = U_{\mathrm{CS}} / U$, $\delta_{\mathrm{CS}} = 0$.

If we know that it is control standard with significantly greater uncertainty caused by the deviation of the control standard values, then we will use the conservative approach. In this case conditions are $\delta_{\mathrm{CS}} = U_{\mathrm{CS}}$, $\lambda_{\mathrm{A}} = 0$.

If we can evaluate both components of uncertainties, we will use the combined approach. Here constants equal $\lambda_{\mathrm{A}} = 2\,u_{\mathrm{ACS}} / U$, $\delta_{\mathrm{CS}} = U_{\mathrm{B_{CS}}}$, $\vartheta = \delta_{\mathrm{CS}} / \sigma$.

If there is no previous case, we will use the consideration for a conservative approach, since the empirical index can always be greater than the real one.

The correction factors where we take into the “worst” case of the impact of the uncertainty of CS on the calculation of eligibility indexes for each approach in Table 2.

The analysis showed that for the C_{gm} index for the relative measurement process deviation from the target value $\xi \leq 0.1$, the CS uncertainty on the capability index is negligible at the relative uncertainty CS $U_{\mathrm{CS}} / U \leq 0.1$. If $\xi = 0.5$ with a total relative uncertainty of CS up to 10%, the empirical capability index may be up to 4%, which is negligible in practice. So, we could see that the parameter h is of immense importance, because in some forms represents the relation between CS's and MP's deflection.

Table 2. Correction factors for different approaches assessment of capability index with impact of CS.

Index	Combined approach	Statistical approach	Conservative approach
C_{gm}	$\sqrt{\frac{\frac{1}{1-\lambda_A^2 c_g^{Y^2}}+(\xi+\vartheta)^2}{1+\xi^2}}$	$\sqrt{\frac{\frac{1}{1-\lambda_A^2 c_g^{Y^2}}+\xi^2}{1+\xi^2}}$	$\sqrt{\frac{1+(\xi+\vartheta)^2}{1+\xi^2}}$
C_{gmk}	$\frac{1}{1+h}\sqrt{\frac{\frac{1}{1-\lambda_A^2 c_g^{Y^2}}+(\xi+\vartheta)^2}{1+\xi^2}}$	$\frac{1}{1+h}\sqrt{\frac{\frac{1}{1-\lambda_A^2 c_g^{Y^2}}+\xi^2}{1+\xi^2}}$	$\frac{1}{1+h}\sqrt{\frac{1+(\xi+\vartheta)^2}{1+\xi^2}}$

An analysis of the impact of the CS uncertainty on the C_{gmk} showed that CS uncertainty affects the value of the empirical index. In a conservative approach, the necessary values of the empirical index C_{gmk}^Y are given so that the actual index reaches $C_{gmk} = 1.33$. Knowing the components of CS uncertainty, a less stringent correction can be made.

Acknowledgements

The authors thank the Faculty of Mechanical Engineering of the Slovak University of Technology in Bratislava and Technical University of Košice.

This work was supported by the Slovak Research and Development Agency under the Contract no. APVV-21-0216 and 21-0195. Grant agency VEGA grants No. 1/0687/21 and 1/0675/22, as well as KEGA grant agency, projects 020STU-4/2023 and 016TUKE-4/2021.

References

1. JCGM 100:2008, Evaluation of measurement data - Guide to the expression of uncertainty in measurement (GUM 1995 with minor corrections), in *JCGM - Joint Committee for Guides in Metrology*, (ISO, BIPM, IEC, IFCC, ILAC, IUPAC, IUPAP and OIML, 2008).
2. W. L. Pearn and S. Kotz, *Encyclopaedia and Handbook of Process Capability Indices* (World Scientific, Singapore, 2006).

3. W. L. Pearn and M. Y. Liao, One-sided process capability assessment in the presence of measurement errors, *Quality and Reliability Engineering International* **22**(7), 771–785 (2006).
4. A. K. Baral and M. Z. Anis, Assessment of Cpm in the presence of measurement errors, *Journal of Statistical Theory and Applications* **14**, 13–27 (2015).
5. D. Grau, Testing capability indices for one-sided processes with measurement errors, *International Journal of Metrology and Quality Engineering* **4**, 71–80 (2013).
6. A. Haq, J. Brown, E. Moltchanova and A. I. Al-Omari, Effect of measurement error on exponentially weighted moving average control charts under ranked set sampling schemes, *Journal of Statistical Computation and Simulation* **85**, 1224–1246 (2015).
7. M. R. Maleki, A. Amiri and P. Castagliola, Measurement errors in statistical process monitoring: A literature review. *Computers & Industrial Engineering* **103**, 316–329 (2017).
8. NIST/SEMATECH, *e-Handbook of Statistical Methods* (National Institute of Standards and Technology, 2012), https://www.itl.nist.gov/div898/handbook/.
9. S. Bordignon and M. Scagliarini, Statistical analysis of process capability indices with measurement errors, *Quality and Reliability Engineering International* **18**(4), 321–332 (2002).
10. E. Kurekova, Measurement Process Capability - Trends and Approaches, *Measurement Science Review* **1**, 43 (2017).
11. J. Palencar, R. Palencar, S. Duris and P. Pavlasek, Correction of the capability index computation based on the uncertainty of the check standard in measurement process, *Journal of Physics: Conference Series* **1065**, (2018).
12. J. Palencar, M. Dovica, R. Palencar, M. Halaj, A. Kovaciková, L. Soos, How can the Check Standard Influence Measurement Process Capability, *Quality Innovation Prosperity* **26**, (2022).

F Pavese, A Bošnjaković, S Eichstädt, A B Forbes, J Alves e Sousa (eds.):
Advanced Mathematical and Computational Tools in Metrology and Testing XIII
Series on Advances in Mathematics for Applied Sciences, Vol. 94

https://doi.org/10.1142/9789819800674_0022

Interpolation of the correction values of a thermometer calibration certificate for a real number of calibration points

Ciro Alberto Sánchez

Instituto Nacional de Metrología - INM,
Bogotá, D.C.ZIP/111321, Colombia
E-mail: csanchez@inm.gov.co
www.inmgov.co

The interpolation of the correction values of a thermometer calibration certificate is presented following literal H3 of JCGM 100:2008 for an actual number of calibration points. From JCGM 100 text is very non usual to have eleven points of calibration, so, its literal H3 methodology is applied for a typical number of points of calibration, up to 3. For each case, the analysis for correlated input quantities: the intercept and slope of the calibration curve is developed to predict the value and standard uncertainty of the correction to be applied to the thermometer for any value t of the temperature.

Keywords: interpolation; measurement uncertainty.

1. Introduction

The interpolation of correction values within a thermometer calibration certificate, as outlined in Clause H.3 of JCGM 100:2008 (Evaluation of Measurement Data - Guide to the Expression of Uncertainty in Measurement, 2008),[1] employs a linear model integrating variance and covariance to ensure precise adjustments. This methodology aims to anticipate the necessary corrections for the thermometer across diverse temperature ranges by scrutinizing interrelated input parameters, notably the calibration curve's intercept and slope. The subsequent exploration delves into this interpolation technique, adhering to the directives of literal H3 within JCGM 100:2008, which are tailored to accommodate the standard calibration point count, typically up to 3. Each calibration point mandates a comprehensive analysis of correlated input quantities, facilitating forecasts of both correction value and standard uncertainty for the thermometer across varying temperature measurements (t).

2. Methodology

2.1. *The measurement problem*

The least squares method is applied to obtain the interpolation of the correction for the linear model. Focusing solely on linear interpolation, the declared measurement uncertainty in the calibration certificate is not considered. For 11 temperature measurement values in the measurement range from 21.521 °C to 26.511 °C, a correction is determined for each temperature of the instrument under calibration (IUC).

$n = 11.$

k: 1…n.

$t_{R,k}$: reference temperature.

t_k : is IUC temperature.

$b_k = t_{R,k} - t_k$: correction.

The correction b_k and the measured temperature t_k are the input quantities for the evaluation. For the linear calibration curve:

$$b(t) = y_1 + y_2(t - t_0) \tag{1}$$

It is adjusted using the least squares method for both the corrections and the measured temperatures. The parameters y_1 and y_2, are respectively the intercept and the slope of the calibration curve, and they represent the two measurands or output quantities to be determined.

The temperature t_0 is chosen by convenience to a convenient exact reference temperature value is chosen; therefore, it is not an independent parameter determined by the least squares fit. Once y_1 and y_2 are determined, along with their estimated variances and covariance, Eq. (1) is used to predict the value and the standard measurement uncertainty of the interpolated correction at any temperature value t and a particular temperature value t_0.

2.1.1. *Results for 11 data*

Utilizing the principles delineated in Sec. 2.2, Least Squares Fitting of Clause H.3 from JCGM 100:2008 (Evaluation of Measurement Data - Guide to the

Expression of Uncertainty in Measurement, 2008), the subsequent results are presented for a set of 11 measurement values.

The measurement values to be fitted are located in the second and third columns of Table 1, as shown in Fig. 1. Using $t_0 = 20\ °C$ as the reference temperature the following results are obtained:

$$y_1 = -0.171\,2\ °C \qquad s(y_1) = 0.002\,9\ °C$$

$$y_2 = 0.002\,18 \qquad s(y_2) = 0.000\,67$$

$$r(y_1, y_2) = -0.930 \qquad s = 0.003\,5\ °C$$

The fact that the slope y_2 is more than three times greater than its standard deviation is an indication that a calibration curve is needed rather than a constant average correction.

Table 1. Measurement values used to obtain a linear fit curve for thermometer correction by the least squares method.

Reading number k	**Thermometer reading** t_k (°C)	**Observed or measured correction** $b_k = t_{R,k} - t_k$ (°C)	**Interpolated correction** $b(t_k)$ (°C)	**Difference between the observed or measured correction and the interpolated correction** $b_k - b(t_k)$ (°C)
1	21.521	−0.171	−0.167 9	−0.003 1
2	22.012	−0.169	−0.166 8	−0.002 2
3	22.512	−0.166	−0.165 7	−0.000 3
4	23.003	−0.159	−0.164 7	0.005 7
5	23.507	−0.164	−0.163 6	−0.000 4
6	23.999	−0.165	−0.162 5	−0.002 5
7	24.513	−0.156	−0.161 4	0.005 4
8	25.002	−0.157	−0.160 3	0.003 3
9	25.503	−0.159 2	−0.159 2	0.000 0
10	26.002	−0.161	−0.158 1	−0.002 9
11	26.511	−0.160	−0.157 0	−0.003 0

The calibration curve is:

$$b(t) = -0.171\,2(29)\ °C + 0.002\,18(67)(t - 20\ °C) \tag{2}$$

Where the numbers in parentheses represent the numerical values of the standard measurement uncertainties corresponding to the last digits of the results cited for the intercept and slope. This equation provides the interpolated value of the correction *b(t)* for a temperature *t*, and in particular, the value $b(t_k)$ at $t = t_k$. These values are given in the fourth column of Table 1, while the last column provides the difference between the observed or measured values and the interpolated ones, $b_k - b(t_k)$, as shown in Figs. 2 and 3.

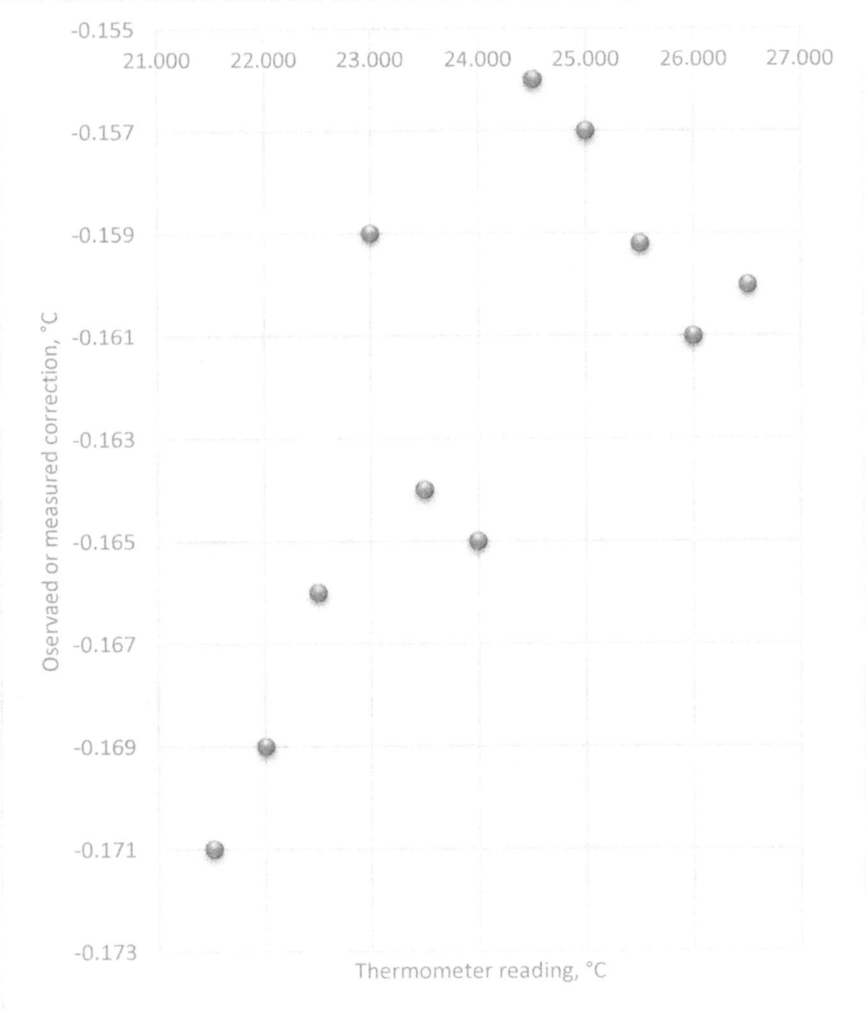

Fig. 1. The 11 measurement values to be fitted are located in the second and third columns of Table 1.

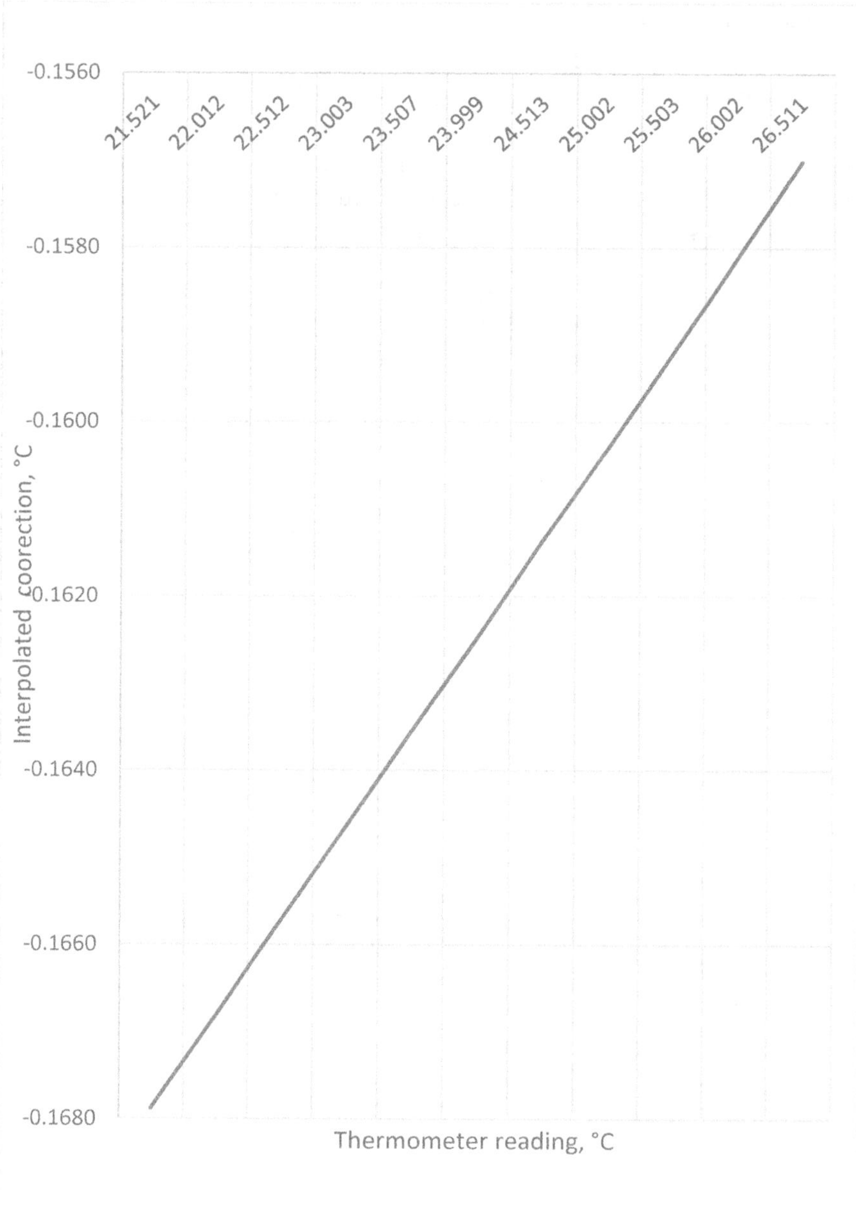

Fig. 2. Calibration curve for b(t) = −0.1712 °C + 0.00218 (t −20 °C). The red curve represents the predicted temperature values.

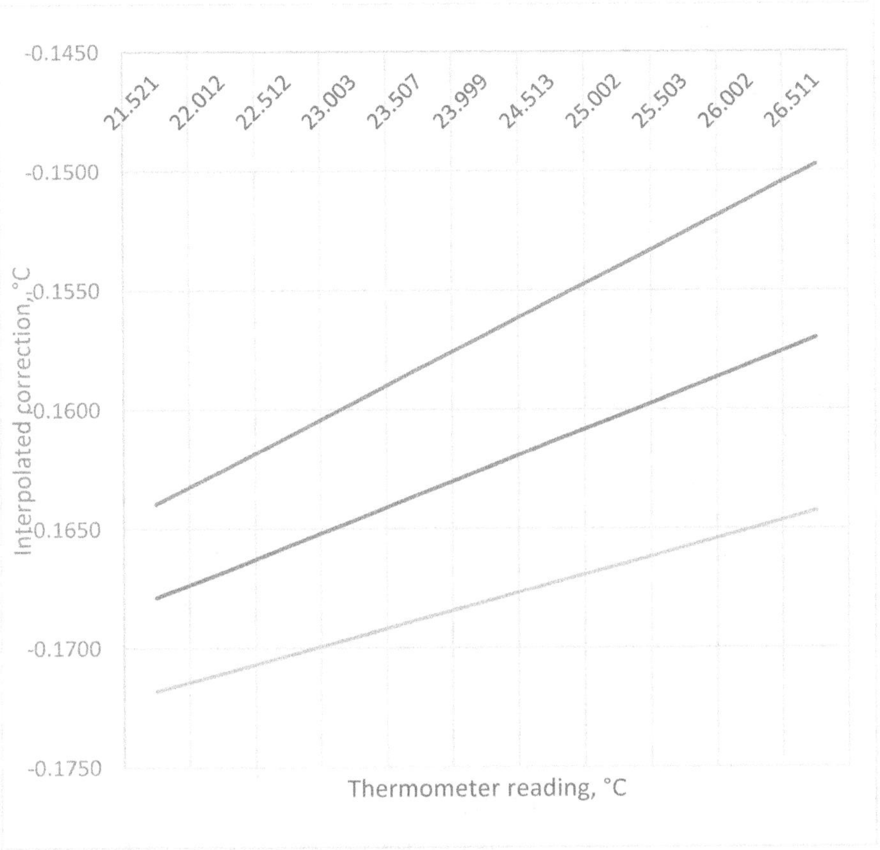

Fig. 3. Calibration curve illustrating the relationship between b(t) and temperature (t) based on the equation b(t) = −0.1712(29) °C + 0.00218(67)(t −20 °C). The blue, red, and green curves respectively represent variations of the equation for different experimental conditions, taking into account measurement uncertainties, the numbers in parentheses indicate the uncertainties associated with the measurements. Specifically, the blue curve corresponds to b(t) = −0.1683 °C + 0.00285(t −20 °C), the red curve to b(t) = −0.1712 °C + 0.00218(t −20 °C), and the green curve to b(t) = −0.1741 °C + 0.00151(t −20 °C).

2.1.1.1. The temperature at which the estimated variance $u_c^2[b(t)]$ is minimized

Equation (3) describes the process of determining the temperature value, denoted as $t_{u_c^2\,min}$, at which the estimated variance $u_c^2[b(t)]$ is minimized. This equation is derived by applying the criterion of setting the first derivative equal to zero and ensuring the second derivative is greater than 0.

The equation is represented as:

$$t_{u_c^2\, min} = t_0 - \frac{u(y_1)\, r(y_1, y_2)}{u(y_2)} \tag{3}$$

Where:

$t_{u_c^2\, min}$ represents the temperature value at which the estimated variance is minimized.
t_0 is a reference temperature.
$u(y_1)$ and $u(y_2)$ denote standard uncertainties associated with y_1 and y_2.
$r(y_1, y_2)$ represents the correlation coefficient between measurements y_1 and y_2.

In the current context, this equation yields a value of $t_{u_c^2\, min} = 24.007\ 7\ °C$, as shown in Fig. 4.

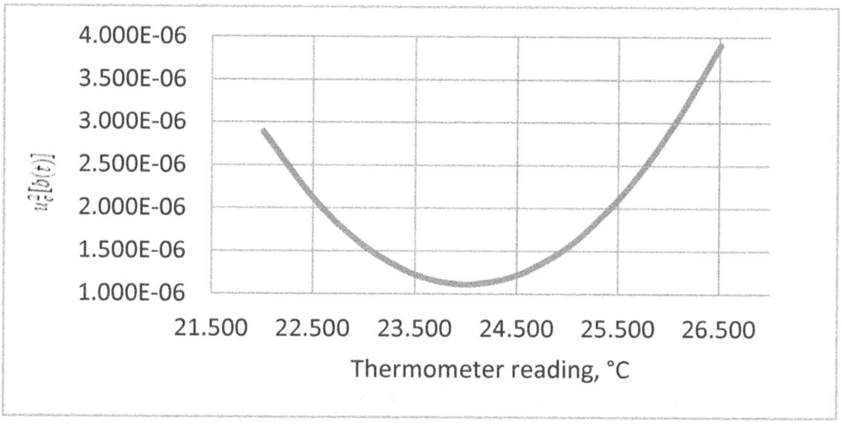

Fig. 4. Illustration showing the relationship between the estimated variance $u_c^2[b(t)]$ and temperature, highlighting its minimum value at $t_{u_c^2\, min} = 24.007\ 7\ °C$, as determined by Eq. (3).

3. Results

In this section, the results will be presented for n = 9, n = 7, n = 5 and n = 3, showcasing tables and values similar to those previously observed for n = 11. This comprehensive analysis aims to highlight the consistency, variations, and similarities across different values of n, providing insights into the trends and comparative aspects among the data and associated values for each n, see Tables 2 and 3.

Table 2. Set of values for n < 11.

n = 9		n = 7	
Thermometer reading, °C	**Observed correction, °C**	**Thermometer reading, °C**	**Observed correction, °C**
21.521	−0.171	21.521	−0.171
22.512	−0.166	23.003	−0.159
23.003	−0.159	23.507	−0.164
23.507	−0.164	23.999	−0.165
23.999	−0.165	24.513	−0.156
24.513	−0.156	25.002	−0.157
25.002	−0.157	26.511	−0.160
25.503	−0.1592		
26.511	−0.160		
n = 5		**n = 3**	
Thermometer reading, °C	**Observed correction, °C**	**Thermometer reading, °C**	**Observed correction, °C**
21.521	−0.171	21.521	−0.171
23.507	−0.164	23.999	−0.165
23.999	−0.165	26.511	−0.160
24.513	−0.156		
26.511	−0.160		

Table 3. Values of y_2, $s(y_2)$, y_1, $s(y_1)$ and $t_{u_c^2\,min}$ for n = 9, n = 7, n = 5 and n = 3.

n	y_2	$s(y_2)$	y_1, °C	$s(y_1)$, °C	$t_{u_c^2\,min}$, °C
9	0.00224	0.00083	−0.1709	0.0035	24.0079
7	0.00223	0.00111	−0.1707	0.0048	24.0080
5	0.00243	0.00114	−0.1729	0.0049	24.0102
3	0.00220	0.00013	−0.1742	0.0006	24.0103

4. Conclusions

This study has effectively demonstrated the application of literal H3 of JCGM 100:2008 in a practical context, where calibration points rarely exceed three, deviating from the atypical scenario of eleven points described in the JCGM 100 text. By adapting this methodology to align with more customary calibration practices, the analysis of correlated input quantities, including the calibration curve's intercept and slope, has facilitated estimations of correction values and their standard uncertainties across varying temperature ranges. This work serves as a bridge between theoretical guidelines and practical industry standards, showcasing the feasibility and effectiveness of applying JCGM 100:2008 to a more realistic calibration point count, essential for enhancing accuracy in temperature measurements within industrial settings.

Acknowledgements

Grateful to the Instituto Nacional de Metrologia – INM for enabling my participation in the Symposium AMCTM 2023. Their support was instrumental.

References

1. JCGM 100:2008. *Evaluation of measurement data-Guide to the expression of uncertainty in measurement. Évaluation des données de mesure - Guide pour l'expression de l'incertitude de mesure.* (2008). https://www.bipm.org/documents/20126/2071204/JCGM_100_2008_E.pdf/cb0ef43f-baa5-11cf-3f85-4dcd86f77bd6.

F Pavese, A Bošnjaković, S Eichstädt, A B Forbes, J Alves e Sousa (eds.):
Advanced Mathematical and Computational Tools in Metrology and Testing XIII
Series on Advances in Mathematics for Applied Sciences, Vol. 94
 (pp. 249–259)
https://doi.org/10.1142/9789819800674_0023

Uncertainty quantification and global sensitivity analysis of a pollutant dispersion problem

J. A. Mackenzie, A. Ramage and T. Shahid*

Department of Mathematics and Statistics,
University of Strathclyde, Glasgow G1 1XH, Scotland
**E-mail: tasnia.shahid@strath.ac.uk*

A. B. Forbes

National Physical Laboratory,
Hampton Road, Teddington TW11 0LW, England

In this paper, we consider a computational modelling approach to gain some insight into the factors that strongly influence atmospheric pollution concentration and how uncertainties associated with these factors contribute to the uncertainties in pollution estimates. We consider a parametric advection-diffusion pollution dispersion model to estimate pollutant concentration over a spatial domain, given values for a set of model parameters and hyper-parameters. A global sensitivity analysis (GSA) is conducted to assess how the variability in the pollutant concentration can be attributed to the variability in the model parameters. We use a variance-based GSA technique based on the estimation of first order and total Sobol indices across the spatial domain using Quasi-Monte Carlo sampling. A possible use for the resulting spatial sensitivity maps is in the design of an optimal placement strategy for sensors.

Keywords: global sensitivity analysis; advection-diffusion model; Sobol indices.

1. Introduction

Urban air quality is a pressing issue in modern society that substantially threatens public health and environmental well-being. Therefore, it is no surprise that the accurate measurement of air quality has become a very important subject in metrology. Recent advances in computational mathematics have made it feasible to model some of the key processes contributing to air quality problems over a range of parameter values in order to gain some insight into the factors that strongly influence pollutant concentration. In particular, parametric differential equations can be used to predict how pollutants behave with respect to parameterised physical processes.[1]

In this paper, we study a parametric differential equation as an uncertainty quantification (UQ) problem by assigning probability distribution functions to model uncertainties associated with the parameters. We are interested in assessing how the variance in the pollutant concentration in a simple pollutant dispersal model can be attributed to the variance in model input parameters such as the wind speed and diffusion coefficient. Variance-based Global Sensitivity Analysis (GSA) measures known as Sobol indices[2,3] are used to carry out a spatial analysis[4] in order to understand better which parameters contribute the most uncertainty to the pollutant concentration estimates throughout the spatial domain. We focus on a simple 1D model of pollutant dispersal which retains the key features of the underlying (much more complex) physical problem and so is still useful for gaining insight into sensitivity analysis of parameters.

2. A simplified pollutant dispersion model

Suppose we have a pollutant on the 1D domain $\Omega = [0, b]$, which originates from sources at points $x_{s_1}, x_{s_2} \in \Omega$ with source rates s_1 and s_2, respectively. We assume the pollutant is transported through advection with a wind speed $a > 0$ and turbulent eddy diffusion at a rate $D > 0$. With these assumptions, steady-state pollutant dispersion can be modelled using the simple, non-dimensional, boundary value problem (BVP)

$$-D\frac{d^2u}{dx^2} + a\frac{du}{dx} = s(x) \text{ in } \Omega, \qquad u(0) = 0, \;\; u(b) = 0, \tag{1}$$

where $s(x) = s_1 \exp\left\{-(x - x_{s_1})^2/\sigma^2\right\} + s_2 \exp\left\{-(x - x_{s_2})^2/\sigma^2\right\}$ is a sum of continuous Gaussian source functions over a length scale σ. We assume that the boundary conditions are implemented sufficiently far upstream and downstream of both source locations such that there is zero pollutant concentration at the boundaries.

In (1), the parameters for the wind speed, diffusivity, and source rates are taken as fixed quantities that quantify physical processes, but these values are unknown in practice. By sampling parameters from probability distribution functions, we can carry out a parametric study that enables us to interpret the model as a UQ problem. We characterise the uncertainty associated with the model parameters by sampling parameters from probability distribution functions as follows:

$$\begin{aligned} a &\sim \text{Weibull}(\lambda, \kappa), & D &\sim \text{Beta}(A, B), \\ s_1 &\sim \text{Uniform}(\alpha_1, \beta_1), & s_2 &\sim \text{Uniform}(\alpha_2, \beta_2), \end{aligned} \tag{2}$$

where λ is a scale parameter and κ, A and B are shape parameters. Note that $A, B, \alpha_1, \alpha_2, \beta_1, \beta_2 > 0$. We use a Weibull distribution for wind speed,[5] and the uncertainty associated with diffusivity is modelled using a Beta distribution so that non-negative diffusion is guaranteed. Source strengths are taken to have a Uniform distribution reflecting a prior set of plausible bounds for source strengths but no further information. More definitive knowledge of the source strength would reduce the variance contribution of this parameter to the overall variance in the pollutant concentration.

A numerical approximation of the solution to the steady-state BVP (1) is obtained using a Streamline Upwind Petrov-Galerkin (SUPG) finite element method.[6] In some situations, standard Galerkin finite element solutions of (1) are known to suffer from spurious oscillations if the mesh Péclet number, as defined by $P_e = ha/2D$ (where h is a measure of element size), is greater than one. Using a SUPG discretisation ensures that our numerical approximation technique is robust to the possibility of advection-dominated problems. We do not present the details of the discretisation here, but simply state that the process leads to the system of linear equations

$$[(D+\alpha)K + aC]\,\boldsymbol{u} = s_1\boldsymbol{l_1} + s_2\boldsymbol{l_2} \tag{3}$$

which can be solved for the vector of unknown concentration values $\boldsymbol{u}$. Here $\alpha \in \mathbb{R}$ is a stabilisation parameter, which we take to be[7]

$$\alpha = \begin{cases} \coth\left(\frac{ah}{2}\right) - \frac{2}{ah} & \text{for } P_e \geq 1, \\ 0 & \text{otherwise.} \end{cases}$$

As we use a uniform linear finite element mesh with N elements, $h = b/N$ and the matrices in (3) are of dimension $(N+1) \times (N+1)$.

To model a transient problem would introduce additional considerations into the numerical simulations as efficient and robust time integration schemes would be necessary to ensure predicted concentrations remained positive.[8] For further details on how sensitivity analysis could be conducted in this case, see Marrel *et al.*[4]

3. Uncertainty quantification of pollutant concentration

The variability in the model parameters, as sampled from the given probability distributions, introduces uncertainty in the solution to the BVP (1). In order to visualise this, we use a Quasi-Monte Carlo (QMC) approach based on Sobol sequences with $m = 10^5$ samples.[9] We set $x_{s_1} = 1.5$ and $x_{s_2} = 3$ in the domain $[0, 5]$ and sample the wind speed as $a \sim \text{Weibull}(4.96, 1.63)$ and diffusivity as $D \sim \text{Beta}(0.7, 2.17)$. The source strengths are sampled as $s_1 \sim \text{Uniform}(0.5, 1.5)$ and $s_2 \sim \text{Uniform}(1, 3)$ with $\sigma = 0.1$.

Figure 1 illustrates the variability in the resulting pollutant concentration over the spatial domain: Fig. 1(a) shows the 2.5 and 97.5 percentiles of the solution profiles, with the mean shown in black, and the variance of the pollutant concentration is shown in Fig. 1(b). We observe that the distribution of pollutant concentration at each x-location is skewed positively, and that some parameter combinations can lead to high pollutant concentration.

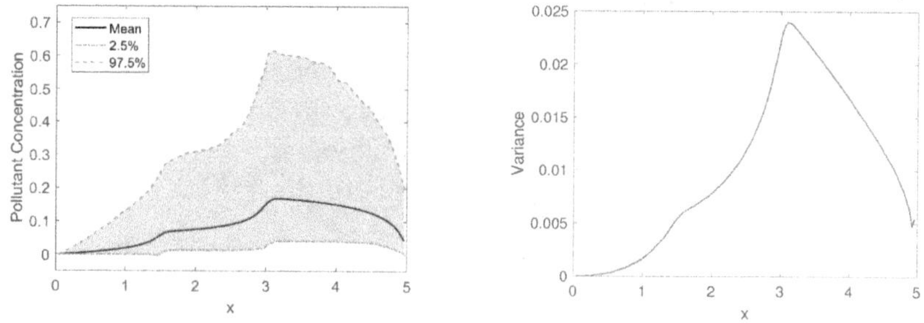

(a) Variability in pollutant concentration.

(b) Variance of pollutant concentration.

Fig. 1. Pollutant concentration using a linear SUPG finite element discretisation with $N = 128$.

In order to study how the distribution of the model parameters affects the pollutant concentration, we now consider the parameter values that give the highest and lowest pollutant concentrations. Figure 2 shows boxplots of the parameters that result in pollutant concentrations below the 2.5 percentile and above the 97.5 percentile at points $x = 1$, $x = 2$ and $x = 4$. We can see that the distribution of wind speed is significant to the level of pollutant concentration. Figure 2(a) shows that large wind speeds correspond to low pollutant concentrations throughout the domain, and conversely, low wind speeds correspond to high pollutant concentrations. Figure 2(b) highlights the importance of the distribution of diffusivity at $x = 1$, where we see that a larger rate of diffusivity results in higher pollutant concentration. Figures 2(c) and 2(d) show that greater source strength also results in high levels of pollutant concentration. These boxplots indicate that the variability in the model parameters affects the level of pollutant concentration, and that the influence of certain parameters change throughout the domain.

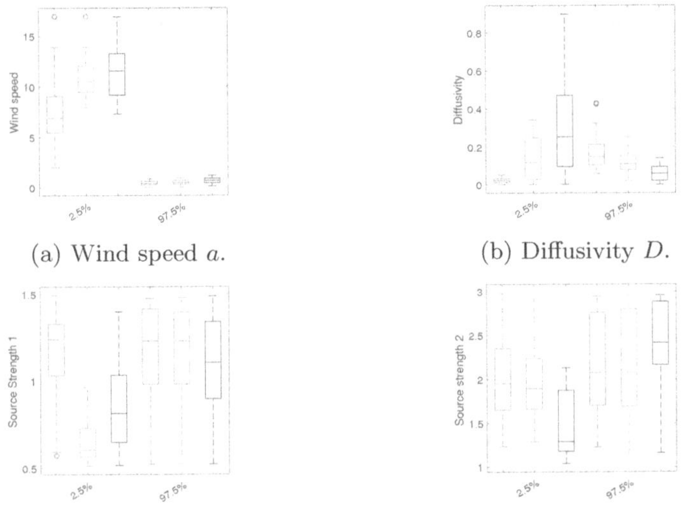

(a) Wind speed a.

(b) Diffusivity D.

(c) Source strength s_1.

(d) Source strength s_2.

Fig. 2. Boxplots of wind speed, diffusivity, and source strength values that result in pollutant concentrations below the 2.5 percentile and above the 97.5 percentile: $x = 1$ (blue), $x = 2$ (red) and $x = 4$ (black).

4. Global sensitivity analysis

We next investigate the relative influence of the model parameters on the variability of the pollutant concentration across the domain. We do this using Sobol indices, which are variance-based GSA measures that assess how sensitive the model output of (1) is to variation in the parameters (2).[2,10]

4.1. *Sobol indices*

Let $\boldsymbol{P} = (P_1, P_2, \ldots, P_k)$ be a vector of parameters, where P_j corresponds to the jth parameter. We may then denote the pollutant concentration at the point x by $Y_x = f_x(\boldsymbol{P})$ (where f_x represents the SUPG solution of the BVP at the point x), and the total variance in the pollutant concentration at the point x by $\mathbb{V}(f_x(\boldsymbol{P}))$. First-order Sobol indices provide a measure of the variance contributed by individual parameters, and can be computed at point x for parameter j by taking the ratio

$$S_{x,j} = \frac{\mathbb{V}_{P_j}(\mathbb{E}_{\boldsymbol{P}_{\bar{j}}}(Y_x|P_j))}{\mathbb{V}(Y_x(\boldsymbol{P}))}, \quad j = 1, \ldots, k, \tag{4}$$

where $\boldsymbol{P}_{\tilde{j}}$ denotes all the parameters except jth parameter. An interpretation[11] of $\mathbb{V}_{P_j}(\mathbb{E}_{\boldsymbol{P}_{\tilde{j}}}(Y_x|P_j))$ is that it is the reduction in the variance that would be obtained if the value of the parameter $\boldsymbol{P}_{\tilde{j}}$ could be fixed.

The total effect measures the total variance contributed by a parameter, including the variance contributed by its interaction with other parameters. The total effect Sobol index at point x for parameter j is given by

$$S^t_{x,j} = \frac{\mathbb{E}_{\boldsymbol{P}_{\tilde{j}}}(\mathbb{V}_{P_j}(Y_x|\boldsymbol{P}_{\tilde{j}}))}{\mathbb{V}(Y_x(\boldsymbol{P}))}, \quad j = 1, \dots, k, \tag{5}$$

where $\mathbb{E}_{\boldsymbol{P}_{\tilde{j}}}(\mathbb{V}_{P_j}(Y_x|\boldsymbol{P}_{\tilde{j}}))$ denote the remaining expected variance given $\boldsymbol{P}_{\tilde{j}}$.

4.2. *Estimation of Sobol indices*

In order to use Sobol indices to analyse the sensitivity of our pollutant dispersion model (1), we require a numerical method of estimating the quantities (4) and (5). We do this using a Quasi-Monte Carlo approach based on Sobol sequences.[11]

Let Z denote an $m \times k$ design matrix, where m is the sample size, and $\boldsymbol{y}_x = f_x(Z)$ be the vector of pollutant concentrations obtained by solving the BVP on a given grid. We generate a resampled matrix $Z_{\sim j}$, where the entries of the jth column are fixed to the values in Z. The BVP is solved such that $\boldsymbol{y}_{x,\sim j} = f_x(Z_{\sim j})$, with mean concentration $\overline{y_{x,\sim j}} = \left[\sum_{i=1}^{m}(\boldsymbol{y}_{x,\sim j})_i\right]/m$. Then we estimate the first-order indices for each parameter as[12]

$$\hat{S}_{x,j} = \frac{\sum_{i=1}^{m}\left[(\boldsymbol{y}_x)_i - \overline{y_x}\right]\left[(\boldsymbol{y}_{x,\sim j})_i - \overline{y_{x,\sim j}}\right]}{\sum_{i=1}^{m}\left[(\boldsymbol{y}_x)_i - \overline{y_x}\right]^2}, \quad j = 1, \dots, k, \tag{6}$$

where $\overline{y_x} = \left[\sum_{i=1}^{m}(\boldsymbol{y}_x)_i\right]/m$ is the sample mean concentration at the point x.

For the estimation of the total effects, we generate a resampled matrix $Z^t_{\sim j}$, where only the entries of the jth column are changed, and all other entries are fixed. Then $\boldsymbol{y}^t_{x,\sim j} = f_x(Z^t_{\sim j})$ is the solution to the BVP corresponding to the resampled input $Z^t_{\sim j}$. We estimate the total effect for each parameter as[11]

$$\hat{S}^t_{x,j} = \frac{1}{2}\frac{\sum_{i=1}^{m}\left[(\boldsymbol{y}_x)_i - (\boldsymbol{y}^t_{x,\sim j})_i\right]^2}{\sum_{i=1}^{m}\left[(\boldsymbol{y}_x)_i - \overline{y_x}\right]^2}, \quad j = 1, \dots, k. \tag{7}$$

4.3. *Spatial dependence of Sobol indices*

Figure 3 shows the spatial dependence of the first-order and total effect Sobol indices across the domain, with the constant wind blowing from left ($x = 0$) to right ($x = 5$). We see that upstream of the first source at $x_{s_1} = 1.5$, the wind speed is the most influential parameter. We also observe that the variance contributed by the source strength s_1 is localised about the source location $x_{s_1} = 1.5$, and there is a similar effect for s_2 about $x_{s_2} = 3$. As we move further downstream, we note that the relative influence of the wind speed begins to decrease. Around $x = x_{s_2}$, the influence of diffusivity begins to increase and downstream of the second source, it becomes the dominant factor contributing uncertainty to the pollutant concentration.

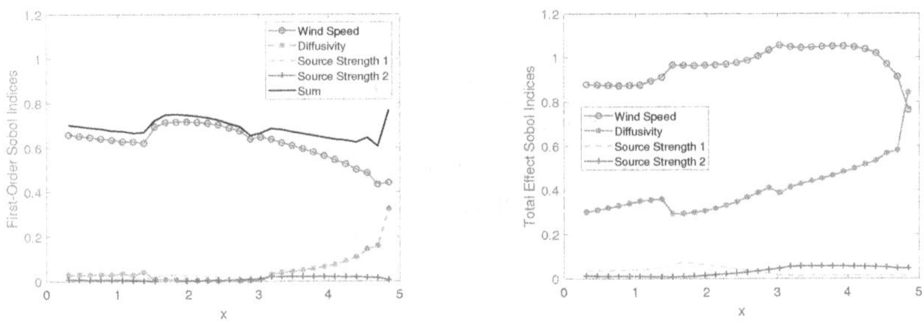

(a) First-order indices. (b) Total effect indices.

Fig. 3. Spatial distribution of Sobol indices.

The total effect indices shown in Fig. 3(b) highlight the interactions between parameters. We notice that while Fig. 3(a) suggests that diffusivity has a small first-order effect upstream of x_{s_1}, the total effect is much larger and hence diffusivity does contribute to the total variance but its impact is through interactions with other parameters. We also note that the wind speed total effect index is noticeably larger than its first-order index, indicating that the variance in the wind speed makes additional contributions to the total variance through interactions with other parameters. There is very little change in the total effect indices for s_1 and s_2 as compared with their first-order indices.

The results presented above indicate that variability in the wind speed is a key factor in the overall variability in the pollutant concentration throughout the domain. Hence, it is desirable to know more about how changes to the wind speed distribution affect the variability in pollutant concentration. We therefore repeat the experiment above with the mean wind speed fixed

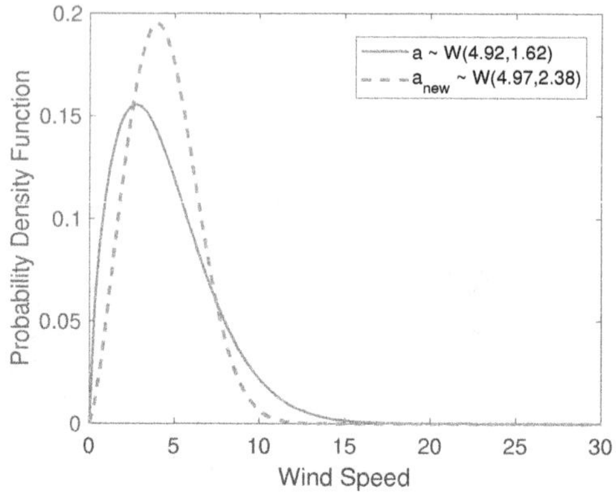

Fig. 4. Probability density functions of Weibull distributions for wind speed.

as before but we reduce the variance by a half by adjusting the shape and scale parameters of the Weibull distribution accordingly. Specifically, we use $a_{new} \sim \text{Weibull}(4.97, 2.38)$ as the new wind speed distribution. Figure 4 compares the probability density functions of the two distributions a and a_{new}.

Figure 5 shows the variability in the pollutant concentration with the new wind speed distribution, with all other parameter distributions unchanged. Comparing with Fig. 1(b), we see that the variance of the pollutant concentration (Fig. 5(b)) has reduced by a half. In addition, Fig. 5(a) shows that the 97.5 percentile value is significantly lower. As seen in Fig. 4, the new wind speed probability density function is more localised towards

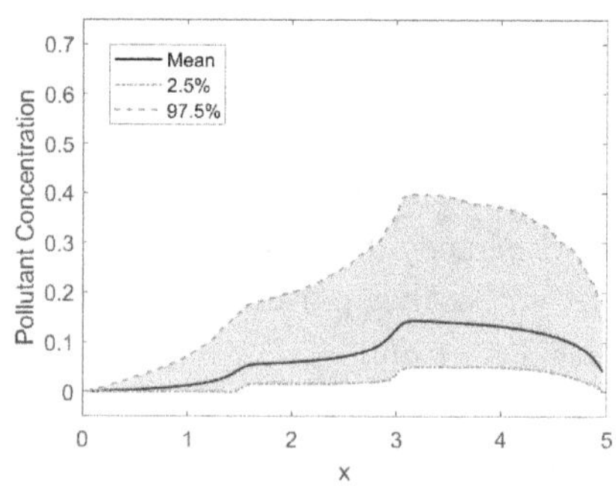

(a) Variability in pollutant concentration.

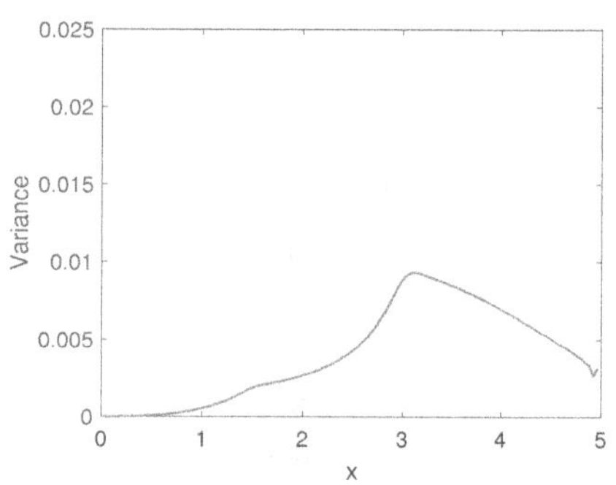

(b) Variance of pollutant concentration.

Fig. 5. Variability in pollutant concentration using $a_{new} \sim \text{Weibull}(4.97, 2.38)$.

its mean and this results in a reduction in the likelihood of low wind speeds. This absence of low wind speeds is responsible for the global reduction in the 97.5 percentile pollution level.

Having seen that reducing variation in the wind speed has resulted in an overall reduction in the variance of the concentration, we now consider whether or not the relative contribution of the parameters to the new variance has also changed. Figure 6 shows the spatial dependence of the new first-order and total effect Sobol indices across the domain. Direct comparison with Fig. 3 shows that the diffusivity and source strengths now play a more prominent role in the overall variance of the concentration. In particular, we now see the influence of diffusivity upstream of the $x_{s_1} = 1.5$ in the first-order indices. However, we note that despite reducing the variance of the wind speed, it is clearly still the most important factor.

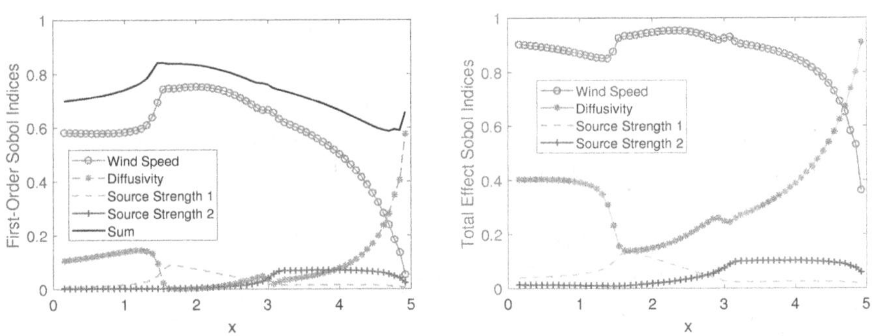

(a) First-order indices.

(b) Total effect indices.

Fig. 6. Spatial distribution of Sobol indices using $a_{new} \sim$ Weibull(4.97, 2.38).

5. Conclusion

In this paper, we have studied how the variability associated with the pollutant concentration for a simple BVP pollutant dispersal model can be attributed to the variability associated with the model input parameters using Sobol indices as a GSA measure. Our results show the significant impact that uncertainty in wind speed has on the variability in pollutant concentration. We have also illustrated how reducing the variance in the wind speed distribution can cause a significant reduction in the variance of pollutant concentration levels. Such results show the potential usefulness of sensitivity measures for pollutant dispersal models.

Acknowledgements

This project is supported by the grant EP/V519777/1 (EPSRC Industrial CASE Account - University of Strathclyde 2020) and the National Physical Laboratory, United Kingdom. For the purposes of open access, the authors have applied a Creative Commons Attribution (CC BY) licence to any Author Accepted Manuscript (AAM) version arising from this submission.

References

1. J. M. Stockie, The mathematics of atmospheric dispersion modeling, *SIAM Review* **53**, 349 (Jan 2011).
2. I. M. Sobol, Global sensitivity indices for nonlinear mathematical models and their Monte Carlo estimates, *Mathematics and Computers in Simulation* **55**, 271 (Feb 2001).
3. A. Allard and N. Fischer, *Sensitivity Analysis in Metrology: Study and Comparison on Different Indices for Measurement Uncertainty*, in *Advanced Mathematical and Computational Tools in Metrology and Testing AMCTM VIII*, (World Scientific Publishing, 2009), pp. 1–6.
4. A. Marrel, N. Saint-Geours and M. D. Lozzo, Sensitivity analysis of spatial and/or temporal phenomena, in *Handbook of Uncertainty Quantification*, eds. R. Ghanem, D. Higdon and H. Owhadi (Springer International Publishing, 2017), pp. 1327–1357.
5. E. C. Morgan, M. Lackner, R. M. Vogel and L. G. Baise, Probability distributions for offshore wind speeds, *Energy Conversion and Management* **52**, 15 (Jan 2011).
6. H. Elman, D. Silvester and A. Wathen, *Finite Elements and Fast Iterative Solvers with Applications in Incompressible Fluid Dynamics*, second edn. (Oxford University Press, Oxford, 2014).
7. A. R. Mitchell and D. F. Griffiths, *The Finite Difference Method in Partial Differential Equations* (Wiley, 1980).
8. W. Hundsdorfer and J. Verwer, *Numerical Solution of Time-Dependent Advection-Diffusion-Reaction Equations* (Springer Berlin Heidelberg, 2003).
9. I. M. Sobol, On quasi-Monte Carlo integrations, *Mathematics and Computers in Simulation* **47**, 103 (Aug 1998).
10. A. Saltelli, M. Ratto, T. Andres, F. Campolongo, J. Cariboni, D. Gatelli, M. Saisana and S. Tarantola, *Global Sensitivity Analysis. The Primer* (Wiley, 2007).

11. A. Saltelli, P. Annoni, I. Azzini, F. Campolongo, M. Ratto and S. Tarantola, Variance based sensitivity analysis of model output. Design and estimator for the total sensitivity index, *Computer Physics Communications* **181**, 259 (Feb 2010).
12. A. Janon, T. Klein, A. Lagnoux, M. Nodet and C. Prieur, Asymptotic normality and efficiency of two Sobol index estimators, *ESAIM: Probability and Statistics* **18**, 342 (2014).

F Pavese, A Bošnjaković, S Eichstädt, A B Forbes, J Alves e Sousa (eds.):
Advanced Mathematical and Computational Tools in Metrology and Testing XIII
Series on Advances in Mathematics for Applied Sciences, Vol. 94

https://doi.org/10.1142/9789819800674_0024

Proof of concept for the digitalized certification of measurement data: An example of vernier-caliper calibration*

K. Shirono†, A. Hirai, G. Matsui, O. Sato, H. Watanabe, T. Takatsuji and T. Fujimoto
National Institute of Advanced Industrial Science and Technology
Tsukuba, Ibaraki, 305-8563, Japan
†*E-mail: k.shirono@aist.go.jp*

We have developed a system in which numerical values can be verified to assure the management and measurement processes in calibrations and testings. The system was developed in accordance with standard operation procedure, and it works with an online application run by a third-party organization to examine data consistency, thus certifying calibration data. An example of the calibration of a vernier caliper is presented.

Keywords: conformity assessment; digitalization; management system; quality assurance; data certification.

1. Introduction

Calibrations and testings require many procedures before measurement. For example, laboratories need to have independent calibrators calibrate their measuring devices and standard references. For this, the standard operation procedure (SOP) of the measurement must be prepared and qualifications of personnel must be verified.

Since this is complicated, third-party accreditation is often used for laboratories to explicitly show the quality of their calibration and testing procedures. We believe that digitalization techniques can support the current form of accreditation. For example, digital calibration certificates (DCCs) can streamline data processing for the management of measuring devices and reference standards. Digital tools to handle the SOP and personnel data will improve the efficiency of the measurement and accreditation processes.

In addition, digitalization can realize a new type of conformity assessment of a laboratory's activities. Currently, laboratories are externally reviewed only once every several years, and no third party can assure a laboratory's procedures during every calibration or testing. However, digital data could be used to do so.

*This work was supported by JSPS KAKENHI Grant Number JP20K04997.

To process data for individual calibrations and testings, the keys seem to be machine-readable data and a programmed SOP. Proposals for a digitalized format of calibration and testing records have been presented, including a DCC developed by the Physikalisch-Technische Bundesanstalt (PTB) [1, 2]. Here, we investigated what data are practically required in assessing laboratory performance and how to handle the data. As a programmed SOP is conceptually important, we assumed that a program would work with an online application run by a third party, which checks the validity of the data in accordance with the predetermined SOP.

We propose *data certification* of the measurement data for every calibration (Fig. 1). This may be interpreted as *Accreditation as a Service* (in the sense that it may be related to the concept of *Software as a Service* (*SaaS*)). We have developed a system in which vernier calipers are calibrated. Section 2 describes the design and Sec. 3 the procedures of the system, and Sec. 4 presents a summary of the study. The hardware configuration is explained in Appendix A.

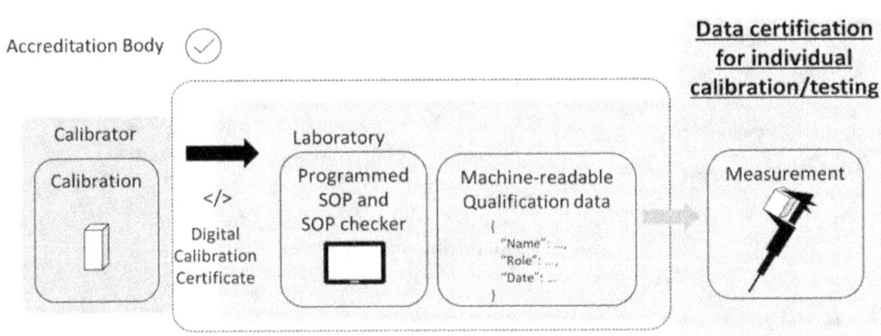

Fig. 1. Data certification with digitalization in this study.

2. Data certification

2.1. *Flow of data certification*

Data processing is implemented in two logically separated environments, managed respectively by the laboratory (on the "laboratory terminal") and the third-party organization (on the "certifier server") (Fig. 2). In the demonstration, the environments were realized by two PCs (Appendix A).

We assume that the laboratory requires an identifier and a password to use the certifier server. Previously [3], we suggested checking external review results, training records, and calibration records. The proposed system could potentially handle them. Here, we explain the steps of qualification (Sec. 3.1), calibration planning, including obtaining the calibration ID (Sec. 3.2), implementation of measurements (Sec. 3.3), and (laboratory, internal) confirmation and (third-party, external) certification of data (Sec. 3.4).

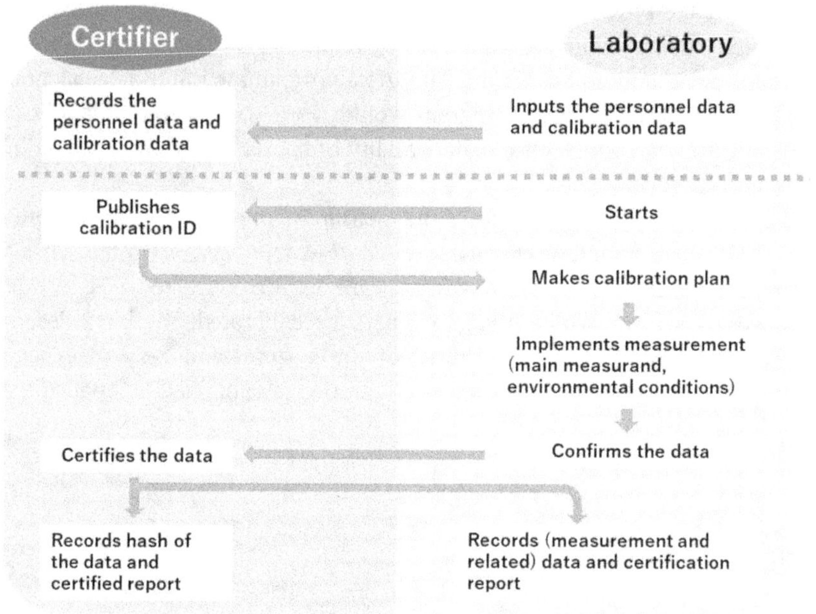

Fig. 2. Flow of data certification.

2.2. *Items to be checked in data certification*

In this example, the following items are checked (the letters in parentheses indicate their software codes):

(1) The calibration certificates (CCs) of the gauge blocks are not expired (CG).
(2) The CCs of the thermometer and hygrometers are not expired (CE).
(3) The expanded uncertainty of the thermometer's calibration value is less than the criterion (TU).
(4) The personnel have the qualifications (PQ).
(5) The temperature is within the predetermined range (TR).
(6) The temperature variation is smaller than the criterion (TV).
(7) The humidity is within the predetermined range (HR).
(8) The offset adjustment is implemented properly (OF).
(9) The differences from the nominal are less than the criteria (BI).
(10) The ranges of the measured values in the repetitions are less than the criteria (DR).

(11) The uncertainties are appropriately computed (UE).
(12) The digital signature is correctly given (DS).

Although item 9 may not be required in typical calibrations, it is included to make the system readily applicable to testing cases. Since items 8, 9, and 10 sometimes cannot be met because of the poor properties of the caliper under calibration, failure to satisfy them may not be due to the calibration system or operation.

3. Demonstration

3.1. *Qualification*

In this example, we consider three user roles in managing the calibration. Those users have the right to execute the events in accordance with the SOP, as shown below. The role of the user is first checked by user ID and password. The user can thus execute only events that he/she is allowed to do.

(a) Administrator:
Qualifying other users (operator and responsible person)
(b) Operator:
Calibration planning and measuring
(c) Responsible person:
Confirming data and requesting data certification.

3.2. *Calibration planning*

Calibration is planned by an operator. In this example, the operator: (1) obtains the calibration ID, (2) determines the calibration date, (3) sets the calibration parameters, and (4) chooses the measuring devices and reference standards.

The measuring operator is not selected in this stage. The calibration ID in step 1 can be obtained from the certifier server; the ID is unique in the system, and all subsequent data are related to this ID. The calibration parameters in step 3 include the levels of the lengths to be measured, the repetition numbers, thermal expansion coefficients, and so on. For step 4, a screenshot of the application shows the data for the thermometer, the hygrometer, and the gauge block (Fig. 3). Data below this screen follow the same format. In this example, the information on the devices and the standards are given in the synthesized DCCs in the PTB developing format, data of which have already been uploaded. Only important data extracted from the DCCs are shown for the operator to check. The devices and standards can be changed as needed.

Fig. 3. Screenshot of the laboratory terminal application for choosing the thermometer, the hygrometer, and the gauge blocks to be used in the measurement.

3.3. *Measurement*

Two types of measurements were implemented in accordance with the SOP and the planning. The first is environmental (temperature and humidity). The second is length, measured against the reference gauge blocks.

In this example, the lengths were set as 25 mm, 50 mm, 100 mm, 125 mm, and 150 mm, with three repetitions (Fig. 4). These values are fixed at the planning stage and cannot be changed at the measurement stage. Thus, the software eliminates any procedures deviating from the SOP.

The colored lines beneath the two-letter codes at the top in Fig. 4 indicate the validity of the data (green, conforming; red, non-conforming; blue, validating) in terms of each item, expressed by its code (Sec. 2.2). For example, the expiration of the calibration certificates of the gauge blocks (code CG) is checked at the end of the measurement process. Since the final assessment cannot be given during measurement, CG gets a blue line (validating), and the operator hence knows what could be problematic in the certification stage even during measurement.

The measurement uncertainties are also computed in this system.

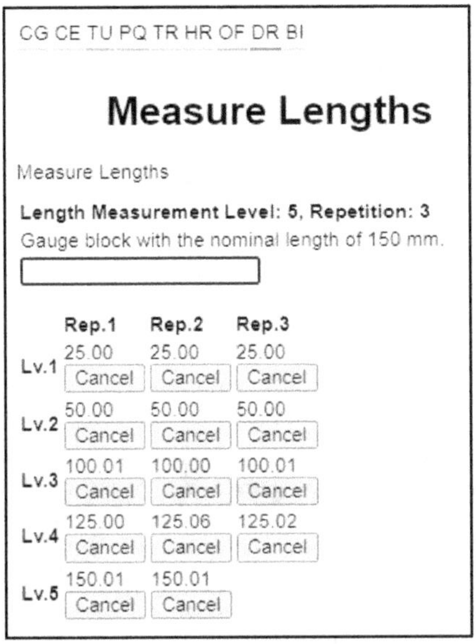

Fig. 4. Screenshot of the application on the laboratory terminal for obtaining indications from the vernier caliper. The indications were input via a Bluetooth interface (see Appendix A). Color codes: green, conforming; red, non-conforming; blue, validating.

3.4. *Confirmation and certification of the data*

In this example, the responsible person confirms the data. XML data are then generated to be checked on the certifier server. A digital signature incorporated within the data validates the origin.

The XML data comprise all data reported by the calibration system. From the relationships among the data, the validity of the programmed SOP is checked on the certifier server, which assesses the consistency of the data.

In this example, the checking report is given in HTML format (Fig. 5). After sending the report to the laboratory terminal, the certifier server deletes the XML file for privacy. (The laboratory holds all the data in the laboratory terminal even after the certifier server deletes the data.) The certifier server stores only the data hash of the calibration data and the checking report.

What is held and what is made publicly available both depend on the characteristics of the calibration and testing data. If the checking report is made publicly available, the caliper user can know that the caliper was properly calibrated by finding the report on the server using the calibration ID as the search key. This is just an example, and other ways to control the data realistically are possible.

In this example, item 10 in the "Check results" failed to be met. The reason could be the calibration system or the calibrated item. This system has a support function, and it does not make any decisions, such as the reimplementation of calibrations or the publication of calibration certificates.

Checking report for the vernier caliper calibration (length) :

Data information:

Test Id: TEST1703225511355
Test data Hash: 3e055c5fdec5de1bb5b546d6b89294c2c78a6237fb9bd802e5a16332b6caa363

Check results:

1 The calibration certificates (CCs) of the gauge blocks were not expired.	true
2 The CCs of the thermometer and hygrometers were not expired.	true
3 The maximum expanded uncertainty of the thermometer was less than the criterion.	true
4 The personnel had the qualification.	true
5 The temperature was within the appropriate range.	true
6 The temperature variation was less than the criterion	true
7 The humidity was within the appropriate range.	true
8 The offset adjustment was implemeted properly.	true
9 The differences from the nominal were less than the criteria.	true
10 The ranges of the measured values in the repetitions were less than the criteria.	false
11 The uncertainties were appropriately computed.	true
12 The digital signature was correctly given.	true

Fig. 5. Data certification report with digitalization.

4. Additional remarks

This paper presents a use case. Data certification can be applied to other cases also. For example, we used a blockchain to ensure data integrity [3]. Other digital techniques [4] will be helpful in developing the system further. We believe that digitalization will provide higher-level quality assurance in the current accreditation system and will allow the development of alternatives for quality assurance. Further investigations and practices are required.

Appendix A. System configuration

We developed the demonstration system with two PCs (Fig. 6) in the roles of a laboratory terminal and a certifier server. Temperature and humidity were measured by a sensor mounted on a microcontroller controlled by the laboratory terminal PC via the local network. A Bluetooth interface on the vernier caliper sent the caliper reading to the laboratory terminal PC as a text string delineated by a carriage return and line feed.

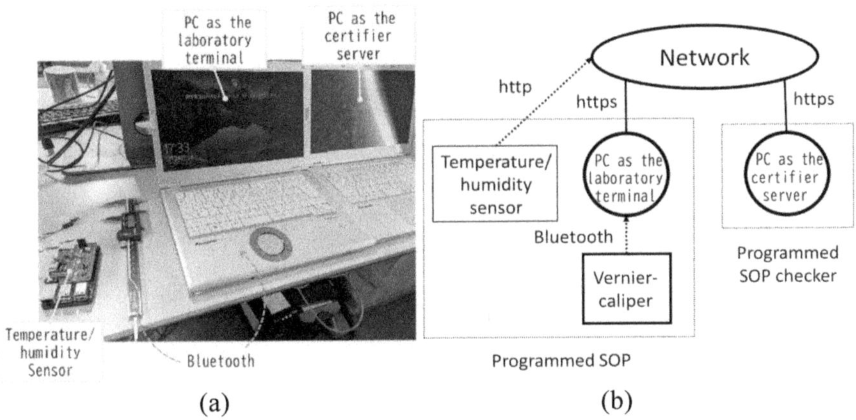

Fig. 6. (a) Photograph and (b) schematic explanation of hardware setup for the demonstration.

References

1. S. G. Hackel, F. Härtig, J. Hornig and T. Wiedenhöfer. Digital Calibration Certificate. PTB-Matriline. **127**, 75–81 (2017).
2. Physikalisch-Technische Bundesanstalt, Digital Calibration Certificate, html https://www.ptb.de/dcc/.
3. Katsuhiro Shirono, Naoki Takegawa, Mahbuba Moni, Daniel Peters, Design of the digitalized conformity assessment for laboratory activities in individual certifications, First International IMEKO TC6 Conference on Metrology and Digital Transformation, Berlin, Germany, 2022.
4. D. Peters, A. Yurchenko, W. Melo, K. Shirono, T. Usuda, J.-P. Seifert and F. Thiel. IT security for measuring instruments: Confidential checking of software functionality. In Advances in Intelligent Systems and Computing, 1129, 701–720. Springer, Cham, 2020.

F Pavese, A Bošnjaković, S Eichstädt, A B Forbes, J Alves e Sousa (eds.):
Advanced Mathematical and Computational Tools in Metrology and Testing XIII
Series on Advances in Mathematics for Applied Sciences, Vol. 94

https://doi.org/10.1142/9789819800674_0025

On choosing two-sided power distribution for measurement data

A. V. Stepanov+ and A. G. Chunovkina
D.I. Mendeleev Institute for Metrology, Russia
+*E-mail: stepanov17@yandex.ru*

The application of the method for choosing a distribution law in relation to a family of two-sided power distributions for further use in metrological applications is considered. An efficiency of the method is assessed by statistical modeling based on the Monte Carlo method. Comparison is made with the maximum likelihood method when estimating a single parameter of distribution law (power). For the considered family of distributions, the coverage factor and coverage interval are estimated.

Keywords: probability density; distribution fitting; Monte Carlo simulation; coverage factor; coverage interval.

1. Introduction

In the practice of scientific metrology the use of Monte Carlo simulations for uncertainty evaluation is very wide [1–4]. An approach, using the propagation of distributions by a Monte Carlo method, is recommended in the Supplement 1 to the Guide to the expression of the uncertainty in measurement (JCGM 101, 2008) [5]. In particular, it allows to avoid the limitations imposed onto the basic GUM uncertainty framework, mentioned in the JCGM 101 (linearization of the model, Gaussian or a scaled and shifted t-distribution for the output quantity, the input quantities should be characterized by best estimates and standard uncertainties etc.). Application of statistical modeling methods in processing measurement data and accuracy assessment opens up opportunities for the widespread use of a variety of distribution laws for modeling measurement errors and describing fuzzy information used in measurements (for example, information about the accuracy of referential data, expert information, etc.). In this regard, the question arises about a reasonable choosing of distribution law. Statistical criteria for testing hypotheses about the type of distribution law turn out to be uninformative, since their power depends on the amount of experimental data, which in practice is often insufficient for these purposes. In addition, in the Bayesian approach, distribution laws are used to describe inaccuracy of the information used, where criteria based on frequency interpretation probabilities turn out to be inapplicable. The latter relates to methods for Type B evaluating of measurement uncertainties. In view of the above, it seems important to develop general principles for the selecting distributions used in processing measurement data. Firstly, one has to ensure continuity with established practice, which focuses mainly on the use of normal and uniform distribution laws having unbounded and bounded domains, correspondingly, which is a fundamental difference between these two distributions. This difference creates difficulties in comparison such metrological characteristics as boundaries (limits of permissible error) used when establishing accuracy of measuring instruments, and confidence interval or expanded uncertainty, when dealing with accuracy of measurement procedures.

Based on the above, two basic principles for choosing distribution laws are proposed when processing measurement results, at least within the framework of solving a single measuring task. Firstly, the distributions should belong to the same family, which provides a wide variety of them, including close approximations for normal and uniform distribution mentioned above, as well as asymmetric distribution laws. Secondly, the distributions should allow a simple modeling procedure which is more convenient in case of bounded domain. The latter property is especially important when calculating, e.g., a confidence interval or expanded uncertainty. Below, we consider a two-sided power (TSP) distributions family, as one of the families of distributions satisfying the listed conditions.

A PDF for a symmetric TSP distribution [6,7] is given by the following expression:

$$f(x) = \begin{cases} \frac{p}{2r}\left(1 + \frac{x-x_0}{r}\right)^{p-1}, & x_0 - r < x < x_0, \\ \frac{p}{2r}\left(1 - \frac{x-x_0}{r}\right)^{p-1}, & x_0 \leq x < x_0 + r, \\ 0, & \text{otherwise}. \end{cases} \tag{1}$$

Here $r > 0$; $p > 0$ is the distribution power parameter ($p = 1$ corresponds to a uniform distribution case, and $p = 2$ corresponds to a triangular one, in case when $p \approx 3$, it approximates the truncated normal distribution quite well). This family of distributions is quite wide and flexible (the authors of [7] demonstrate it using the moment ratio diagram and comparing the family with a Beta distribution family, suggesting to use it as a convenient alternative to the Beta family in some cases). At the same time it has an uncomplicated mathematical description; a simple expression for its CDF allows to generate the corresponding random values using the inverse CDF method. This, in particular, simplifies the efficient implementation of the Monte Carlo method. The above properties encourage the use of the TSP distribution family for modeling and practical application (in particular, it was used by authors in metrological applications [10]).

In more general case, an asymmetric TSP distribution may be considered:

$$f(x) = \begin{cases} \frac{p}{r_1+r_2}\left(1+\frac{x-x_0}{r_1}\right)^{p-1}, & x_0 - r_1 < x < x_0, \\ \frac{p}{r_1+r_2}\left(1-\frac{x-x_0}{r_2}\right)^{p-1}, & x_0 \le x < x_0 + r_2, \\ 0, & \text{otherwise}; \end{cases} \tag{2}$$

here $r_1 + r_2 > 0$, $r_{1,2} \ge 0$.

2. Choosing of distribution law

Consider a method of choosing a distribution law (CDL) proposed in [8], providing the choosing of a distribution law from a given set of distributions that best fits the experimental data. The method is based on use of inverse CDF to obtain a mapping of an empirical sampling distribution onto a reference line. For each distribution from a given set (family) one obtains a value of the functional equal to the RMS value of error when mapping onto the reference line, and then choose the most probable law for the initial sample as one minimizing the value of the specified functional.

Of course, this method is not the only possible for fitting the distribution of experimental data, please see the bibliography of [8,9].

The procedure of the method is quite simple. Denote $\mathcal{X}_0$, $\mathcal{R}$, $\mathcal{P}$ some **finite** subsets of possible values for the TSP distribution family (1) parameters x_0, r, p, respectively. Then, for any sample x_i of length n from the considered TSP family, the set of parameters estimates

$$(\hat{x}_0, \hat{r}, \hat{p}) = \hat{s} \in \mathcal{S} \subset \mathcal{X}_0 \times \mathcal{R} \times \mathcal{P}$$

is selected as follows [8]: $(\hat{x}_0, \hat{r}, \hat{p}) = \arg\min_{s\in\mathcal{S}} \sum_{i=1}^{n} \left(F^{-1}\left(\frac{i}{n}\right) - x_{(i)}\right)^2$, here $x_{(i)}$ is a variational series composed from the initial sample values x_i, and $F^{-1}(y)$ is an inverse CDF for the considered TSP distribution. The CDF for (1) is given by formula

$$F(x) = \begin{cases} \frac{1}{2}\left(1+\frac{x-x_0}{r}\right)^p, & x_0 - r < x < x_0, \\ 1-\frac{1}{2}\left(1-\frac{x-x_0}{r}\right)^p, & x_0 \le x < x_0 + r, \end{cases}$$

so the inverse CDF may be easily calculated as

$$F^{-1}(y) = \begin{cases} x_0 - r + r\sqrt[p]{2y}, & y \in [0, 1/2], \\ x_0 + r - r\sqrt[p]{2(1-y)}, & y \in [1/2, 1]. \end{cases}$$

The inverse CDF for (2) looks as follows

$$F^{-1}(y) = \begin{cases} x_0 - r_1 + \sqrt[p]{r_1^{p-1}(r_1 + r_2)y}, & y \leq \frac{r_1}{r_1+r_2}, \\ x_0 + r_2 - \sqrt[p]{r_2^{p-1}(r_1 + r_2)(1-y)}, & y > \frac{r_1}{r_1+r_2}. \end{cases}$$

The rest of constructions remain the same, except that instead of the set $\mathcal{R}$ one have a pair $\mathcal{R}_{1,2}$, respectively. In the same manner, one can also consider a pair of power parameters $p_{1,2}$ instead of single p [7].

Note that before performing the calculations, it may be worth to check the hypothesis about the symmetry of the distribution. One possible way to do this is to calculate a sample skewness. The hypothesis about the symmetry of the TSP distribution is rejected if its absolute value is greater than the corresponding critical value calculated for a given sample length n and value of the parameter p. Table 1 contains some critical values for $n = 100, 200$ and the confidence probability level P of 99%.

Table 1. Critical values for the sample skewness.

p	$n = 100$	$n = 200$
1.00	0.37	0.26
2.00	0.40	0.28
3.00	0.51	0.36
4.00	0.62	0.44
5.00	0.73	0.52

The advocated method for evaluating parameters of TSP distribution can be considered as a relatively simple alternative to using the maximum likelihood estimation (MLE) method [7], which in case of TSP family is quite complex and requires the numerical solution of systems of nonlinear equations to estimate all the distribution parameters. The values $F^{-1}(i/n)$ can be calculated in advance and tabulated for the given set of F and n to simplify the further calculations.

The other strength of this method is that when selecting the distribution law that best fits the experimental data, it allows to arbitrarily expand the set of alternatives. For example, one can combine some set of TPS distributions with some set of truncated normal distributions in order to select the most suitable one of them (or consider mixtures etc.). The method can also be used to select the best approximation of an arbitrary distribution by the TSP one.

The efficiency of the CDL method was assessed using a Monte Carlo method. Unfortunately, there is not enough space here to provide specific examples [11], nevertheless it shows quite high efficiency, the measure of which is a probability of choosing the proper distribution from a finite number of alternatives. Of course, this probability depends on the length of the original sample and the number of alternatives, as well as on the parameters of the original distribution. Note that the distribution of this probability is asymmetric with respect to distribution parameter p (TSP distributions corresponding to smaller p values are better detected).

While the parameter p, which is responsible for the type of distribution, is purely exhaustive, the boundaries of the sets $\mathcal{X}_0$, $\mathcal{R}$ can be estimated basing on metrological considerations (for example, one may know the limit of permissible error of the measuring instrument), problem statement or statistical considerations. Regarding the last point, having sample statistics calculated (sample mean m, standard deviation S and sample range R), one can use the estimates based on the assumption of the type of distribution:

$$m - CS \leq x_0 \leq m + CS, \quad C_1 R \leq r \leq C_2 R$$

(here we are talking mostly about the symmetrical distribution (1), but the reasoning in the asymmetric case is similar). Critical values for C_1, C_2 for $p \in [1, 5]$) and the coverage probability level P of 99% are given in Table 2. Critical values for C are 0.262 and 0.184 for the same parameters and $n = 100$ and 200, correspondingly.

Table 2. Critical values for $C_{1,2}$.

p	$n = 100$	$n = 200$
1.0	1.001, 1.078	1.001, 1.038
1.5	1.010, 1.200	1.006, 1.118
2.0	1.031, 1.346	1.022, 1.225
2.5	1.063, 1.506	1.047, 1.345
3.0	1.104, 1.673	1.080, 1.472
3.5	1.150, 1.843	1.120, 1.606
4.0	1.200, 2.016	1.163, 1.740
4.5	1.254, 2.191	1.210, 1.878
5.0	1.310, 2.367	1.259, 2.017

For a rough estimate of the $\mathcal{P}$ bounds one can use the sample kurtosis coefficient confidence bands. Similarly, in case of asymmetrical distribution one can use confidence bands for the sample skewness to roughly estimate the bounds of $\mathcal{R}_i$. Some confidence bands for the mentioned sample estimators are given in [11].

3. Sensitivity of the method: estimating of power parameter *p*

As a trivial measure of the sensitivity of the method, consider the probability value P to choose the correct distribution TSP(p) from two, when $\mathcal{P} = \{p, p + \delta p\}$. Figure 1 contains plots for the sensitivity measure for some values of p. It should be noted that the plots are asymmetric, as the estimate for the parameter p is biased (as shown below). It is noticeable that in order for the two distributions to be distinguishable (separable), a sufficiently large value of dp should be taken. From a common sense point of view, this confirms the idea that when choosing the distribution law, one should not consider too close alternatives, since this will only waste (reduce) the probability of choosing the correct (the nearest) distribution law.

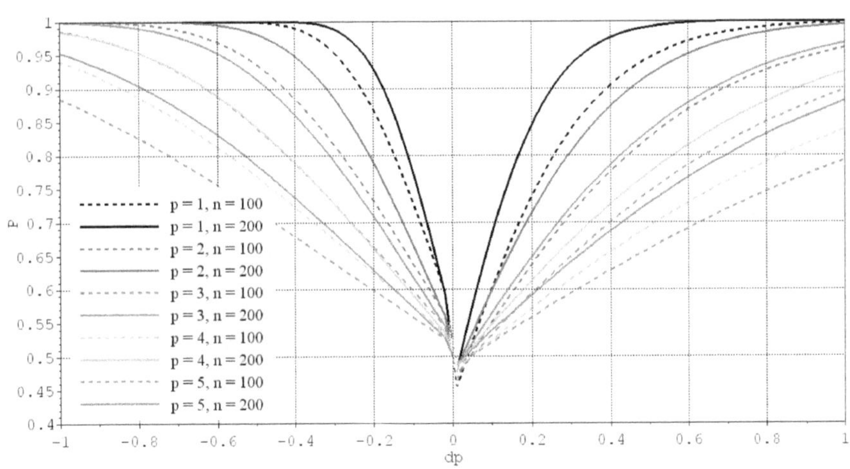

Fig. 1. CDL method sensitivity with respect to p.

The sensitivity of the CDL method is visibly decreases with increasing value of parameter p.

4. Comparison with MLE when estimating p

Let us consider the limiting case when the parameters x_0, r of the distribution are assumed to be known, and the unknown power parameter p belongs to the set $\mathcal{P}$ consisting of very close values (for example, the distance between its adjacent values is constant and equals to dp). In such a case, applying the CDL method one can obtain a histogram for $\hat{p}$ close enough to its distribution function, and the sample statistics (such as mean, standard deviation and skewness) can be estimated.

A MLE for parameter p may be obtained by formula [7]:

$$\hat{p} = -\frac{n}{\log M}, \quad M = \max_{1 \le k \le n} \prod_{i=1}^{k-1} \frac{x_{(i)} - x_0 + r}{x_{(k)} - x_0 + r} \prod_{i=k+1}^{n} \frac{x_0 + r - x_{(i)}}{x_0 + r - x_{(k)}}.$$

Consider its distribution as well as the distribution of CDL estimate obtained for some values of p (a value of 0.001 for dp was used when applying Monte Carlo method). Corresponding PDF plots are given on the Fig. 2. Note that both the methods provide biased estimates of p.

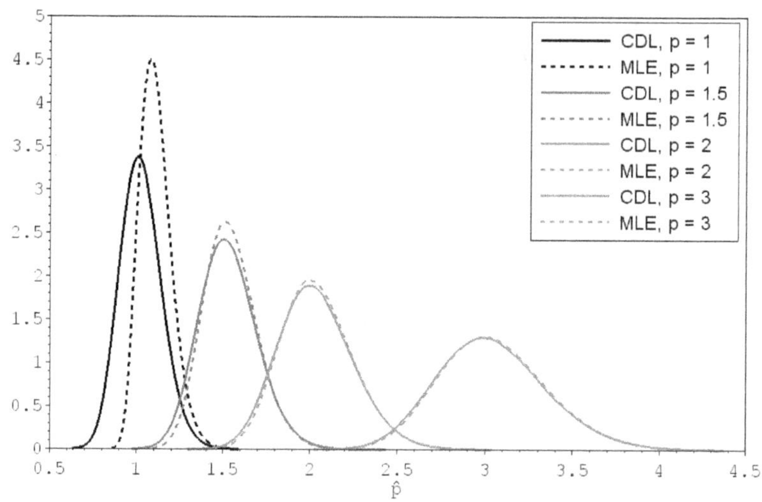

Fig. 2. PDFs for $\hat{p}$, CDL and MLE methods.

Some sample statistics values are given in Table 3. It should be noted that for relatively low values of p CDL provides less value of the estimate $\hat{p}$ bias, and its distribution has lower skewness but higher deviation value. As the value of the parameter p increases, the variance values for CDL and MLE estimates become comparable, as well as the bias.

Table 3. Sample statistics: CDL and MLE methods.

Sample parameters	Statistic	CDL	MLE
$p = 1, n = 100$	mean	1.024	1.108
	standard deviation	0.118	0.091
	skewness	0.33	0.67
$p = 1, n = 200$	mean	1.012	1.073
	standard deviation	0.082	0.061
	skewness	0.23	0.57
$p = 1.5, n = 100$	mean	1.529	1.542
	standard deviation	0.165	0.154
	skewness	0.35	0.42
$p = 1.5, n = 200$	mean	1.515	1.520
	standard deviation	0.115	0.107
	skewness	0.25	0.29
$p = 2, n = 100$	mean	2.035	2.042
	standard deviation	0.213	0.206
	skewness	0.38	0.41
$p = 2, n = 200$	mean	2.018	2.021
	standard deviation	0.149	0.144
	skewness	0.26	0.29
$p = 3, n = 100$	mean	3.044	3.055
	standard deviation	0.313	0.309
	skewness	0.40	0.40
$p = 3, n = 200$	mean	3.024	3.027
	standard deviation	0.218	0.216
	skewness	0.28	0.29

5. Coverage factor and coverage interval

Finally, consider the calculation of coverage factors and coverage intervals for the TSP distribution; a trivial expression for the distribution PDF makes this task very simple.

In case of symmetric TSP distribution (1), a symmetric coverage interval $[x_0 - c, x_0 + c]$ for the confidence level of probability P_0 can be found from condition $F(c + x_0) - F(x_0 - c) = P_0$, so $c = r\left(1 - \sqrt[p]{1 - P_0}\right)$, and the coverage interval looks like:

$$\left[x_0 - r\left(1 - \sqrt[p]{1 - P_0}\right),\ x_0 + r\left(1 - \sqrt[p]{1 - P_0}\right)\right].$$

So, as the standard deviation for (1) is given by $\sigma = r\sqrt{\frac{2}{(p+1)(p+2)}}$, then the coverage factor is described by the following expression:

$$K_{P_0} = \left(1 - \sqrt[p]{1 - P_0}\right)\sqrt{\frac{1}{2}(p+1)(p+2)}.$$

For example, $K_{0.95} = 1.645, 1.902, 1.997, 2.065$ for $P_0 = 0.95$ and $p = 1, 2, 3, 5$, correspondingly.

In case of (2), the coverage factor cannot be calculated, and the coverage interval can be obtained in the following ways. Firstly, one can impose the requirement that the coverage interval $[c_1, c_2]$ provides equal probability tailings on the left and right boundaries, that is $F(c_1) = 1 - F(c_2) = \frac{1-P_0}{2}$ (consider $c_1 < x_0 < c_2$). Some transformations will lead to the following expression for the coverage interval $[c_1, c_2]$ boundaries: $c_1 = x_0 - r_1\left(1 - \sqrt[p]{\frac{r_1+r_2}{2r_1}(1 - P_0)}\right)$, $c_2 = x_0 + r_2\left(1 - \sqrt[p]{\frac{r_1+r_2}{2r_2}(1 - P_0)}\right)$. Secondly, one can calculate the shortest coverage interval $[d_1, d_2]$ such that $F(d_2) - F(d_1) = P_0$, and $d_2 - d_1 \to \min$, which will require the use of numerical methods. Note that quite often it gives a result comparable in terms of interval length (the difference is a few percent), while the intervals boundaries may differ significantly; for some examples please see [11]. So generally there probably is not much point in trying to estimate the shortest coverage interval since this is a more complex computational task and only gives an approximate result (although, of course, there may be cases where this is desirable).

6. Conclusions

Monte Carlo simulation method becomes more and more popular in measurement data processing and uncertainty evaluation. The choice of distribution law is an important initial step in this process. In this paper, use of quite wide and simple TSP distribution family is advocated. Use of the asymmetric TSP distribution can be meaningful when measurand value is close to its physical limit, as 100%, for an example. Application of CDL algorithm for choosing a best TSP distribution for measurement data modeling and further use in metrological applications is discussed and compared with MLE method. CDL method can be used as a simplified alternative of MLE when dealing with TSP distribution family. Finally, coverage intervals and coverage factors are calculated for symmetric and asymmetric TSP distribution. The use of Chi-square test to check Goodness-of-Fit as a final step of CDL was also investigated, including an assessment of its power (unfortunately, not reflected in the text of the article). A confidence bands for the CDL method functional were considered as well.

References

1. P. Harris and M. Cox, On a Monte Carlo method for measurement uncertainty evaluation and its implementation, *Metrologia* **51** (2014), pp. S176–S182.
2. A. van der Veen and M. Cox, Getting started with uncertainty evaluation using the Monte Carlo method in R, *Accreditation and Quality Assurance* **26** (2021), pp. 129–141.
3. S. Crowder, C. Delker, E. Forrest, N. Martin, Monte Carlo Methods for the Propagation of Uncertainties, In: *Introduction to Statistics in Metrology* (Springer, 2020), pp. 153–180.
4. J. Zhang, Modern Monte Carlo methods for efficient uncertainty quantification and propagation: A survey, *WIREs Computational Statistics* **15(5)** (2020).
5. BIPM, IEC, IFCC, ILAC, ISO, IUPAC, IUPAP and OIML, *Supplement 1 to the "Guide to the Expression of Uncertainty in Measurement" — Propagation of distributions using a Monte Carlo method* (JCGM 101:2008, 2008).
6. S. Kotz and J. R. Van Dorp, The Standard Two-Sided Power Distribution and its Properties, *The American Statistician*, **56(2)** (2002), pp. 90–99.
7. S. Kotz and J. R. Van Dorp, *Beyond Beta: Other Continuous Families of Distributions with Bounded Support and Applications* (World Scientific Publishing, 2004).
8. A. N. Tyrsin, The method of selecting the best distribution law for continuous random variables on the basis of inverse mapping, *Bulletin of the South Ural State University. Series "Mathematics. Mechanics. Physics"* **9(1)** (2017), pp. 31–38.
9. I. A. Klyavin and A. N. Tyrsin, Method of choosing the best distribution law for a random variable based on experimental data, *Optoelectronics, Instrumentation and Data Processing* **49** (2013), pp. 14–20.
10. A. V. Stepanov, A. G. Chunovkina, On testing of the homogeneity of variances for two-sided power distribution family, *Accreditation and Quality Assurance* **28** (2023), pp. 129–137.
11. https://github.com/stepanov17/TSP_fit_appendix.

F Pavese, A Bošnjaković, S Eichstädt, A B Forbes, J Alves e Sousa (eds.):
Advanced Mathematical and Computational Tools in Metrology and Testing XIII
Series on Advances in Mathematics for Applied Sciences, Vol. 94
 (pp. 278–287)
https://doi.org/10.1142/9789819800674_0026

Towards an explainable GNN based approach for lifetime prediction using multivariate time-series data

A. P. Vedurmudi*, S. M. H. Zaidi and S. Eichstädt

Department 9.4 - Metrology for the Digital Transformation,
Physikalisch-Technische Bundesanstalt,
Berlin, 10587, Berlin
**E-mail: anupam.vedurmudi@ptb.de*
www.ptb.de

Deep learning is a versatile tool for condition monitoring that provides accuracy at the cost of explainability. In this contribution we explore the use of graph neural networks (GNN) for predicting the remaining useful lifetime (RUL) of an electromechanical cylinder. Through a pre-processing scheme based on *a priori* knowledge of the constituent sensors and the corresponding dataset, the time-series data is embedded into a graph structure. The preprocessed data is used to train a GNN to correctly predict the RUL of the cylinder. Explainable AI (xAI) methods are then used to assess the importance of nodes and edges in making predictions for the specific case of a wrongly predicted value.

Keywords: condition monitoring; explainable AI; graph neural networks.

1. Introduction

A graph neural network (GNN)[1] is a type of artificial neural network designed to operate on and process data that can be represented by graphs. Representative examples of such data with an inherent graph-like structure can be found in social networks and biomolecules. GNNs improve transparency and explainability by allowing a graph-based visualization of their decision-making process and of relationships between its constituent elements. The present work explores the application of GNNs on multivariate time-series data for a condition-monitoring task. The component being monitored is an electromechanical cylinder (EMC) with a spindle drive[2] and the associated experiments were conducted at the Zentrum für Mechatronik und Automatisierungstechnik gGmbH (ZeMA) in Saarbrücken, Germany. The EMC is based on a ball screw drive and serves as a mechanical linear actuator that enables a rotational to linear translation of motion. The main purpose of the experiments was to develop methods to assess the behavior of the EMC while operating under high-loads and high-speeds for the purposes of condition monitoring. In previous research,[3] a series of machine learning methods for the estimation of the remaining useful lifetime (RUL) of the EMC were implemented as part of an automated toolbox. In the present work, we motivate the use of GNNs by the desire to improve the human interpretability in condition monitoring tasks using time-series data by exploiting the inherent transparency and explainability of graph-based visualizations. In particular, the advantage of the graph representation will be emphasized by combining GNNs with two common explainable AI (xAI) methods namely, saliency maps and integrated gradients in order to attribute the results of a prediction to the input data.

2. Methods

2.1. *Experimental setup and dataset*

The setup in question, illustrated in Fig. 1, consists of an EMC subject to a constant axial force of 7 kN via a pneumatic cylinder and lateral forces by means of a fluidic muscle. The tests were carried out until the operational failure of the EMC. Furthermore, the setup was outfitted with eleven sensors continuously measuring various physical properties during the operation of the EMC. The operational EMC velocity is around 200 mms^{-1} with a stroke range of 100–350 mm. Moreover, it accelerates and decelerates at the rate of around 5 ms^{-2}. The combination of a high axial load, a high motion speed, and a high acceleration leads to a fast wear progression.[3] The drag error was used as failure criterion, i.e. the lifetime test was considered to have failed when a deviation larger than 30 mm between the set-position and the actual EMC end position was recorded.[4] A schematic illustration of the setup, showing the three-phase motor driven by a external power source as well as the approximate placement of the sensors, is illustrated in Fig. 2. The physical attributes measured by the sensors during the experiments are as follows:

(1) The active current (ACu) input to the EMC,
(2) The three motor currents (MC1-3) corresponding to the phases of the electric motor driving the EMC,
(3) The acoustic output near the EMC measured by a microphone (Mic),
(4) The pneumatic pressure exerted by the EMC (Pre),
(5) The velocity (Vel) of the EMC,
(6) The accelerations of the plane bearing (APB), ball bearing (ABB) and the plane ring (APR) and,
(7) The axial force (AxF) exerted by the EMC.

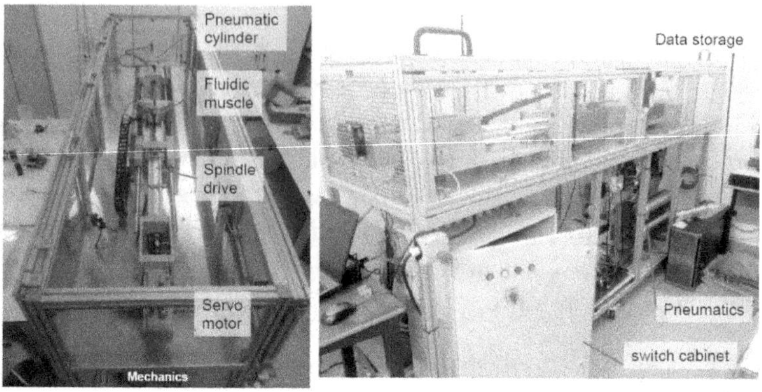

Fig. 1. The test-bed used to carry out the experiment and generate the dataset used in this work at the Zentrum für Mechatronik und Automatisierungstechnik gGmbH (ZeMA) in Saarbrücken, Germany.

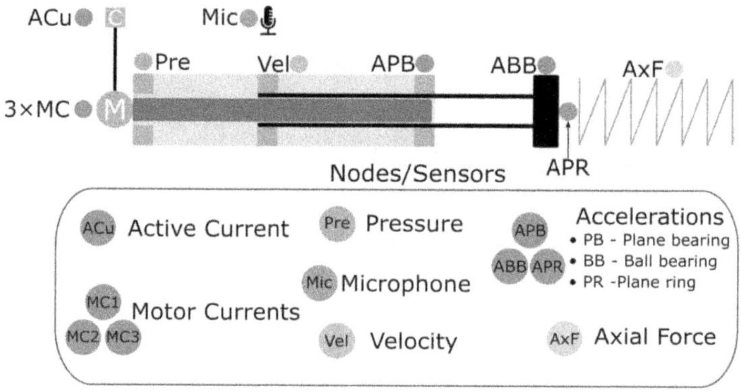

Fig. 2. A schematic representation of the EMC testbed and the associated sensors shown in Fig. 1. Adapted from Dorst *et al.*[5]

The dataset under consideration[6] contains the outputs of eleven sensors monitoring the EMC and consists of 4766 samples where each sample corresponds to 11 time-series with 2000 elements. Corresponding to a sampling rate of 2 kHz, each time-series is one second in duration and represents the measurements taken during one return stroke of the EMC. Moreover, each sample is provided with an associated remaining useful lifetime (RUL) estimate as a percentage. A sample from the aforementioned dataset given by the time-series output of the eleven sensors with corresponding to an RUL of 70% is shown in Fig. 3.

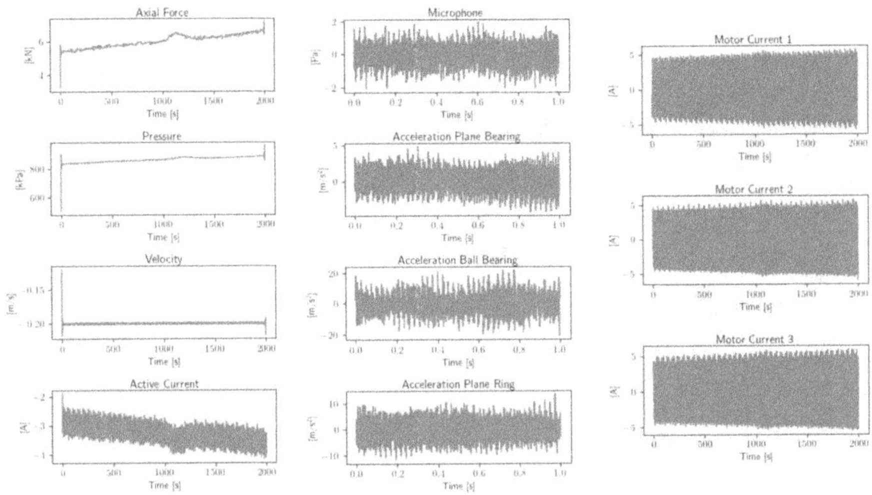

Fig. 3. A single sample from the dataset corresponding to the time-series output from the eleven sensors with an associated RUL of 70%.

2.2. *Preprocessing and graph embedding*

In order to be able to the use GNNs on time-series, an appropriate graph representation of the data needs to be chosen. One way to do so is by interpreting the individual sensors as nodes and choosing an appropriate interrelation between them as the basis to set edges.[7] Both the resulting edges and nodes can be provided with data attributes. We do so by embedding the multivariate time-series into a graph-like structure such that each sample results in a graph consisting of eleven nodes representing the sensors with their individual (normalized) time-series as node-attributes. Furthermore, the edges are weighted by the correlation between the time-series of connected nodes which in turn correspond to the edge attributes. Edges with a weight below the average of the absolute correlations between all nodes (55 possible edges) are discarded. In doing so, we have effectively used the *a priori* information about the setup and the sensors to pre-process the data. An example of this procedure for the time-series dataset from Fig. 3 is illustrated in Fig. 4. In the figure, the nodes have been labeled according to their corresponding sensors as shown in Fig. 2. The edges have been labeled by their weights, i.e. the correlation between the sensor nodes. The thickness of the edges in the figure has been scaled in the same manner.

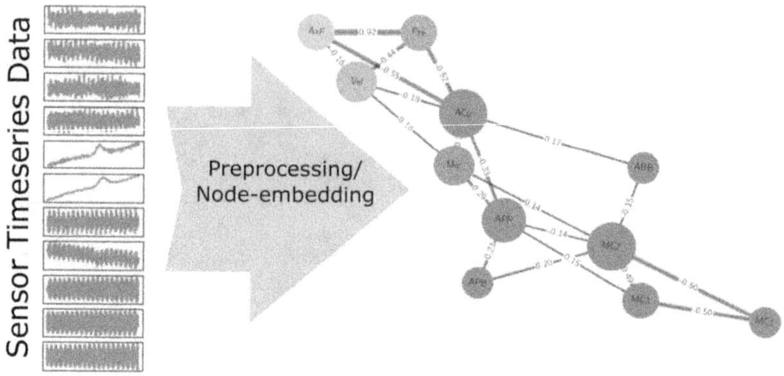

Fig. 4. Illustration of the preprocessing step in which the time-series data is restructured as a graph with eleven nodes.

2.3. *GNN structure and training*

The GNN consists of three GraphConv convolutional layers[8] followed by one pooling, one dropout and two alternating linear layers. The pooling layer returns batch-wise graph-level-outputs by adding node features across the node dimension. The algorithm was implemented using the PyTorch Geometric[9] library. Furthermore, the target value was modified such that the percentage of lifetime passed was binned into steps of 10%. In doing so, the RUL prediction task was reformulated as the classification of a given data sample into one of 10 categories. The negative log-likelihood (NLL) function was chosen as the target loss for the classification problem. From the resulting set of 4766 graphs, 4290 are used to train the GNN, while the remaining are set aside for testing.

3. Results

The performance of the GNN classification is shown in Fig. 5 in terms of a training and test accuracy as well as the training loss as a function of the training epochs. The NLL loss after 30 epochs was around 0.33%. Furthermore, after the same number of epochs a training accuracy of 97% and a test accuracy of 92% was achieved. As an example, the case of a correct prediction made by the GNN for the graph from Fig. 4 is illustrated in Fig. 6. In order to demonstrate the advantage of the combination of xAI and GNNs, we consider another sample from the test dataset where the network makes a false prediction. In Fig. 7, we see that for the given graph

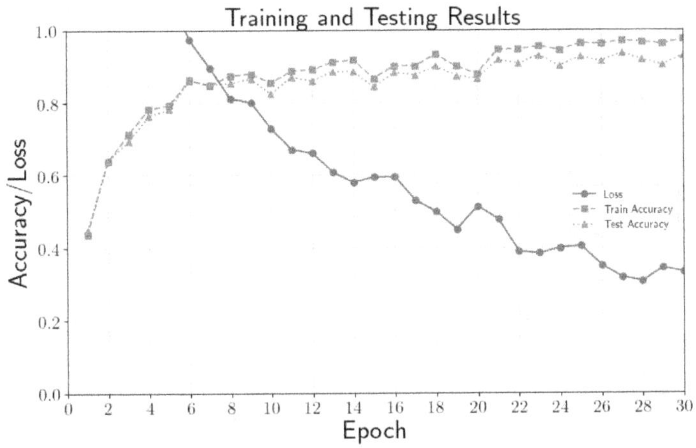

Fig. 5. The training and test accuracy as well as the loss of the GNN against the training epochs.

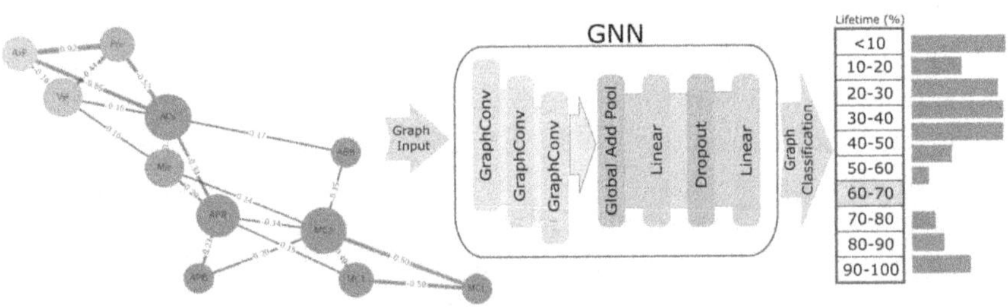

Fig. 6. An example from the test dataset where the GNN makes a correct prediction (blue). The bar plot indicates the negative log-likelihood associated with each prediction.

input, the GNN predicts an RUL of 50–60% instead of the true value of 60–70%. In this case, it would be advantageous to attribute this discrepancy to the sensor inputs.

3.1. *Explainability*

Explainable AI or xAI refers to techniques used to improve the human-interpretability of machine learning methods.[11] Deep learning (DL) algorithms in particular have a tendency to behave as 'black boxes', i.e., the explanation for a specific decision or result taken by the algorithm is opaque even to the designers. In some cases, explainability is achieved by using ML

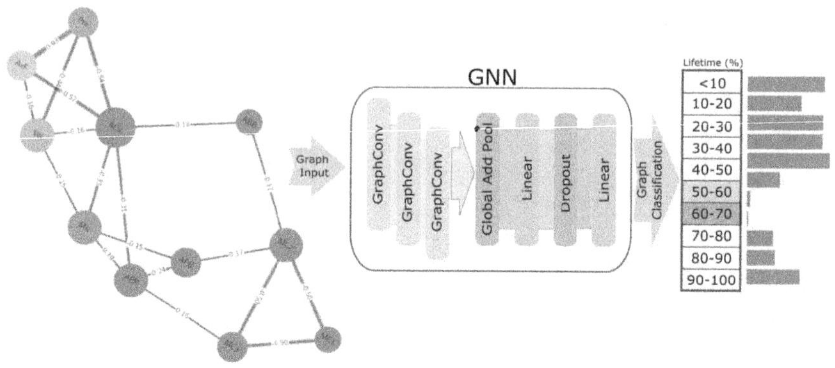

Fig. 7. An example from the test dataset where the GNN makes a false prediction (red) instead of the true prediction (blue). The bar plot indicates the negative log-likelihood associated with each prediction.

methods that are inherently human-interpretable, such as linear regression or decision trees, while in other cases supplementary methods need to be used in addition to the implemented ML algorithm. Our goal is to use the graph structure to identify the nodes (sensors) or interactions (edges) that significantly contributed to the wrong decision taken by the GNN as shown in Fig. 7. To do so we apply two commonly used methods to improve the interpretability of DL algorithms by providing numerical attributions to the input features for a given output — saliency maps[12] and integrated gradients.[13] Both methods require a knowledge of the implemented algorithm, which in our case refers to the weights and parameters of the GNN learned via training. Saliency maps correspond to a simple gradient of the output of a neural network with respect to a particular feature. For an output C and an input feature x_i, the saliency E_{grad} is expressed as

$$E_{\text{grad}} = \left.\frac{\partial F_C}{\partial x}\right|_{x=x_i} . \tag{1}$$

The attribution can be attained at both the edge and node levels. The result of this computation for the classification shown in Fig. 6 is illustrated in Fig. 8. Typically, calculating saliency maps is computationally intensive and not feasible for large datasets. On the other hand, integrated gradients require the computation of the path integral of the gradient of the output with respect to a given feature and a baseline input value x'_i. In this case, the baseline corresponds to the GNN receiving a zero input for all nodes

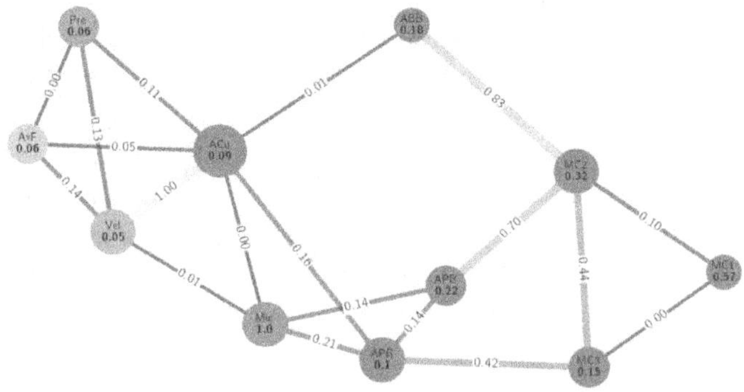

Fig. 8. Graph depicting node- and edge-level attributions for a false prediction generated using saliency maps.

and edges. The path integral computed in this way is given by

$$I_{\text{grad}} = (x_i - x_i') \int_{\alpha=0}^{1} \left. \frac{\partial F_C(x_i' + \alpha(x_i - x_i'))}{\partial x} \right|_{x=x_i} d\alpha. \quad (2)$$

The corresponding attributions at a node and edge level obtained via integrated gradients is shown in Fig. 9. In contrast to saliency maps, the computation of integrated gradients in this manner requires fewer computational resources.

In both cases, the accelerations of the plane ring (APR) and plane bearing (APB) seem to contribute significantly to the decision taken by the

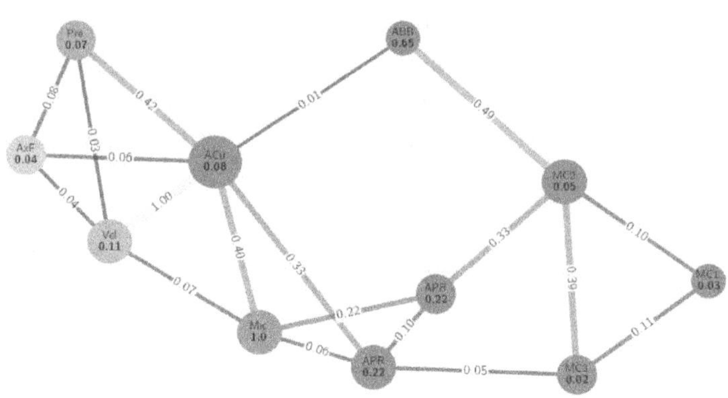

Fig. 9. Graph depicting the node- and edge-level attributions generated using integrated gradients for a false prediction.

GNN. In the case of the saliency map, the correlation between APB and the second motor current (MC2) also has a high contribution, while the use of integrated gradients identifies a high attribution of the interaction between APR and APB. In contrast to integrated gradients, a high impact of the motor currents can be seen in the explanation via saliency maps, while the microphone output has a significant contribution in both cases. While neither the active current (ACu) nor its interaction with other sensors was given a high numerical attribution for the given example, the interaction of ACu and the velocity (Vel) seem to significantly contribute to the output in both cases.

4. Conclusions

In the present work, we have demonstrated the use of graph neural networks for a condition monitoring task involving the prediction of the remaining useful lifetime (RUL) of an electromechanical cylinder. Time-series data collected by means of a series of high-load and high-speed driving tests was first embedded into a graph structure using the time-series data and the correlation between the sensors as *a priori* information. A graph neural network was trained using this preprocessed data to predict the RUL, which was reformulated as a simplified classification task. Two common xAI methods — integrated gradients and saliency maps — were used in conjunction with the GNN to provide node- and edge-level attributions for a given false prediction. In doing so, GNNs were shown to offer increased explainability/interpretability for condition monitoring, while the graph structure enabled us to account for the effects of both individual sensors and their interactions. Future research will explore the applicability of model agnostic attribution methods like SHAP and LIME to graph neural networks and consider the integration of semantic information about the sensors into the graph structure.

References

1. J. Zhou, G. Cui, S. Hu, Z. Zhang, C. Yang, Z. Liu, L. Wang, C. Li and M. Sun, Graph neural networks: A review of methods and applications, *AI Open* 1, 57 (2021).
2. Festo S. E. & Co. K. G., *Electric cylinders ESBF, with spindle drive*. Available: https://www.festo.com/cat/en-gb_gb/data/doc_ENUS/PDF/US/ESBF_ENUS.PDF. [Accessed: 16 Nov 2023].

3. T. Dorst, Measurement Uncertainty in Machine Learning — Uncertainty Propagation and Influence on Performance (Doctoral Dissertation, Universität des Saarlandes), August 2023. doi:10.22028/D291-40173.
4. N. Helwig, Zustandsbewertung industrieller Prozesse mittels multivariater Sensordatenanalyse am Beispiel hydraulischer und elektromechanischer Antriebssysteme. (Doctoral Dissertation, Universität des Saarlandes), December 2018. doi:10.22028/D291-27896.
5. T. Dorst, M. Gruber, B. Seeger, A. P. Vedurmudi, T. Schneider, S. Eichstädt and A Schütze, Uncertainty-aware Data Pipeline of Calibrated MEMS Sensors Used for Machine Learning. *Measurement: Sensors* **22**, 100376 (2022).
6. T. Dorst, *Sensor data set of 3 electromechanical cylinder at ZeMA testbed (2 kHz)*, 5, Zenodo, 2019. [HDF5]. Available: https://doi.org/10.5281/zenodo.3929385. [Accessed: 26 June 2023].
7. Wang, Y., Duan, Z., Huang, Y., Xu, H., Feng, J. and Ren, A. MTHetGNN: A heterogeneous graph embedding framework for multivariate time series forecasting. *Pattern Recognition Letters* **153**, 151 (2022),
8. C. Morris, M. Ritzert, M. Fey, W. Hamilton, J. Lenssen, G. Rattan and M. Grohe, Weisfeiler and Leman Go Neural: Higher-order Graph Neural Networks. *AAAI Conference On Artificial Intelligence*, (New Orleans, USA, 2018).
9. M. Fey and J. Lenssen, Fast Graph Representation Learning with PyTorch Geometric. *ICLR 2019 Workshop On Representation Learning On Graphs And Manifolds.* (New Orleans, USA, 2019).
10. H. Yuan, H. Yu, S. Gui and S. Ji, Explainability in Graph Neural Networks: A Taxonomic Survey, *IEEE Transactions on Pattern Analysis & Machine Intelligence* **45**, 5782 (2023).
11. C. Molnar, *Interpretable Machine Learning: A Guide for Making Black Box Models Explainable* (2nd ed.), (2022). Available: https://christophm.github.io/interpretable-ml-book/cite.html.
12. K. Simonyan, A. Vedaldi and A. Zisserman, Deep Inside Convolutional Networks: Visualising Image Classification Models and Saliency Maps, in *Proc. 2nd International Conference On Learning Representations (ICLR'14)* (Banff, AB, Canada, 2014).
13. M. Sundararajan, A. Taly and Q. Yan, Axiomatic Attribution for Deep Networks, in *Proc. 34th International Conference On Machine Learning (ICML'17)*, (Sydney, Australia, 2017).

F Pavese, A Bošnjaković, S Eichstädt, A B Forbes, J Alves e Sousa (eds.):
Advanced Mathematical and Computational Tools in Metrology and Testing XIII
Series on Advances in Mathematics for Applied Sciences, Vol. 94

https://doi.org/10.1142/9789819800674_0027

The method of fitting a non-linear function and its uncertainty band to data of measured points

J. Puchalski[1] and Z. L. Warsza[2]

[1]*Central Office of Measures,*
Warsaw, 00-139, Poland
E-mail: jacek.puchalski@gum.gov.pl

[2]*Assoc. Professor (emeritus)*
Polish Metrological Society
E-mail: zlw1936@gmail.com

This paper presents a proposed new method of determining parameters and uncertainty bands of a specific non-linear function fitted to given measured data of examined points. One or both of the variables of this non-linear function are changed so as to linearize it. Using the linear regression method, fined are the most favorable parameters of this straight line for its adjustment to the measured values of the coordinates of points tested according to the weighted total mean square WTLS criterion. Their autocorrelation and cross-correlation coefficients as well as uncertainties estimated according to the rules of the GUM guide [1] are considered. The parameters and the uncertainty band of the non-linear function result from the parameters of this straight line and its uncertainty band. Numerical examples of determining the parameters and uncertainty bands for the branch of a 2nd degree parabola (two methods) and for the complex exponential function are given.

Keywords: linear regression of non-linear functions; fit; uncertainty and its band; correlation; weighted total least squares method.

Part 1. Theoretical basis

1. Introduction

The nominal characteristics of many sensors, transducers and measuring instruments, as well as the dependencies of the parameters of studied phenomena, material properties, characteristics of devices and their components are described by nonlinear functions. In metrological practice (e.g., calibration and periodic inspection of instruments), the actual values of the parameters of the tested characteristics are determined. Their mileage and uncertainties are also compared with the nominal ones to determine whether they meet specific requirements.

Linear regression is the basic mathematical tool used in fitting parameters of various linear and nonlinear functions to the measured values of one or more coordinates of the studied points. For two variables, it concerns the determination of the function $y = f(P_i, x)$ with a linear dependence on m parameters P_i, $i = 1, \ldots, m$. For example, for the parabola $y = ax^2 + bx + c$, the parameters a, b, c are determined. Various variants of the least squares method are used as a criterion, including its most general form with the English acronym WTLS (Weighted Total Least Squares). It considers all the data of the examined points, i.e., measured coordinate values, distributions, and uncertainties, as well as autocorrelations and mutual correlations of the results of their measurements. The equation of the tested function and the limits of its extended uncertainty band (also known as the uncertainty corridor) as the interval of occurrence of errors with a certain probability are determined. Linear regression is used in all scientific, natural, and social scientific disciplines, in medicine and health care, in the economy and technical fields, including instrumentation branch and the service of measures [23]. Analytical methods and numerical algorithms were developed to determine the parameters of linear and many nonlinear functions fitted to the measurement data [2–8].

There are many linear regression methods for fitting different nonlinear functions to measurement data. They are numerically complex and usually named after their creators. The oldest is the Gauss–Newton method published at 1809, then the Quasi-Newtonian methods include the Davidson [9], Fletcher and Powell methods [10], the BFGS [11], Levenberg–Marquardt methods [12–14], and the general numerical method for solving Raphson–Newton nonlinear equations. The Monte Carlo method is also used.

In the basic mathematical papers on regression methods, the uncertainty of measurements is usually not assessed strictly according to the principles of the internationally used Guide to the to the expression of Uncertainty in Measurement with the acronym GUM [1], commonly now used in measurement practice. This is also due to the fact that the first version of GUM, initiating the concept of measurement uncertainty, was published in 1993, but many regression methods were developed earlier. Even the most recent works usually do not consider the type B uncertainty (u_B), which is estimated by the combined randomized effect of a set of unknown biases in the instruments used in a measurement experiment. This is done on the basis of the maximum permissible instrument error given by the manufacturers. If the measurements of the coordinates of the studied points are repeated, then their random complex components u_A are also determined. The parameters of the searched function $y = f(x)$ are determined by linear regression according to the least squares criterion for the coordinates of the studied points, considering their uncertainty $u = \sqrt{u_A^2 + u_B^2}$, autocorrelation, and mutual correlation [20]. The expanded uncertainty band $U_P(y) = k_P\, u(y)$ with a given probability p is also estimated.

2. Proposed method of fitting a nonlinear function to the measurement data

To determine the parameters of mathematical models describing the characteristics of devices and their components, including nonlinear functions, computer calculations and the method of least squares are now commonly used. Previously, this method was avoided as it was very computationally cumbersome. Among other things, visual adjustment of curve diagrams to measure points marked on graph paper was used. Nonlinear curves were linearized by selecting an appropriate scale on this paper for the function sought, e.g., square, logarithmic, or exponential scale. Such method was not very precise.

A similar numerical method for determining the parameters of a specific nonlinear function $y = f(x)$ fitted to experimental data, but implemented by computer, is proposed below. In this method, the variables determining the coordinates of the measurement points are changed so that a straight line is obtained in the new Cartesian system. The new variables ψ and ξ are different functions of one of the variables, y or x. It is also assumed that they are parameterized by one common parameter. The equation of a straight line β in new coordinates is then of the form:

$$\psi(y, \beta) = \theta_1 \xi(x, \beta) + \theta_0. \tag{1}$$

In this method the parameters of Eq. (1) will be adjusted by the linear regression to the measurement data according to the generalized weighted least squares criterion WTLS. From the obtained parameter values and the parameters θ_1, θ_0, β of the nonlinear function will be determined. In this approach, due to the inaccuracy of the measurements, both the slope coefficient θ_1 and the free term θ_0 will have uncertainties $u(\theta_1) \neq 0$ and $u(\theta_0) \neq 0$.

In solving the problems of measurement technique using regression methods, a linear approximation of the values of errors and uncertainty by the term with the first derivative will be used. This is permissible when their values do not exceed a few percent of the range and the deviations of the position of the points from the determined function (residual variable) are not too large. That is why we have called the method proposed here as approximate. In the new variables ψ and ξ, there is a β parameter. For each set of measurement data, it has the most favorable constant value, i.e., of uncertainty $u(\beta) \approx 0$. Limiting the dispersion to the term with the first derivative allows the criterion function to be minimized $\phi_{\psi\xi}$ as two-dimensional θ_1 and β. The solution requires the use of complex numerical algorithms. A larger number of selectable parameters would greatly complicate the procedure of minimizing the function $\phi_{\xi\psi}$. There could be local minimums or even a criterion function without a global minimum. The time of numerical calculations would also increase significantly.

The accuracy of the result obtained by the method discussed here is assessed by the width of the uncertainty band of the determined function, and in the case of a known rated waveform, by the discrepancy of both curves.

As computational examples of the proposed method, two variants of determining the parameters of the second-degree parabola and the complex exponential function, considered in the classic monograph of N. Draper and H. Smith [23] as impossible to linearize by changing variables, will be considered. The parameters of both nonlinear functions will be adjusted to the measurement data of the examined points for different uncertainties and correlations of random components of their coordinates. The nonlinear function $y = f(x)$, after converting the variables to ξ and ψ, becomes a straight line $\psi(y,\beta) = \theta_1\xi(x,\beta) + \theta_0$ fitted to the points M_i (ξ_i, ψ_i) by linear regression method. It is shown in Fig. 1. For this purpose, we will use our considerations of linear straight-line regression for measurements of both variables and different cases of coordinate dispersion, their uncertainty, autocorrelation, and mutual correlation of variables. They are contained in our earlier works [15–24], including Polish-language ones [15–22].

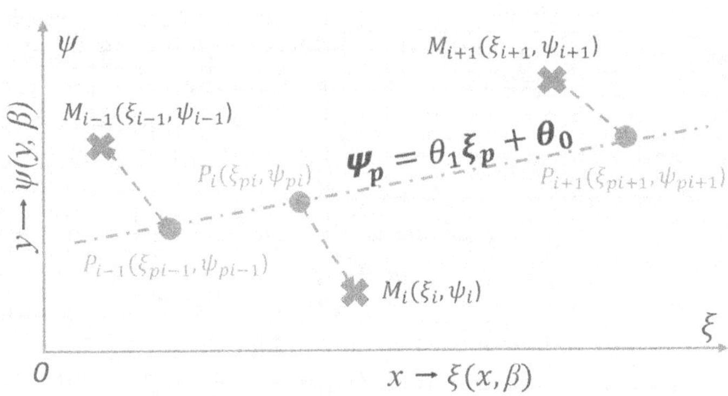

Fig. 1. The result of transforming the non-linear function $y = f(x)$ into a straight line.

The uncertainties of the parameters θ_1 and θ_0 of the line (1) depend on the uncertainties of the coordinates of measured points $M_i(x_i, y_i)$ for $i = (1, ..., n)$ and on all autocorrelations and correlations of the mutual sets of coordinates that may occur in these measurements. These influences are generally described by the covariance matrix $\boldsymbol{U_Z} = \begin{bmatrix} \boldsymbol{U_X} & \boldsymbol{U_{XY}} \\ \boldsymbol{U_{XY}^T} & \boldsymbol{U_Y} \end{bmatrix}$. It consists of four matrices and is used in the criterion function $\phi(\boldsymbol{\Delta X}, \boldsymbol{\Delta Y})$:

$$\phi(\boldsymbol{\Delta X}, \boldsymbol{\Delta Y}) = [\boldsymbol{\Delta X}, \boldsymbol{\Delta Y}] \begin{bmatrix} \boldsymbol{U_X} & \boldsymbol{U_{XY}} \\ \boldsymbol{U_{XY}^T} & \boldsymbol{U_Y} \end{bmatrix}^{-1} \begin{bmatrix} \boldsymbol{\Delta X} \\ \boldsymbol{\Delta Y} \end{bmatrix} \rightarrow min, \tag{2}$$

where: $\boldsymbol{\Delta X} = \boldsymbol{X} - \boldsymbol{X_p}$ i $\boldsymbol{\Delta Y} = \boldsymbol{Y} - \boldsymbol{Y_p}$ are the deviation vectors as differences between the coordinate vectors of measured points $\boldsymbol{X} = [x_1, \dots, x_n]^{\mathrm{T}}$, $\boldsymbol{Y} = [y_1, \dots, y_n]^{\mathrm{T}}$ and vectors $\boldsymbol{X_p} = [x_{p1}, \dots, x_{pn}]^{\mathrm{T}}$, $\boldsymbol{Y_p} = [y_{p1}, \dots, y_{pn}]^{\mathrm{T}}$ of the corresponding coordinates of points on the curve; $\boldsymbol{U_X}$, $\boldsymbol{U_Y}$ and $\boldsymbol{U_{XY}}$ are covariance matrices of coordinates x and y of the measurement points and their cross-correlations.

Fitting a nonlinear function $y = f(x)$ to the measured coordinates (x_i, y_i) test points results from the straight-line regression fit $\psi = \theta_1\xi + \theta_0$ from Fig. 1 to the new point coordinates M_i (ξ_i, ψ_i). These points correspond to the points of the P_i (ξ_{pi}, ψ_{pi}) for $i = 1, \dots, n$, lying on this straight line. The segments connecting the two types of points, in the general case, have different lengths and slopes. They are respectively dependent on the slope of the simple regression and on the elements of the covariance matrix of the measured coordinates. In the absence of correlation, this matrix becomes diagonal and the slopes of the segments connecting the points depend only on the slope of the simple regression and the ratio of the squares of the uncertainty of the two coordinates at each measurement point. In the special case, for equal uncertainty of both coordinates of the points, these segments are perpendicular to the simple regression. On this line, the measuring points M_i correspond to the P_i points with a position dependent on the M_i coordinates and their uncertainties and correlation coefficients. If there are only y_i uncertainties, then the P_i points are projections of the M_i points in the vertical direction y_0.

Relations describing the propagation of deviations when changing variables can be presented in the form of Taylor series expansion. Small deviations of the residual variable together with the measurement errors of the coordinates of the points are approximated with sufficient accuracy by the first terms of this expansion. The first derivative is used to determine the deviations ψ and ξ as $\Delta\xi \approx \xi'(x, \beta)\Delta x$ and $\Delta\psi \approx \psi'(y, \beta)\Delta y$. They are stored together in a diagonal matrix $\boldsymbol{L}$ about the size $[2n \times 2n]$:

$$\boldsymbol{L}=\begin{bmatrix} \xi'(x_1,\beta) & 0 & \dots & 0 & \dots & 0 \\ \dots & \dots & \dots & \dots & \dots & \dots \\ 0 & \dots & \xi'(x_n,\beta) & 0 & \dots & 0 \\ 0 & \dots & 0 & \psi'(y_1,\beta) & 0 & 0 \\ \dots & \dots & \dots & \dots & \dots & \dots \\ 0 & \dots & 0 & 0 & \dots & \psi'(y_n,\beta) \end{bmatrix}, \tag{3}$$

for which $\boldsymbol{L}\boldsymbol{L}^{-1}=\boldsymbol{L}^{-1}\boldsymbol{L}=\mathbf{I}$, where $\mathbf{I}$ is an identity matrix.

The uncertainties in the new Cartesian coordinates result indirectly from this linear approximation. For Gaussian density distributions: $u(\psi)\approx|\psi'(y,\beta)|u(y)$ and $u(\xi)\approx|\xi'(x,\beta)|u(x)$ and Eq. (2) takes the form:

$$\phi(\boldsymbol{\Delta X},\boldsymbol{\Delta Y})=[\boldsymbol{\Delta X},\boldsymbol{\Delta Y}]\boldsymbol{L}\boldsymbol{L}^{-1}\begin{bmatrix}\boldsymbol{U_X} & \boldsymbol{U_{XY}}\\ \boldsymbol{U_{XY}^T} & \boldsymbol{U_Y}\end{bmatrix}^{-1}\boldsymbol{L}^{-1}\boldsymbol{L}\begin{bmatrix}\boldsymbol{\Delta X}\\ \boldsymbol{\Delta Y}\end{bmatrix}$$

$$\approx[\boldsymbol{\Delta\xi},\boldsymbol{\Delta\Psi}]\begin{bmatrix}\boldsymbol{U_\xi} & \boldsymbol{U_{\xi\Psi}}\\ \boldsymbol{U_{\xi\Psi}^T} & \boldsymbol{U_\Psi}\end{bmatrix}^{-1}\begin{bmatrix}\boldsymbol{\Delta\Xi}\\ \boldsymbol{\Delta\Psi}\end{bmatrix}=\phi_{\xi\psi}(\boldsymbol{\Delta\xi},\boldsymbol{\Delta\Psi},\beta). \tag{4}$$

In the new coordinates the measurement points are given by vectors respectively $\boldsymbol{\xi}$ and $\boldsymbol{\Psi}$. Their covariance matrices are $\boldsymbol{U_\xi}$, $\boldsymbol{U_\Psi}$ i $\boldsymbol{U_{\xi\Psi}}$. Only the β is unknown parameter. Its constant value should be chosen in order to minimize the criterion function $\phi_{\xi\psi}$. It is assumed that the uncertainty of this operation is negligibly small, i.e., $u(\beta)\approx 0$. When transforming vectors $\boldsymbol{X}$, $\boldsymbol{X_p}$ and $\boldsymbol{Y}$, $\boldsymbol{Y_p}$ respectively into vectors $\boldsymbol{\xi}$, $\boldsymbol{\xi_p}$ and $\boldsymbol{\Psi}$, $\boldsymbol{\Psi_p}$, with (2) and (4) in order to fit a nonlinear function, a system of two equations is obtained: describing the minimization of the criterial function $\phi_{\xi\psi}$ and the equation of a straight line. These are given below as n-dimensional vector:

$$\begin{cases}\phi_{\xi\psi}(\boldsymbol{\Delta\xi},\boldsymbol{\Delta\Psi},\beta)\rightarrow min\\ \boldsymbol{\Psi_p}=\theta_1\boldsymbol{\xi_p}+\boldsymbol{\theta_0}.\end{cases} \tag{5a,b}$$

In the general case, considering all possible correlations described in [14, 19, 22], solution of this system in the new Cartesian coordinates ξ and ψ is the same as for simple regression using the WTLS together with specific cases [4–14, 17]. Criterion function $\phi_{\xi\psi}$ shall be determined numerically. It depends on the slope of the straight line θ_1 and its offsets linked unambiguously as a relation $\theta_0=(S_y-\theta_1 S_x)/S$ with auxiliary parameters S, S_x and S_y, which are different functions of coordinates x_i, y_i and uncertainties defined and described in the [17, 19, 21]. Using the characteristics calculated at several points of the variable θ_1, parameter is selected β and finds a minimum of features $\phi_{\xi\psi}(\theta_1)$, to obtain a straight line aligned with the measuring points. In the new coordinates ψ and ξ uncertainty are related only to the parameters of the θ_1 and θ_0. When they are correlated with each other, the correlation coefficient is an important parameter $\rho_{\theta_0\theta_1}$. These elements can be determined from the matrix specified in (4) using the uncertainty propagation and approximate formulas for sensitivity coefficients described in [17, 19]. The boundaries of the uncertainty band, as the area around the function to be sought, are determined numerically in the software environment, e.g., *R*, Matlab, and Monte Carlo Method.

In the generation of samples from the covariance matrix (4), the Choleski decomposition method is used [24]. This method allows you to generate a multidimensional variable $\boldsymbol{Z}$ with correlated components defined by the covariance matrix $\boldsymbol{U_Z}$. Cholesky's distribution is to represent the covariance matrix $\boldsymbol{U_Z}$ as the product of two triangular matrices: $\boldsymbol{H}$ and the upper one as its transposition $\boldsymbol{H}^{\mathrm{T}}$. Then $\boldsymbol{U_Z} = \boldsymbol{H}\,\boldsymbol{H}^{\boldsymbol{T}}$. In this way, random multidimensional vectors are obtained $\boldsymbol{Z}$ with components correlated by matrix $\boldsymbol{U_Z}$. This is obtained by generating vectors $\boldsymbol{R}$ as uncorrelated samples from the normal distribution $N(0,1)$, and then multiplying the matrix $\boldsymbol{H}^{T}$ by these vectors, it is $\boldsymbol{Z} = \boldsymbol{H}^{\boldsymbol{T}}\boldsymbol{R} + \boldsymbol{E}$. Vector $\boldsymbol{E} = [\boldsymbol{\xi}, \boldsymbol{\Psi}]$ is the multivariate expected value for the original distribution. The generated correlated random vectors are used to determine from Eq. (5b) directional coefficient distributions θ_1 and intercept θ_0, to find the extension interval $\pm U_\psi$ about probability 0,95 for any selected value of the variable ξ. The values of the parameters of the line are determined numerically as marked with indices '*min*', they are: θ_{1min}, θ_{0min}, β_{min}, $u(\theta_{1min})$, $u(\theta_{0min})$ and coefficient of correlation $\rho_{\theta_0\theta_1}$. An inverse function is assumed to exist ψ^{-1}. The fitted nonlinear function can then be written as:

$$y = \psi^{-1}(\theta_{1\min}\,\xi(x, \beta_{min}) + \theta_{0\min}). \tag{6}$$

The limits of the uncertainty band of the extended output quantity are described by:

$$U_{y+/-} = |\psi^{-1}(\theta_{1min}\xi(x, \beta_{min}) + \theta_{0min} \pm U) - \psi^{-1}(\theta_{1min}\xi(x, \beta_{min}) + \theta_{0min})| \tag{7}$$

or approximate

$$U_y \approx U/\psi', \tag{8}$$

where

$$U = t_{1-\frac{\alpha}{2}, n-2}\sqrt{u^2(\theta_{1min})\xi^2 + 2\rho_{\theta_0\theta_1}\xi\, u(\theta_{1min})u(\theta_{0min}) + u^2(\theta_{0min})} \tag{8a}$$

is the extended uncertainty of the argument of the inverse function for correlated distributions θ_{1min} and θ_{0min}, and the coefficient $t_{1-\frac{\alpha}{2},\, n-2}$ corresponds to the inverted distribution of the *t*-Student's distribution with *n* - 2 degrees of freedom and coverage with probability of $1 - \alpha = 0{,}95$ ($\alpha = 0{,}05$).

Part 2. Numerical examples

3. Adjusting the second-degree function to the measurement points

The characteristics of many types of measuring instruments, sensors, and transducers, as well as the properties of the tested materials, are determined from the measurements of the coordinates of the appropriate number of test points. When they are described by nonlinear functions, the parameters of the equation of this function are determined using the linear regression method. When calibrating or calibrating instruments, the results are compared with the rating data and their accuracy is estimated. We will consider an example of fitting a characteristic described by a second-degree function [17, 22] as a branch of a parabola:

$$y = Ax^2 + Bx + C \tag{9}$$

or in canonical form

$$y = Ax^2 + Bx + C = A\left(x + \frac{B}{2A}\right)^2 + C - \frac{B^2}{4A}. \tag{10}$$

By changing the variables accordingly, this equation can be transformed into a linear equation and adjusted to the new coordinates of the test points [22]. We will consider two methods of fitting a parabolic curve.

3.1. *Parabola fitting - method I*

Method I is to convert the x coordinate into a new Cartesian coordinate $\xi(x, \beta) = (x + \beta)^2$ and substitution $\theta_1 = A$, $\beta = \frac{B}{2A}$ and $\theta_0 = C - \frac{B^2}{4A}$. And the ordinate does not change, i.e., $\psi = y$. Equation (10) will be transformed to the form:

$$y = \theta_1 \xi + \theta_0. \tag{11}$$

Measurement errors are determined by approximations:

$$\Delta\xi_i \approx \frac{\partial \xi}{\partial x_i} \Delta x_i = 2(x_i + \beta)\Delta x_i. \tag{11a}$$

Uncertainty of the new coordinate ξ is determined by the law of uncertainty propagation. For small values, only words with the first derivative are considered:

$$u(\xi_i) \approx \left|\frac{\partial \xi}{\partial x_i}\right| u(x_i) = 2|x_i + \beta|\, u(x_i),\ i = 1, \ldots, n. \tag{11b}$$

As an example, we will consider simulated tests of the actual processing characteristics of a voltage converter with a rated parabolic characteristic generally described by formula (10). For the value x_i of its input voltage x, the y_i values of the output voltage y were measured. Measurements of both voltages were made with a meter of standard relative uncertainty $\delta = 2$ %. Table 1 shows the values of the two measurement point coordinates and the simulated deviations Δy from the nominal characteristics $y \approx 0.133x^2 + 0.04x + 1.003$.

Table 1. Coordinates of the examined points of the 2nd degree parabola with parameters rated $A = 0.13333 \approx 0.133$; $B = 0.039999 \approx 0.04$; $C = 1.003$ and their deviations Δy.

No.	Coordinates of measurement points		Matching errors to nominal parabolic characteristic Δy [V]
	x_i [V]	y_i [V]	
1	1	1.14	−0.034
2	2	1.72	0.10
3	3	2.14	−0.18
4	4	3.26	0.04
5	5	4.66	0.12
6	6	5.71	−0.33
7	7	6.83	−0.99
8	8	10.19	0.33
9	9	12.50	0.34
10	10	13.64	−1.10

Using an EXCEL workbook (Fit, parabola I.xlsx ver. 1.01), based on measurement data from ten points, the values of the function were determined $\phi_{\xi\psi}(\theta_1)$. The formulas for the effective inverse matrix to the covariance matrix and auxiliary parameters were used here [19, 22]. Minimization was made by means of a parameter by β monitoring the determined characteristics $\phi_{\xi\psi}(\theta_1)$ containing a local minimum. For parameter matching $\beta = \beta_{min}$ a global minimum is obtained. This is illustrated in Fig. 2(a).

For the same data of sampling points, the characteristics of the $\phi_{\xi\psi}(\theta_0)$ using the unambiguous relationship between the value of the free term and the value of direction coefficient at the point of global minimum of the function $\phi_{\xi\psi}$ – Fig. 2(b).

For the minimum of the criterion function, the following formulas for the parameters of the parabola are obtained:

$$A = \theta_{1min},\ B = 2A\beta_{min} = 2\beta_{min}\theta_{1min},\ C = \theta_{0min} + \beta_{min}^2\theta_{1min}. \quad (12a,b,c)$$

Matched second-stage function with uncertainty band $\pm U_y(x)$ describes the equation:

$$y = \theta_{1min}\,(x + \beta_{min})^2 + \theta_{0min} \pm U_y(x) \quad (13)$$

where the expanded uncertainty U_y formulas (7, 8).

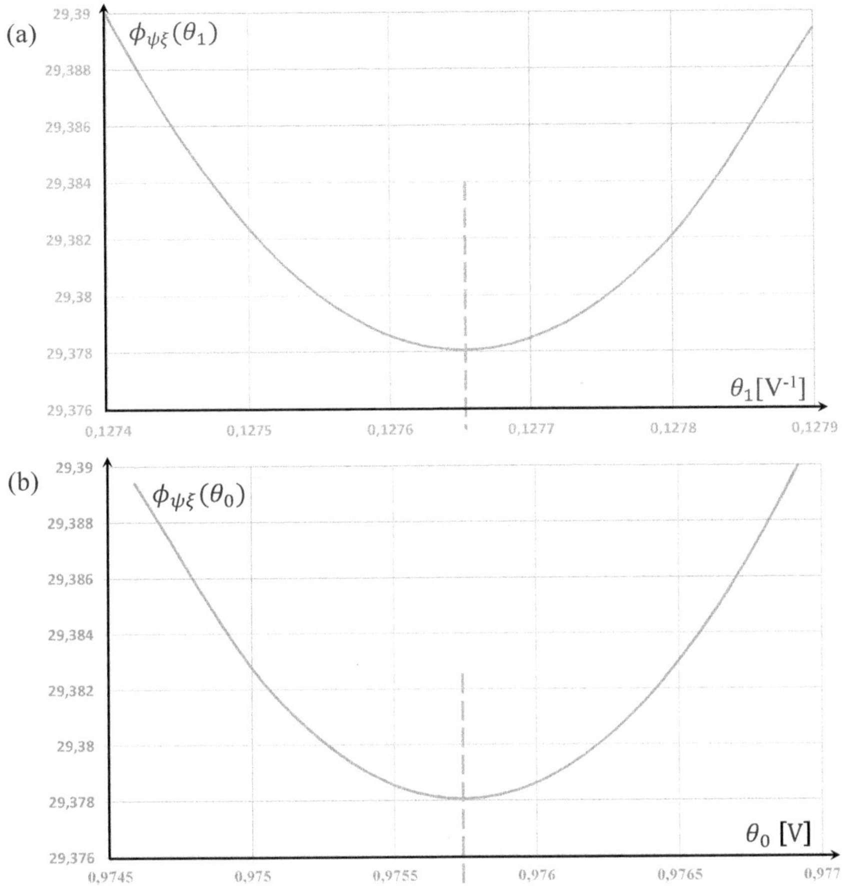

Fig. 2. Quasi-parabolic numerical characteristics (a) $\phi_{\xi\psi}(\theta_1)$, (b) $\phi_{\xi\psi}(\theta_0)$ for β_{min} = 0.21, $\phi_{\xi\psi min} \approx 29.38$.

From coordinate values of the global minimum point in Figs. 2(a) and 2(b), the following parameters of the fitted second-degree parabola were obtained: $A = 0.1276$, $B = 0.0534$, $C = 0.9812$. Together with the uncertainty band, it is shown in Fig. 3(a).

After assuming the same autocorrelation coefficients +0,2 for matrix elements $\boldsymbol{U_X}$, $\boldsymbol{U_Y}$ and negative coefficients −0,2 cross-correlations of x_i and y_j for $i, j = 1, \ldots, n$, i.e., for matrixes $\boldsymbol{U_{XY}}$. The following parameters were obtained: $\beta_{min} = 0.24$, $\theta_{1min} = 0.1275$, $\theta_{0min} = 0.968$, the global minimum $\phi_{\psi\xi min} \approx$ 36,49 and another equation of the fitted parabola $y = 0{,}1275x^2 + 0.0612x + 0.975$. They are given together with the uncertainty band in Fig. 3(b).

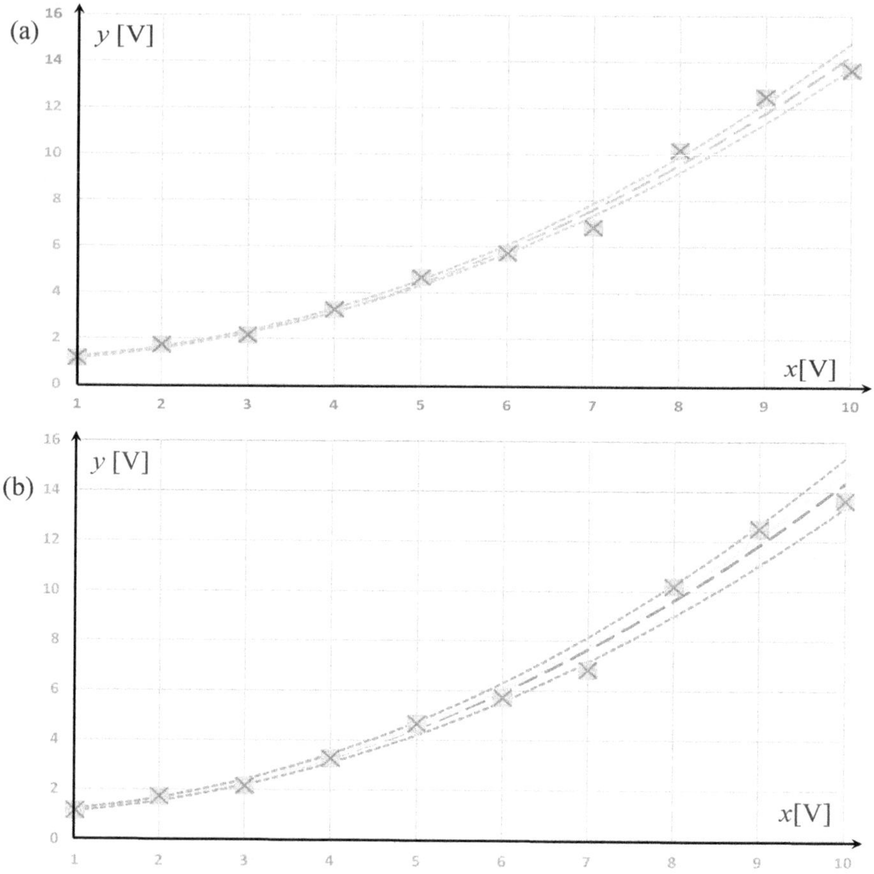

Fig. 3. Parabola fitted to the data of the test points X from Table 1 — dashed (blue) line and its uncertainty band for coverage 0.95 — finely dashed (red) lines: (a) with correlation, (b) without correlation.

The influence of correlation on the uncertainty band boundaries is in Fig. 4.

For coordinates correlated in the entire range of the parabolic characteristic, an almost twice as wide as uncertainty band is in the absence of correlation.

Equations of both parabolas: the obtained and the nominal one $y \approx 0.133x^2 + 0.04x + 1.003$ differ little. This parabola runs inside their uncertainty bands.

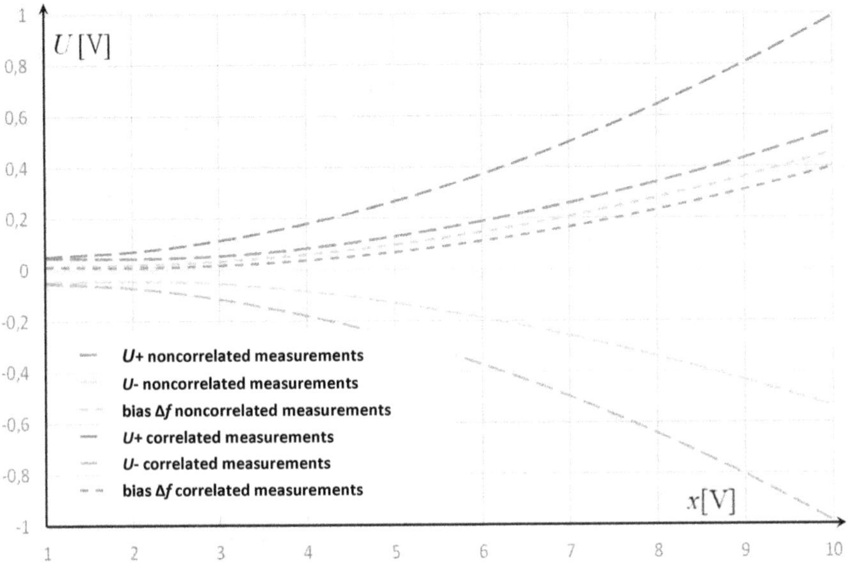

Fig. 4. Expanded uncertainties U and deviation Δf of the fitted characteristic from the nominal one, for correlated and uncorrelated quantities measured as functions of the x coordinate

3.2. *Parabola fitting – method II*

Method II consists of the root of Eq. (10) performed after the free term has been transferred $\beta = C - \frac{B^2}{4A}$ and converts the y coordinate to the new coordinate $\psi(y,\beta) = \sqrt{y-\beta}$ for $y > \beta$ and for substitution $\theta_1 = \sqrt{A}, (A > 0)$, and $\theta_0 = \frac{B}{2\sqrt{A}}$. At that time $C = \beta + \frac{B^2}{4A}$. It is assumed that, over the entire measurement range, the value of $x + \beta/2A > 0$ and that the modulus of squared after taking the square root in Eq. (10) is equal to this value. In contrast, the x coordinate remains unchanged, i.e., $\xi(x,\beta) = x$ and Eq. (10) is transformed to the form:

$$\psi = \theta_1 x + \theta_0. \tag{14}$$

The uncertainties for the new coordinate ψ are determined by the uncertainty propagation law, considering terms with only the first derivative $u(\psi_i) \approx \left|\frac{\partial\psi}{\partial y_i}\right| u(y_i) = \frac{u(y_i)}{2\sqrt{y-\beta}}$ for $i = 1, \ldots, n$. Errors are due to their propagation, i.e. $\Delta\psi_i \approx \frac{\partial\psi}{\partial y_i}\Delta y_i = \frac{\Delta y_i}{2\sqrt{y-\beta}}$. The standard uncertainty $u(\beta) \approx 0$. Figure 5 shows the absolute uncertainty bands and deviations of the characteristics adjusted from the nominal for method II. Obtained using an EXCEL workbook (Fitting - parabola II.xlsx ver. 1.01). Table 2 shows the y-values of the second-degree parabola and extended uncertainties of the correlated measurements ($k = t_{1-\frac{0,05}{2},10-2} = 2.31$) and uncorrelated, determined by methods I and II for x_i test points.

Figure 6 shows the relative uncertainties δ [%] obtained for I and II methods If measurement results are uncorrelated, these values are close to each other.

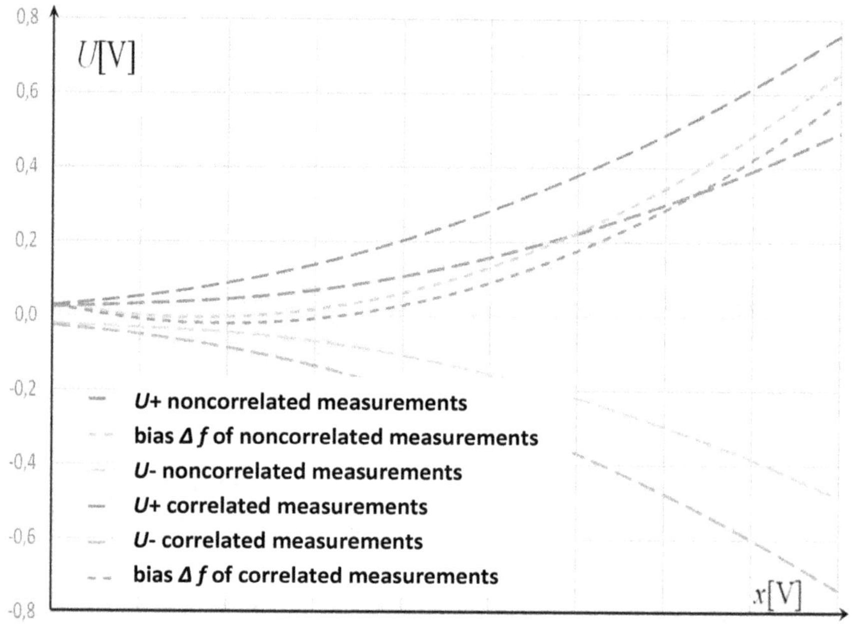

Fig. 5. Expanded uncertainties U_y of the parabola and deviation Δf of the fitted characteristic from the nominal in method II for correlated and uncorrelated measurand, as a function of x.

Table 2. The y-values of the second-order curve fit, and the expanded uncertainties of uncorrelated and correlated measurements determined by methods I and II.

Lp.	Method I				Method II			
	Match y [V]		Expanded uncertainty U_y [V]		Match y [V]		Expanded uncertainty U_y [V]	
	Non correlated	Cor-related	Non Correlated	Cor-related	Non correlated	Cor-related	Non correlated $U_{y-} \approx U_{y+}$	Correlated $U_{y-} \approx U_{y+}$
1	1.16	1.16	0.05	0.05	1.15	1.15	0.02	0.03
2	1.60	1.61	0.04	0.07	1.62	1.63	0.03	0.05
3	2.29	2.31	0.05	0.11	2.33	2.34	0.05	0.09
4	3.24	3.26	0.08	0.18	3.28	3.30	0.07	0.14
5	4.44	4.47	0.13	0.27	4.47	4.51	0.11	0.20
6	5.90	5.93	0.19	0.37	5.91	5.95	0.16	0.28
7	7.61	7.65	0.26	0.50	7.59	7.64	0.22	0.38
8	9.58	9.63	0.34	0.64	9.51	9.57	0.30	0.49
9	11.80	11.85	0.43	0.81	11.67	11.74	0.39	0.62
10	14.28	14.34	0.54	0.99	14.08	14.15	0.50	0.76

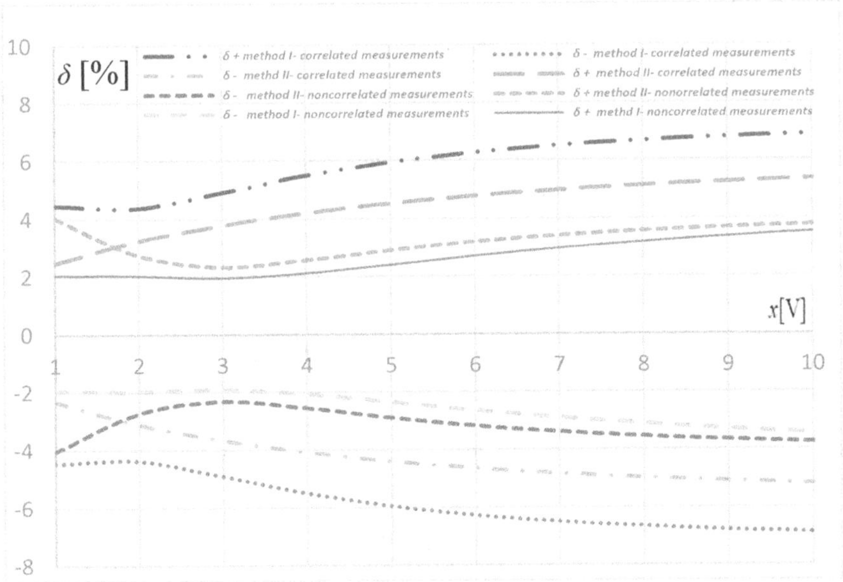

Fig. 6. Expanded relative uncertainty in methods I and II for correlated ones, and uncorrelated measured quantities as a function of the x.

4. Fitting a complex exponential function

The fitting a complex exponential curve with two parameters will be considered α_1 i α_2, described by the formula:

$$y = \frac{\alpha_1}{\alpha_1 - \alpha_2}[\exp(-\alpha_2 x) - \exp(-\alpha_1 x)] + \alpha_3. \quad (15)$$

The parameters of Eq. (15) are adjusted using the weighted method of least squares (WTLS) with values of

$$\alpha_1 = 0.999 \approx 1;\ \alpha_2 = 0.0999 \approx 0.1;\ \alpha_3 = 0. \quad (15a)$$

Fitting the curve (15) to the measurement points in Table 2 approximate method is to use the new variables ξ and ψ described as

$$\xi(x, \alpha_1, \alpha_2) = \exp(-\alpha_2 x) - \exp(-\alpha_1 x) \text{ and } \psi(y) = y.$$

Function (15) in Cartesian coordinates ψ and ξ takes the form of:

$$\psi(y) = \theta_1\, \xi(x, \alpha_1, \alpha_2) + \theta_0, \quad (15b)$$

where the additionally entered parameters are unambiguously defined as $\theta_1 = \frac{\alpha_1}{\alpha_1 - \alpha_2}$ and $\theta_0 = \alpha_3$. In this case, the match parameter β is one of the parameters α_1 or α_2 for example $\beta = \alpha_1$.

Numerical matches were made with the use of a prepared workbook EXCEL-a (Matching, exponential. xlsx ver. 1.01) and the environment R and also repeated in the Matlab environment (script.*Uab*.txt).

In order to minimize the criterion function, the characteristics of the $\phi(\theta_1)$. The fitting is done by a parameter $\beta_{min} = \alpha_1$. Then $\alpha_2 = \beta_{min}(1 - 1/\theta_{1min})$, $\alpha_3 = \theta_{0min}$. From the propagation of errors and the propagation of uncertainty, the following are obtained:

$$\Delta\xi \approx \xi'(x)\Delta x = [-\alpha_2 \exp(-\alpha_2 x) + \alpha_1 \exp(-\alpha_1 x)]\Delta x,$$

$$u(\xi) = |\xi'(x)|u(x) = |-\alpha_2 \exp(-\alpha_2 x) + \alpha_1 \exp(-\alpha_1 x)|\, u(x). \quad (16a,b)$$

In this example, it is assumed that the uncertainty of the parameters θ_1 and θ_o, it is $u(\theta_1) \neq 0$ and $u(\theta_0) \neq 0$, due to the uncertainty of the measuring points ($\delta =$ 2 %) and allow the boundaries of the uncertainty band to be determined. The effect of the uncertainty of the β is negligible.

As a result of fitting the function for uncorrelated data, the function and its uncertainty bands are shown in Fig. 7(a). For comparison with the correlated data series, it was assumed that these data have the same correlation coefficients for the covariance matrix $\boldsymbol{U}_X$ and $\boldsymbol{U}_Y$ equal 0,2 for the variable x and y. A fitted curve with an uncertainty band for mutual correlations (covariance matrix $\boldsymbol{U_{XY}}$ with the same correlation coefficients –0,2) given in Fig. 7(b). Both figures also show the nominal exponential curve indicated by a solid (yellow) line.

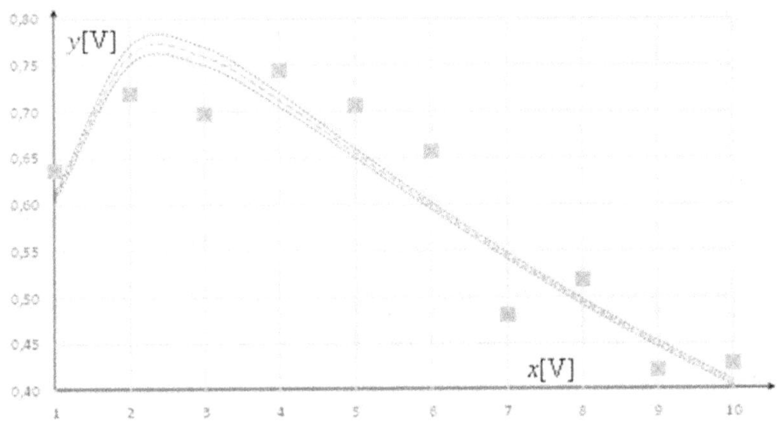

Fig. 7(a). Fitted curve with parameters θ_{1min} = 1.086 and θ_{omin} = −0.032, $\alpha_1 = \beta_{min}$ = 1.15, $\alpha_2 = \beta_{min}(1 - 1/\theta_{1min}) = 0.091$ at the level of $\phi_{\xi\psi min}$ ≈103.5 and uncertainty band – non correlated measurements.

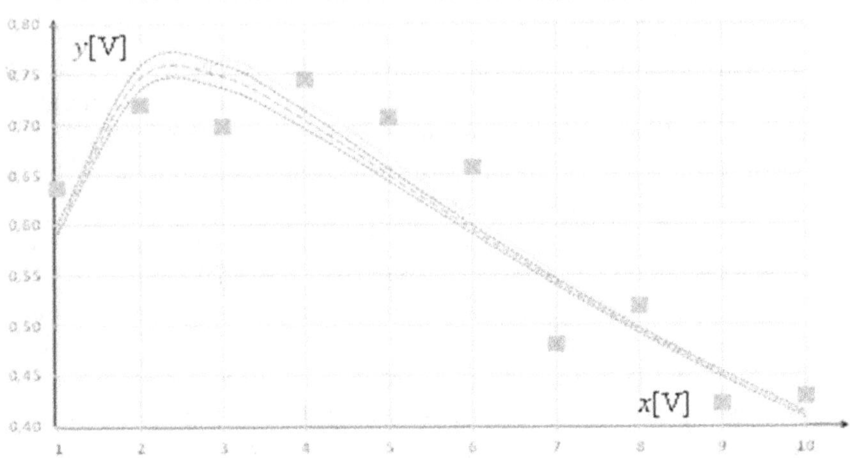

Fig. 7(b). Fitted curve with parameters θ_{1min} = 1.05 and θ_{omin} = −0.013, $\alpha_1 = \beta_{min}$ = 1.1, $\alpha_2 = \beta_{min}(1 - 1/\theta_{1min}) = 0.052$ for $\phi_{\xi\psi min}$ ≈ 128 and uncertainty band - correlated measurements.

A comparison of the waveforms of the upper limit of the uncertainty bands of two cases from Figs. 7(a) and 7(b) is shown in Fig. 7(c).

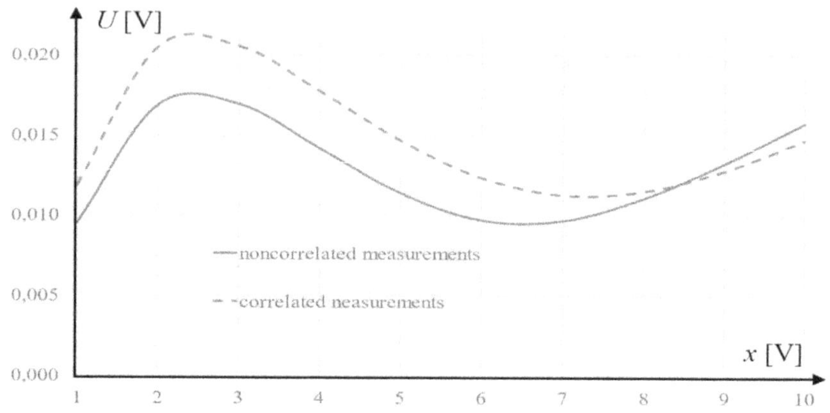

Fig. 7(c). Changes of expanded uncertainty U of coverage $p = 0.95$ (upper of the uncertainty band) of uncorrelated and correlated results of bi-exponential function measurements (15).

5. Summary

A method of approximate adjustment of the parameters of nonlinear functions to the measured coordinates of the measured points is presented and checked on several examples. This method is based on the application of linear regression for the transformed forms of these functions by changing the variables accordingly. In the new Cartesian coordinate system, they become linear. This method uses the laws of error propagation and uncertainty for small deviations from the nominal value. This is an approximate method, as the value of the first derivative of the variable transforming function was used for the calculation.

The fit was realized for two examples of a nonlinear function, i.e., a branch of a second-degree parabola, and for a complex exponential function, which in the monograph of N. Draper and H. Smith [23, p. 163] was considered to be irreducible to a linear function by changing variables. For the quadratic function, two variants, I and II, of the discussed method were used. Different coordinate systems and different results of fitting the two functions and unequal bandwidths of uncertainty were obtained. This also applies to the sets of correlated and uncorrelated coordinates of measured points.

For both methods I and II, the obtained uncertainty band of the function widens with the increase in the measurement uncertainty of each of the coordinates of the test points. Changes in the uncertainty bandwidth along with x for both methods are greater when correlated. With an increase in positive correlation coefficients for coordinates x and y and an increase in the value of the modulus of negative correlation coefficients between coordinates x and y. A widening of the uncertainty band is observed at each point.

The accuracy of the approximate method depends on the type of nonlinear function being fitted. It should be verified whether it is satisfactory for a particular task in which the values of the regression function are used.

According to the GUM guide for the assessment of the complex uncertainty u_i coordinates x_i, y_i of the individual points studied, geometric summation of uncertainties of type A and B, e.g., $u_i = \sqrt{u_{Ai}^2 + u_{Bi}^2}$. Uncertainty $u_{B\,i}$ is due to the maximum relative error E instrument manufacturer or in the calibration certificate. When different measurand values are measured with the same instrument with a constant E, each will have a different component of uncertainty u_B, proportionate to the value to be measured. In measurements with low uncertainty values u_A, these differences need to be considered u_B. In the uncertainty assessment *of* mathematical operations on sets of values measured with the same instrument with constant E, their correlation coefficient $k_{ij} \approx 1$. From the $u_{Bij} = \sqrt{u_{Bi}^2 + 2k_{ij}u_{Bi}u_{Bj} + u_{Bj}^2}$ for the sum follows $u_{Bij} = u_{Bi}+u_{Bj}$ and for the difference $u_{Bij} = |u_{Bi}-u_{Bj}|$. For multiplication and division, the same relationships apply to relative errors.

Acknowledgements

The authors would like to thank the organizers of the AMCTM XIII symposium from Bosnia and Herzegovina NMI in Sarajevo for making it possible to present proposed linear regression method for nonlinear functions and their uncertainties.

References

1. Joint Committee for Guides in Metrology (BIPM, IEC, IFCC, ILAC, ISO, IUPAC, IUPAP & OIML) JCGM 100:2008 Evaluation of Measurement Data — Guide to the Expression of Uncertainty in Measurement. Sevres, France: International Bureau of Weights and Measures BIPM.
2. Reed B. C. (1992). Linear least-squares fits with errors in both coordinates. II: Comments on parameter variances *Am. J. Phys.* **60**, 59–62.
3. Nielssen, Lars, Evaluation of Measurements by the Method of Least Squares. Reprint from Algorithms for Approximation IV, Levesley J., Anderson I.J., Mason J.C. (Eds.) University of Huddersfield, (2002), 170–86.

4. York D., Evensen M. N., Lopez Martines M., DE Basabe Delgato J., Unified equations for the slope, intercept, and standard errors for the best straight line. AM. J. Phys. 72(3) March (2004) pp. 367–375.
5. Lira I., Grientschnig D., Error-in-variables models in calibration. Metrologia. (2017);54: p. 133.
6. Krystek M., Mathias A., A least-squares algorithm for fitting data points with mutually correlated coordinates to a straight line. Meas. Sci. Technol. 22 (2011) 035101 DOI:10.1088/0957-0233/3/035101.
7. Amiri-Simkooei, A.R. *et al.*, Estimation of straight-line parameters with fully correlated coordinates. Measurement 48 (2014) 378–386.
8. Davidson W.C., Variable Metric Method for Minimization. AEC Research and Development Report ANL 5990. DOI: 10.2172/4252678.
9. Dennis J.E. Jr., Schnabel R.B., Numerical Methods for Unconstrained Optimization and Nonlinear Equations. SIAM 1996 reproduction of Prentice-Hall 1983 edition DOI: 10.1137/1.9781611971200.
10. Fletcher R., Practical methods of optimization (2nd ed.) New York John Wiley and Sons, 1987, ISBN 978-0-471-91547-8.
11. Mascarenhas W.F., The divergence of the BFGS and Gauss Newton Methods, "Mathematical Programming", 2013, arXiv:1309.7922, DOI: 10.48550/arXiv.1309.7922.
12. Levenberg K., A Method for the Solution of Certain Non-Linear Problems in Least Squares. "Quarterly of Applied Mathematics", Vol. 2, No. 2, 1944, 164–168, DOI: 10.1090/qum/10666.
13. Marquardt D., An Algorithm for Least-Squares Estimation *of Nonlinear Parameters*. "SIAM Journal on Applied Mathematics", Vol. 11, No. 2, 1963, 431–441, DOI: 10.1137/0111030.
14. Kanzow C. H., Yamashita N., Fukushima M., Levenberg–Marquardt methods with strong local convergence properties for solving nonlinear equations with convex constraints. "Journal of Computational and Applied Mathematics", Vol. 172, No. 2, 2004, 375–397, DOI: 10.1016/j.cam.2004.02.013.
15. Warsza Z.L., Puchalski J., Method of estimation uncertainties of indirect multivariable measurement including accuracy of processing function as extension of GUM S2. Advanced Mathematical and Computational Tools in Metrology and Testing AMCTM XII. Pavese F., Forbes A.B., Zhang N.F., Chunovkina A.G. (Eds.): Series on Advances in Mathematics for Applied Sciences. Vol. 90, (2021), 451–63, World Scientific New Jersey • London • Singapore ISBN 978-981-124-37-3 DOI: 0I:10.1142/9789811242380_0029.

16. Warsza Z.L., Puchalski J., Uncertainty bands of the regression line for autocorrelated data of dependent variable Y. Proceedings of 22nd Conference on Automation 2021, Vol. 1390, 364–386, doi:10.1007/978-3-03074893-7_33.
17. Puchalski, J., A new algorithm for generalization of least square method for straight line regression in cartesian system for fully correlated both coordinates. Int. J. of Automation AI Mach. Learn. (2021); 2(2) pp. 20–54.
18. Puchalski J., Warsza Z.L.: Regression and uncertainty of a straight line for measurements of *x* and *y* variables with all correlations. (in Polish) PAR (Pomiary Automatyka Robotyka) R.26. Nr 2/2022 s.47-58 ISSN 427-9126.
19. Puchalski J., Warsza Z.L.: Determining a straight line by the linear regression method and its uncertainty band with reference to the GUM guide from measurements of both variables *x* and *y* with their autocorrelation and cross-correlation. Monography in Polish: METROLOGIA badania i zastosowania. Monografia M152, TU Politechnika Świętokrzyska Kielce s.235-256.
20. Warsza Z.L., Puchalski J., Uncertainty Bands of the Regression Line for Data with Type A and Type B Uncertainties of Dependent Variable Y. Proceedings of 22nd Conference on Automation 2021, Vol. 1390, 342–363, DOI: 10.1007/978-3-03074893-7_32.
21. Puchalski J., Warsza Z.L., Linear regression method of matching the parabolic curve to tested points of both correlated coordinates Abstract 2p. Paris MathMet 2022 EMN Conference.
22. Puchalski J., Warsza Z. L.: The method of fitting a non-linear function to data of measured points and its uncertainty band. (in Polish) PAR R.26. Nr 3/2023 s.45-55, ISSN 427-9126, DOI:10.14313/PAR_249/45.
23. Draper R.D., Smith H. (1998) Applied Regression Analysis. 3rd Edition Willey, New York. (Polish translation of ed. 1 1966).
24. Grafarend E.W., Awange J. L., Applications of Linear and Nonlinear Models–Fixed Effects, Random Effects, and Total Least Squares. "Journal of Geodetic Science", Vol. 3, No. 1, 2013, 77–78 DOI: 10.2478/jogs-2013-0009.

F Pavese, A Bošnjaković, S Eichstädt, A B Forbes, J Alves e Sousa (eds.):
Advanced Mathematical and Computational Tools in Metrology and Testing XIII
Series on Advances in Mathematics for Applied Sciences, Vol. 94

https://doi.org/10.1142/9789819800674_0028

On testing uncertainty statements and estimating biases in a measurement comparison

R. Willink[1,2]

[1]*Measurement Standards Laboratory of New Zealand, Lower Hutt, New Zealand*
[2]*University of Otago (Wellington), New Zealand*
E-mail: robin.willink@otago.ac.nz

Two possible objectives for a laboratory participating in a measurement comparison are (i) to test the reliability of its submission (x, u) and (ii) to estimate the systematic, i.e., enduring, error or bias in the technique that generated the estimate x. The first of these objectives relates to a statistical hypothesis test and the second relates to a statistical estimation procedure. With regard to the first objective, the usual way in which the results are graphed appears to confuse the two statistical tasks, but the conduct of the test itself remains appropriate. However, with regard to the second objective, careful consideration shows that — if the measurand has true value θ and reference estimate $\hat{\theta}$ — the best estimate of the bias is *not* the conventional difference $d = x - \hat{\theta}$ that forms part of the *degree of equivalence*. In fact, it is numerically smaller than d, meaning that — on average — the laboratory is measuring more accurately than might previously have been thought. Similarly, the standard uncertainty of the best estimate of the bias is generally smaller than the figure often denoted $u(d)$.

Keywords: hypothesis test; systematic error; degree of equivalence.

1. Introduction

The way in which the data from a measurement comparison are to be utilized depends on the objectives of the laboratory. Two reasonable objectives are (i) to test the reliability of the submission (x, u) and (ii) to estimate the systematic, i.e., enduring, error or bias in the technique that generated the estimate x. The first of these objectives relates to a hypothesis test and the second relates to an estimation procedure. The concepts of these two types of statistical analysis differ, and so these two tasks should not be conflated. Here we consider them in turn, and we obtain some results that might be surprising. In the type of comparison studied, the measurand has (unknown) true value θ and the comparison reference value (CRV) is the combined estimate $\hat{\theta}$. The logic of a hypothesis test suggests that the usual graphical way of summarizing the results is misleading but not erroneous, while a standard result from multivariate statistics implies that the conventional difference or deviation $d = x - \hat{\theta}$ is not the best estimate of the tendency of the laboratory to produce measured values that depart from the SI values.

We will consider the overall problem from the point of view of one participating laboratory. Hence the pair (x, u) will often be referred to as *our* submission to the comparison, and (in general) we do not need to use a subscript to identify our laboratory.

2. First objective: testing the uncertainty statement

A correctly conducted statistical test involves the pre-specification of (a) a null hypothesis, (b) an intended means of measuring a relevant departure from that hypothesis, and (c) a maximum acceptable Type I error probability, which is the 'significance level' α and which is taken here to be 0.05. In the test relevant to our laboratory, the null hypothesis H_0 is

> H_0: x is the realization of a random variable X with mean θ and standard deviation $u(x)$.

(So $u(x)$ could be written as σ_X.) The associated test is a 'goodness of fit' test because, unless there is sufficient data-based evidence against it, the participants in the comparison are being asked to accept the model, i.e., are being asked to accept that $u(x)$ is a correct figure of standard uncertainty for x. Thus, the intention of our laboratory is that hypothesis H_0 will be considered true. This hypothesis is to be rejected if the test statistic is extreme enough to lie in the 'rejection region' of conditional probabilistic size α. Therefore, our laboratory hopes that the test statistic lies in the complementary 'non-rejection region', which here is an interval.

In accordance with H_0, our laboratory is initially claiming that

$$X = \theta + E \qquad\qquad E \sim (0, u^2(x)), \tag{1}$$

where E is the error variable realized in the measurement. We can always write

$$X = \theta + E_{\mathrm{u}} + \Delta \qquad\qquad E_{\mathrm{u}} \sim (0, u^2(x)),$$

where E_{u} is error properly assessed and Δ is other error. So, in effect, the claim is that $\Delta = 0$. Likewise, the model for the combined measurement of θ is

$$\widehat{\Theta} = \theta + E_\theta \qquad\qquad E_\theta \sim (0, u^2(\hat{\theta})),$$

where E_θ is the random variable for the corresponding error. The realizations of the random variables X and $\widehat{\Theta}$ are the datum x and the combined best estimate $\hat{\theta}$, which acts as the 'comparison reference value' (CRV).

The random test statistic is $D \equiv X - \widehat{\Theta}$. If the null hypothesis is true, the mean of D is zero and the standard deviation of D is

$$\sigma_D \equiv \sqrt{u^2(x) + u^2(\hat{\theta}) - 2u(x, \hat{\theta})},$$

where $u(x, \hat{\theta})$ is the covariance of X and $\widehat{\Theta}$, which is non-zero if the value taken by X contributes to the value taken by $\widehat{\Theta}$. The realization of D is the observed test statistic $d \equiv x - \hat{\theta}$, and the interval $0 \pm 1.96\sigma_D$ is the non-rejection region for the test. If d lies in this non-rejection region then we accept that $\Delta = 0$.

So the logical graphical representation of this test involves the display of the relationship of the test statistic d to the non-rejection region $0 \pm 1.96\sigma_D$, as in Fig. 1(a). This may be compared with the more familiar representation in Fig. 1(b), which compares the constant zero with the interval $d \pm 1.96\sigma_D$. It is apparent that d lies in the non-rejection region in Fig. 1(a) if and only if the point zero lies inside the interval $d \pm 1.96\sigma_D$ in Fig. 1(b), so the conclusions to be drawn do not depend on which representation is used. However, the representation of the test using Fig. 1(b) is misleading, for it seems to imply that the interval $d \pm 1.96\sigma_D$ is an interval of uncertainty to be associated with the derived statistic d. This misrepresents the test as an estimation procedure.

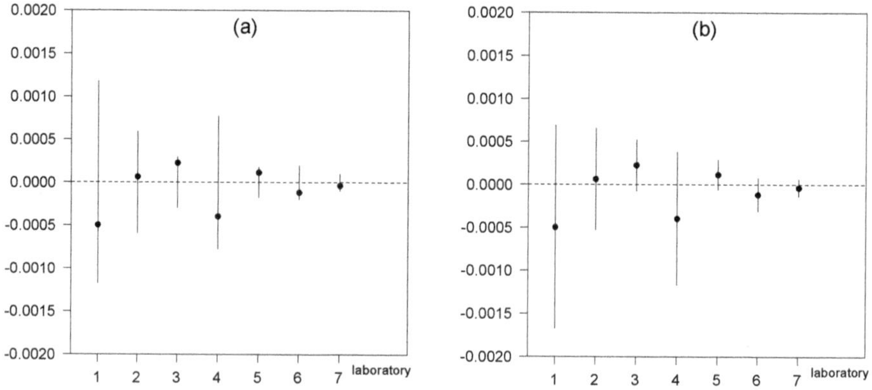

Fig. 1. Example results in a measurement comparison with seven participants. (a) Test statistics d and non-rejection regions $0 \pm 1.96\sigma_D$. (b) Estimates d and intervals $d \pm 1.96\sigma_D$.

Thus, our conclusion is that, *for this important objective*, the form of Fig. 1(a) is appropriate in displaying the results of a measurement comparison but that the form of Fig. 1(b) is misleading.

3. The 'uncertainty of the deviation'

The quantity d is the deviation of the submitted result x from the CRV $\hat{\theta}$. If d lies in the non-rejection region then the participants accept that $\Delta = 0$. But if it does not then the best point estimate of Δ is d and the quality of this estimate of Δ is summarized by the standard deviation of the associated random variable D, which is σ_D.

The notation σ_D is to be preferred to the notation $u(d)$, which would imply that the quantity in question was the 'standard uncertainty' of d. The term 'standard uncertainty' is the abbreviated form of 'standard uncertainty of measurement', so if d were to have a standard uncertainty then d must be interpretable as the measured value of some measurand. But what is that measurand? If a laboratory passes the test then the measurand Δ is deemed to be zero, in which case our best estimate of it is zero, not d, and the notation $u(d)$ seems misleading.

This issue is illustrated in the text of the Mutual Recognition Arrangement[2] (MRA), where we read in paragraph T.2:

> The degree of equivalence of each national measurement standard is expressed quantitatively by two terms: its deviation from the key comparison reference value and the uncertainty of this deviation (at a 95% level of confidence).

Thus, the degree of equivalence is expressed by d and $1.96u(d)$, i.e., $1.96\sigma_D$. But what does the phrase "uncertainty of this deviation" actually mean? If the deviation d is a measured value with a standard uncertainty then what is the corresponding measurand?

Here, we appear to have an example of the clash of two understandings of 'uncertainty'. At this point, the text of the MRA seems based on the idea that the standard uncertainty is just the standard deviation of the procedure that generated the deviation d, (i.e., the standard deviation of D), so that d does not need to be an estimate of anything. Yet, in the "uncertainty approach" that is said to underlie modern metrology, the phrase 'uncertainty of d' does not make sense unless d is recognizable as an estimate of a relevant measurand. This further illustrates the misleading nature of Fig. 1(b).

It also follows that d and σ_D are figures of no intrinsic interest. So far, their roles are only as intermediate figures in the hypothesis test: the relevance of σ_D is in the formation of the limits $0 \pm 1.96\sigma_D$ and the relevance of d is found in its relationship to those limits. Beyond this, the familiar interval with limits $d \pm 1.96u(d)$ where $u(d) \equiv \sigma_D$ has no clear scientific meaning. We conclude that the notation $u(d)$ for σ_D is not appropriate.

4. Defining 'bias'

Before proceeding to discuss the second objective, we will seek to clarify the statistical meaning of the term 'bias.' In a traditional view, the process of measurement obeys

$$X = \theta + E_{\text{ran}} + e_{\text{sys}} \qquad\qquad \mathcal{E}(E_{\text{ran}}) = 0,$$

where E_{ran} is a random variable for the error that changes from occasion to occasion, i.e., for the "random(ized)" error; e_{sys} is the constant "systematic" error; and $\mathcal{E}$ denotes the mathematical expectation or expected value. Under this view of measurement, the statistical bias is e_{sys}, because this is the expected value of the overall error. However, in a modern view of measurement, the process obeys

$$X = \theta + E_{\text{A}} + E_{\text{B}} \qquad\qquad \mathcal{E}(E_{\text{A}}) = 0 \qquad \mathcal{E}(E_{\text{B}}) = 0,$$

where E_{A} is the random variable for the component of error whose variance is assessed by statistical means (in Type A evaluation) and E_{B} is the random variable for the component of error with a variance *imputed* by non-statistical means (in Type B evaluation). According to this equation, the total error has expected value 0, so the bias is now zero! Thus there is a difficulty in communicating the meaning of 'bias' in modern measurement.

Although there is not always a one-to-one correspondence, Type B evaluations are associated with systematic errors, so there is interest in estimating the value e_{B} realized by E_{B} in the modern measurement. We shall call this value, $\beta \equiv e_{\text{B}}$, the 'laboratory bias'. It can be regarded as an unknown parameter of the routine measurement procedure of the laboratory, hence our use of a new symbol, β. Even if our laboratory procedure passes the test, we surely wish to know the value taken by the non-statistical error during our measurement, on the understanding that this persists in our measurement procedure. The remainder of this chapter considers the estimation of this unknown laboratory bias β on the condition that our laboratory has passed the hypothesis test, i.e., on the condition that Eq. (1) is accepted as a model of the generation of our measured value, x.

5. Second objective: estimation of the laboratory bias

After the test is passed, we can write Eq. (1) as

$$X = \theta + E_{\mathrm{B}} + E_{\mathrm{A}} \qquad E_{\mathrm{A}} \sim \left(0, u_{\mathrm{A}}^2\right) \qquad E_{\mathrm{B}} \sim \left(0, u_{\mathrm{B}}^2\right),$$

where $u_{\mathrm{A}}^2 \equiv u_{\mathrm{A}}^2(x)$ and $u_{\mathrm{B}}^2 \equiv u_{\mathrm{B}}^2(x)$ are the total variances associated with the components of uncertainty evaluated by Type A and Type B methods. So $u_{\mathrm{A}}^2 + u_{\mathrm{B}}^2 = u^2(x)$.

If two random variables are related then, on the condition that we know the value taken by one, we have some information about the value taken by the other. In this context a standard result is (e.g., Ref. 1, pp. 98, 100):

> *If $(X_1, X_2)'$ has the bivariate normal distribution with mean $(0,0)'$ and covariance matrix $\left[\begin{smallmatrix} a & c \\ c & b \end{smallmatrix}\right]$ then the distribution of X_1 on the condition that $X_2 = x_2$ is normal with mean cx_2/b and variance $a - (c^2/b)$.*

We will use this result in what follows, first in the idealized case where $u(\hat{\theta}) = 0$ and second in the practical case where $u(\hat{\theta}) \neq 0$.

5.1. *The idealized case of $u(\hat{\theta}) = 0$*

Consider the situation where there is no uncertainty in $\hat{\theta}$, so we know that $\theta = \hat{\theta}$. The result above can be applied directly to the vector $(E_{\mathrm{B}}, E_{\mathrm{B}}+E_{\mathrm{A}})'$, which has a bivariate distribution with mean $(0,0)'$ and covariance matrix

$$\begin{bmatrix} u_{\mathrm{B}}^2 & u_{\mathrm{B}}^2 \\ u_{\mathrm{B}}^2 & u^2(x) \end{bmatrix}.$$

Let us assume normality, which is a common step, and let us define f to be the fraction of total variance associated with Type B evaluation, i.e.,

$$f \equiv u_{\mathrm{B}}^2 / u^2(x).$$

Then, on the condition that $E_{\mathrm{B}} + E_{\mathrm{A}}$ takes the value z, the variable E_{B} is normally distributed with mean fz and variance $f(1-f)u^2(x)$. Now $E_{\mathrm{B}} + E_{\mathrm{A}}$ is known to take the value $d = x - \hat{\theta}$, and the laboratory bias β is the outcome of E_{B}. So the best estimate of β is

$$\hat{\beta} \equiv fd$$

and the standard uncertainty of this is

$$u(\hat{\beta}) \equiv \sqrt{f(1-f)}\, u(x).$$

The quantity f is positive and generally less than 1, so regarding d as an estimate of bias would be conservative or pessimistic. The estimate of bias is numerically smaller than d and the associated standard uncertainty is smaller than σ_D. The gain is found in the idea that, when there is random error, i.e., when $u_{\mathrm{A}}^2 > 0$ and so $f < 1$, the bias that we are estimating is not responsible for all of the error.

5.1.1. *An alternative demonstration*

The idea behind this result can be understood in a different way via some natural reasoning. Suppose that we want to estimate the value taken by a random variable Z when we know (i) that there is another random variable Y with expected value $\mathcal{E}[Y] = 0$ and (ii) that $Y + Z = d$, where d is a known number. If that is all we know then, by ignorance, our best estimate will be d. But suppose we also know that $\mathcal{E}[Z] = 0$. Then it is natural to conclude that our best estimate depends on our knowledge of the ratio $\mathrm{var}Y/\mathrm{var}Z$. Substitute E_{A} for Y and E_{B} for Z.

5.2. *The practical case of $u(\hat{\theta}) \neq 0$*

In practice, $u(\hat{\theta}) \neq 0$. This complicates the analysis a little, but the principles remain the same. Recall that the combined estimator of θ is the random variable $D \equiv E_{\mathrm{B}} + E_{\mathrm{A}} - E_{\theta}$. The vector $(E_{\mathrm{B}}, D)'$ has a bivariate distribution with mean $(0, 0)'$ and covariance matrix

$$\begin{bmatrix} u_{\mathrm{B}}^2(x) & u_{\mathrm{B}}^2(x) - u_{\mathrm{B}}(x, \hat{\theta}) \\ u_{\mathrm{B}}^2(x) - u_{\mathrm{B}}(x, \hat{\theta}) & \sigma_D^2 \end{bmatrix},$$

where $u_{\mathrm{B}}(x, \hat{\theta})$ is the relevant covariance. Also, D is known to take the value $d = x - \hat{\theta}$. The relative contribution from Type B is

$$f_{\mathrm{B}} \equiv \frac{u_{\mathrm{B}}^2(x) - u_{\mathrm{B}}(x, \hat{\theta})}{\sigma_D^2}.$$

As above, $\beta \equiv e_{\mathrm{B}}$ is the outcome of the random variable E_{B}. So, assuming bivariate normality, the best estimate of β is

$$\hat{\beta} \equiv f_{\mathrm{B}} d,$$

and the standard uncertainty of this estimate is

$$u(\hat{\beta}) \equiv \sqrt{f_{\mathrm{B}}(1 - f_{\mathrm{B}})\sigma_D^2 + u_{\mathrm{B}}(x, \hat{\theta})}.$$

Because $f_{\mathrm{B}} < 1$, it would again be pessimistic to use d as the estimate of bias. Provided that $u_{\mathrm{B}}(x, \hat{\theta})$ is not large, $u(\hat{\beta})$ will also be smaller than σ_D. Also, whatever the value of $u_{\mathrm{B}}(x, \hat{\theta})$, we have $u(\hat{\beta}) \leq u_{\mathrm{B}}(x)$, in accordance with the idea that the comparison has provided some information about our laboratory bias, e_{B}. If $u_{\mathrm{B}}(x, \hat{\theta}) = 0$ then $f_{\mathrm{B}} = f$ and we recover the simpler result of Sec. 5.1.

We see that an estimate of laboratory bias can be obtained in a measurement comparison from decomposing the total uncertainty into Type A and Type B sub-totals. The equations obtained are similar in form to those obtained by Chunovkina *et al.*[3] in a Bayesian analysis.

6. Example

Our example involves fictitious data, though the figures of standard uncertainty are closely based on actual figures provided during the analysis of key comparison CCPR-K6.2010[4] (personal communication, A. Koo), where measurements were made of optical transmittance. The figures are dimensionless because transmittance is a dimensionless ratio between 0 and 1. The data are shown in the first three columns of Table 1, where seven participating laboratories are numbered in inverse order of f_{B}.

Table 1. Results in fictitious comparison, based on key comparison CCPR-K6.2010.

i	x_i	$u_{\mathrm{A}}(x_i)$	$u_{\mathrm{B}}(x_i)$	$f_{\mathrm{B}i}$	d_i	$u(d_i)$	$\hat{\beta}_i$	$u(\hat{\beta}_i)$
1	0.61705	0.00005	0.0006	0.99	-0.000495	0.000600	-0.000492	0.000068
2	0.61761	0.00005	0.0003	0.97	0.000065	0.000301	0.000063	0.000067
3	0.61777	0.00007	0.00014	0.80	0.000225	0.000150	0.000180	0.000073
4	0.61715	0.00022	0.00033	0.69	-0.000395	0.000394	-0.000274	0.000186
5	0.61766	0.00007	0.00007	0.50	0.000115	0.000088	0.000057	0.000055
6	0.61743	0.00009	0.00006	0.31	-0.000115	0.000098	-0.000035	0.000052
7	0.61751	0.00006	0.00003	0.20	-0.000035	0.000049	-0.000007	0.000028

$\hat{\theta} = 0.6175453$, $u(\hat{\theta}) = 0.0000461$.

The CRV $\hat{\theta}$ was calculated to be the weighted mean of the seven measured values, with the weights being proportional to the inverse-square of the overall standard uncertainty, as usual. That is, $\hat{\theta} = \sum w_i x_i / \sum w_i$ with $w_i \equiv \{u_{\mathrm{A}}^2(x_i) + u_{\mathrm{B}}^2(x_i)\}^{-1}$. We obtain the results $\hat{\theta} = 0.6175453$ and $u(\hat{\theta}) = 0.0000461$. The corresponding random variable is $\widehat{\Theta} = \sum w_i X_i / \sum w_i$, and the covariance of E_i and $\hat{\Theta}$ is $u(x_i, \hat{\theta}) = 1/\sum w_i$. Also, $u_{\mathrm{A}}(x_i, \hat{\theta}) = w_i u_{\mathrm{A}}^2(x_i)/\sum w_i$, $u_{\mathrm{B}}(x_i, \hat{\theta}) = w_i u_{\mathrm{B}}^2(x_i)/\sum w_i$, and $u(x_i, \hat{\theta}) = u_{\mathrm{A}}(x_i, \hat{\theta}) + u_{\mathrm{B}}(x_i, \hat{\theta})$, as required. These expressions are used in the calculation of the figures in the remaining columns of Table 1. The laboratories with small values of f_B are seen to have much smaller estimates of bias than d.

Figure 2 gives the results for both objectives in graphical form. The results for the hypothesis tests are presented in the style of Fig. 1(a), using solid markers and solid lines. None of the seven measured values lies outside its non-rejection region, so none of these estimates is deemed inconsistent with its uncertainty statement. (We have already seen these results in Fig. 1(a).) Subsequently, the estimates of 'laboratory bias' and the standard uncertainties of these estimates are calculated. The results are presented in the style of Fig. 1(b), using hollow markers and dashed lines. For example, the analysis suggests that the measurement procedure of laboratory 3 involves a positive systematic error β_3 with a value in the interval $[0.000046, 0.000325]$.

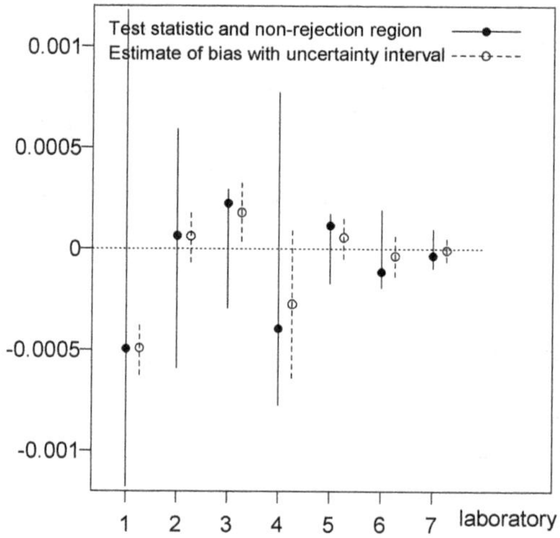

Fig. 2. Results of the comparison for the seven laboratories of Table 1.

In accordance with the theory, Table 1 and Fig. 2 show that $\hat{\beta}/d < 1$ for each laboratory, so that the estimate of bias obtained by this analysis is numerically smaller than the estimate implied in the traditional analysis. The proportional reduction is least for laboratory 1 and greatest for laboratory 7, following the ordering of the ratio f_B. Also, Fig. 2 shows that $u(\hat{\beta}) < \sigma_D$ for each laboratory and that the sizes of the relative reductions behave in a different way. The smallest relative reduction is observed for laboratory 5, which has $f_B = 0.50$ and $u(\hat{\beta})/\sigma_D = 0.62$.

7. Discussion

This paper focuses on the *scientific*, i.e., logical, meanings of the results for a laboratory participating in a measurement comparison. It does not claim to discuss the *administrative* role of a measurement comparison in the system of CIPM Key Comparisons, in which the participants are National Metrology Institutes (NMIs). The literature on measurement comparisons indicates that there are different understandings about the goals and techniques for the associated data analysis. For example, Refs. 5–10 are a selection of papers in which the authors have discussed, perhaps implicitly, either the role of a comparison or the role of a 'degree of equivalence' in a Key Comparison. So it seems important to draw a distinction between the logical and administrative meanings of a comparison dataset.

Here, we have considered two potential scientific objectives for a participating laboratory. The first objective (testing the uncertainty statement) is one that participating NMIs acknowledge, but the second objective (estimation of the laboratory bias) does not seem to be regarded as highly. Moreover, the question of how to use any resulting estimate of bias also does not seem to have an agreed answer. In this section, we will briefly discuss some of these issues using helpful comments from a reviewer of this paper.

In regard to the second objective, the reviewer wrote:

> While $\hat{\beta}$ is numerically no larger than d in magnitude, the random component has not been eliminated and a laboratory cannot ignore its uncertainty contribution. (If there is no random component then $\hat{\beta} = d$.) However, on the basis of the results from the measurement comparison, a laboratory has more information about its systematic effect and can use this information to apply a correction and reduce its uncertainty statement. ... (If the laboratory does not apply this correction, then its original uncertainty statement applies.)

In effect, the reviewer identifies a third possible objective for a laboratory that participates in a comparison: if there is no change made to the measurement procedure after the comparison then the best estimate of the bias $\hat{\beta}$ and its uncertainty $u(\hat{\beta})$ stand, and they can be used to apply a correction and reduce the uncertainty quoted in future measurements. (See Ref. 11 for a non-classical analysis emphasizing the estimation of biases and the use of these estimates with their uncertainties as corrections.)

As was implied by the reviewer, the role of randomness in a measurement procedure cannot be ignored. The idea that we can reduce its effect by making more replications and averaging is associated with a change to the title of this paper. The earlier title finished with the informal phrase "your laboratory procedure is more accurate than you think!", which referred to the fact that $\hat{\beta}$ is, legitimately, a smaller estimate of bias than d. Taken literally, this phrase would invite confusion, because it implies that a laboratory procedure is well-defined and has a fixed level of accuracy. However, the random component of error can be reduced by increasing the number of replications — so for the phrase to have a literal interpretation, our procedure must involve a fixed number of replicated observations. Thus, the text in the body of the paper no longer refers to 'greater accuracy in the measurement procedure' but to 'avoiding conservatism and pessimism in the estimation of bias'.

The fact that the amount of random error depends on the number of replications diminishes the importance of d as a recorded figure of measurement quality. This relates to the important question of how the quality of a measurement made in a comparison relates to the quality of a contemporary measurement made for a client in practice. With the number of replicated measurements not being fixed, the focus seems better placed on the idea of bias. If we regard the accuracy of our procedure as being described by the bias β and its estimate $\hat{\beta}$ then the accuracy is indeed better than would be implied by d.

8. Conclusion

There are at least two potential goals for a laboratory that participates in a measurement comparison. One is to promote confidence that the laboratory is measuring as accurately as implied in the statement of measurement uncertainty, and another is to assess the size of the bias in the laboratory's measurement procedure, i.e., the mean error in the measurement of many similar measurands.

The first goal involves a hypothesis test, where we examine the relationship of a test statistic d to a non-rejection region $0 \pm 1.96\sigma_D$. The logical way to present this test graphically is to show the test statistic and non-rejection region, as in Fig. 1(a), not to compare zero with the interval $d \pm 1.96\sigma_D$, as in Fig. 1(b). The symbol σ_D is appropriate for the standard deviation of the random test statistic (under the null hypothesis). The conventional symbol $u(d)$ is inappropriate because d is not the measured value of any stated measurand. Similarly, the interval $d \pm 1.96u(d)$ has no clear scientific meaning.

The second goal involves the separation of the components of uncertainty associated with Type A (statistical) and Type B (non-statistical) evaluations, on the understanding that Type B evaluation is strongly linked to the existence of systematic error. Analysis shows that the deviation, d, from the reference value and the parent standard deviation of this deviation, σ_D, are not the appropriate figures to report when estimating the laboratory bias β. Instead, the best estimate of the bias, $\hat{\beta}$, is numerically less than d. The key idea is that when some of the total uncertainty is associated with random error, the bias cannot be as large as implied when ignoring the different implications of Type A evaluation and Type B evaluation. Similarly, the standard uncertainty of $\hat{\beta}$ is generally less than $u(d) \equiv \sigma_D$.

The analysis here implies that the enduring meaning of the deviation d and its uncertainty '$u(d)$' can be questioned. This has implications for the understanding of the 'degree of equivalence' described in the Mutual Recognition Arrangement.[2]

References

1. C. Chatfield and A. J. Collins, *Introduction to Multivariate Analysis* (Chapman and Hall, 1980).
2. International Committee for Weights and Measures, *Mutual Recognition of National Measurement Standards and of Calibration and Measurement Certificates Issued by National Metrology Institutes (Paris: Comité International des Poids et Mesures)* (2003).
3. A. G. Chunovkina, C. Elster, I. Lira and W. Wöger, Analysis of key comparison data and laboratory biases, *Metrologia* **45** (2008) 211–216.
4. A. Koo, Report on the consultative committee for photometry and radiometry key comparison of regular spectral transmittance 2010 (CCPR-K6.2010), *Metrologia* **54** (2017) 02001.
5. A. G. Chunovkina, *Measurement Techniques* **46** (2003) 409–414.
6. R. Willink, *Metrologia* **40** (2003) 9–17.
7. R. Willink, *Metrologia* **43** (2006) S220–S230.
8. C. M. Wang and H. K. Iyer, *Metrologia* **47** (2010) 435–443.
9. C. Elster and B. Toman, *Metrologia* **50** (2013) 549–555.
10. A. L. Rukhin, *Metrologia* **50** (2013) 45–59.
11. R. N. Kacker, R. U. Datla and A. C. Parr, *Metrologia* **41** (2004) 340–352.

F Pavese, A Bošnjaković, S Eichstädt, A B Forbes, J Alves e Sousa (eds.):
Advanced Mathematical and Computational Tools in Metrology and Testing XIII
Series on Advances in Mathematics for Applied Sciences, Vol. 94

https://doi.org/10.1142/9789819800674_0029

Two methods for approximating percentiles from moments

R. Willink[1,2]

[1] *Measurement Standards Laboratory of New Zealand, Lower Hutt, New Zealand*
[2] *University of Otago (Wellington), New Zealand*
E-mail: robin.willink@otago.ac.nz

Consider the estimation of a tail percentile or tail quantile from the moments of a continuous random variable. If a normal approximation to the distribution of the variable is applicable then an adequate result is found easily using the mean and variance. However, the use of higher moments would be expected to give a better result. We describe and evaluate two different procedures for estimating tail quantiles, the first of which uses the Pearson system of distributions and moments up to order 4, and the second of which uses the Cornish–Fisher expansion and moments up to order 8. Neither method is uniformly superior.

Keywords: Pearson distribution; Cornish–Fisher expansion; quantile.

1. Introduction

The tails of the distribution of a random variable can be represented by the inverse of the distribution function at small and large values of the variable. The inverse of the distribution function is known as the *quantile function*, and it produces the value of the random variable corresponding to a chosen value of the lower probability integral. Thus, if a random variable X has distribution function $F(x) = \Pr(X \leq x)$ with $F(q_p) = p$ then $q_p = F^{-1}(p)$ and q_p is called the p-quantile or $100p$-percentile of (the distribution of) X. For example, 1.96 is the 0.975-quantile of the standard normal distribution.

A statistical inference involves the study of one or both tails of the distribution. However, the distribution function is often unknown, and the variable might instead be known through its lower order moments, e.g., mean (first moment) and variance (second central moment). So it is common to find that we wish to evaluate tail quantiles of a variable, say the 0.025 and 0.975 quantiles, from a set of low order moments. This paper describes and compares two methods for this task. Section 2 first describes various forms in which the moments of a random variable can be represented. Section 3 then considers the use of a distribution from the Pearson system as an approximating distribution, which involves knowledge of the first four moments. Section 4 then describes the Cornish–Fisher expansion, giving explicit equations for use when the lowest m moments are known, for $m = 2, \ldots, 8$. Section 5 compares the methods by example.

2. Moments and cumulants

Let a continuous random variable X have probability density function $f(x)$. The kth moment of X about the origin is defined to be $m_k \equiv \int_{-\infty}^{\infty} x^k f_X(x)\,dx$, (if the integral exists). The mean of X is $\mu \equiv m_1$. Also, the kth central moment of X is the kth moment of X about the mean, i.e., $\mu_k \equiv \int_{-\infty}^{\infty} (x-\mu)^k f_X(x)\,dx$. The first eight central moments of X are

$$
\begin{aligned}
\mu_1 &= 0 \\
\mu_2 &= m_2 - m_1^2 \\
\mu_3 &= m_3 - 3m_2m_1 + 2m_1^3 \\
\mu_4 &= m_4 - 4m_3m_1 + 6m_2m_1^2 - 3m_1^4 \\
\mu_5 &= m_5 - 5m_4m_1 + 10m_3m_1^2 - 10m_2m_1^3 + 4m_1^5 \\
\mu_6 &= m_6 - 6m_5m_1 + 15m_4m_1^2 - 20m_3m_1^3 + 15m_2m_1^4 - 5m_1^6 \\
\mu_7 &= m_7 - 7m_6m_1 + 21m_5m_1^2 - 35m_4m_1^3 + 35m_3m_1^4 - 21m_2m_1^5 + 6m_1^7 \\
\mu_8 &= m_8 - 8m_7m_1 + 28m_6m_1^2 - 56m_5m_1^3 + 70m_4m_1^4 - 56m_3m_1^5 + 28m_2m_1^6 \\
&\quad -7m_1^8.
\end{aligned}
$$

The variance of X is $\sigma^2 \equiv \mu_2$ and the standard deviation of X is σ.

The quantity $\phi \equiv \mu_3/\mu_2^{3/2}$ is the *coefficient of skewness*, and the quantity $\gamma \equiv \mu_4/\mu_2^2 - 3$ is the *coefficient of excess kurtosis*. These are dimensionless indicators of the shape of the distribution of X. If X is symmetric then $\phi = 0$, and if X is normal then $\phi = 0$ and $\gamma = 0$.

It follows that

$$
\begin{aligned}
\mu &= m_1 \\
\sigma &= \sqrt{(m_2 - m_1^2)} \\
\phi &= (m_3 - 3m_2m_1 + 2m_1^3)/\sigma^3 \\
\gamma &= (m_4 - 4m_3m_1 + 6m_2m_1^2 - 3m_1^4)/\sigma^4 - 3
\end{aligned}
$$

and

$$
\begin{aligned}
m_2 &= \mu^2 + \sigma^2 \\
m_3 &= \mu^3 + 3\mu\sigma^2 + \phi\sigma^3 \\
m_4 &= \mu^4 + 6\mu^2\sigma^2 + 4\phi\mu\sigma^3 + (\gamma + 3)\sigma^4.
\end{aligned}
$$

Therefore, the set $\{\mu, \sigma, \phi, \gamma\}$ is a transform of the set of first four moments.

The first eight *cumulants* of X are (Ref. 1, Eqs. (3.41), (3.43))

$$\kappa_1 = m_1$$
$$\kappa_2 = \mu_2$$
$$\kappa_3 = \mu_3$$
$$\kappa_4 = \mu_4 - 3\mu_2^2$$
$$\kappa_5 = \mu_5 - 10\mu_3\mu_2$$
$$\kappa_6 = \mu_6 - 15\mu_4\mu_2 - 10\mu_3^2 + 30\mu_2^3$$
$$\kappa_7 = \mu_7 - 21\mu_5\mu_2 - 35\mu_4\mu_3 + 210\mu_3\mu_2^2$$
$$\kappa_8 = \mu_8 - 28\mu_6\mu_2 - 56\mu_5\mu_3 - 35\mu_4^2 + 420\mu_4\mu_2^2 + 560\mu_3^2\mu_2 - 630\mu_2^4.$$

Also (Ref. 1, Eq. (3.38))

$$\mu_4 = \kappa_4 + 3\kappa_2^2$$
$$\mu_5 = \kappa_5 + 10\kappa_3\kappa_2$$
$$\mu_6 = \kappa_6 + 15\kappa_4\kappa_2 + 10\kappa_3^2 + 15\kappa_2^3$$
$$\mu_7 = \kappa_7 + 21\kappa_5\kappa_2 + 35\kappa_4\kappa_3 + 105\kappa_3\kappa_2^2$$
$$\mu_8 = \kappa_8 + 28\kappa_6\kappa_2 + 56\kappa_5\kappa_3 + 35\kappa_4^2 + 210\kappa_4\kappa_2^2 + 280\kappa_3^2\kappa_2 + 105\kappa_2^4.$$

So κ_1 is the mean and κ_2 is the variance. The cumulants are important because cumulants of the same order add when independent variables are added, i.e., if Y and Z are independent then $\kappa_r(Y+Z) = \kappa_r(Y) + \kappa_r(Z)$.

The kth-order cumulant ratio for $k \geq 3$ is the dimensionless figure $\eta_k \equiv \kappa_k/\kappa_2^{k/2}$. So $\eta_3 = \phi$ and $\eta_4 = \gamma$. The ratios $\{\eta_k\}$ are indicators of the shape of the distribution.

(The kth moment about the origin is often denoted μ'_k, e.g., Ref. 1. The coefficient of skewness might be denoted γ_1 or $\sqrt{\beta_1}$ (even if negative) and the coefficient of excess might be denoted γ_2 or $\beta_2 - 3$. The notations m_k, ϕ and γ used here are simpler and are consistent with other work.[4,5])

3. Approximation using a Pearson distribution

The first method for approximating a tail quantile uses the Pearson system of distributions. In this system, there is a unique (untruncated) distribution for every possible set of first four moments and, hence, for every set of the figures μ, σ, ϕ and γ. From knowledge of these figures, a number of quantiles of the corresponding Pearson distribution can be found to good accuracy using formulae on the next page. Thus, on the basis that the distribution of X will be similar to the Pearson distribution with the same first four moments, we can estimate various tail quantiles of X.[4,5]

□ Let P be the random variable in the Pearson system with mean μ, variance σ^2, coefficient of skewness ϕ and coefficient of excess kurtosis γ. If $-1 \le \phi \le 1$ and $-1.2 + 1.8\phi^2 \le \gamma \le 6$ then the p-quantile of P is $\mu + q\sigma$, where

$$q = \begin{cases} r_p & \phi \ge 0 \\ -r_{1-p} & \phi < 0 \end{cases} \tag{1}$$

and r_p or r_{1-p} is closely approximated in the appropriate equation below.

$$r_{0.005} \approx \frac{-(2.5758 + 2.6736\,\gamma + 0.7685\,\gamma^2 - 0.817|\phi| - 0.720|\phi|\gamma + a_{0.005})}{1 + 0.8864\,\gamma + 0.2362\,\gamma^2 - 0.001|\phi| - 0.067|\phi|\gamma + b_{0.005}}$$

$$a_{0.005} \equiv -0.113|\phi|\gamma^2 - 0.675\phi^2 - 0.673\phi^2\gamma$$

$$b_{0.005} \equiv -0.009|\phi|\gamma^2 + 0.172\phi^2 - 0.150\phi^2\gamma$$

$$r_{0.010} \approx \frac{-(2.3263 + 2.2846\,\gamma + 0.6161\,\gamma^2 - 0.493|\phi| - 0.506|\phi|\gamma + a_{0.010})}{1 + 0.8809\,\gamma + 0.2328\,\gamma^2 + 0.074|\phi| - 0.066|\phi|\gamma + b_{0.010}}$$

$$a_{0.010} \equiv -0.060|\phi|\gamma^2 - 0.941\phi^2 - 0.511\phi^2\gamma$$

$$b_{0.010} \equiv 0.001|\phi|\gamma^2 - 0.093\phi^2 - 0.125\phi^2\gamma$$

$$r_{0.025} \approx \frac{-(1.960 + 1.845\,\gamma + 0.470\,\gamma^2 - 1.572|\phi| - 1.136|\phi|\gamma - 0.110|\phi|\gamma^2)}{1 + 0.906\,\gamma + 0.239\,\gamma^2 - 0.573|\phi| - 0.453|\phi|\gamma - 0.041|\phi|\gamma^2}$$

$$r_{0.050} \approx \frac{-(1.645 + 1.466\,\gamma + 0.338\,\gamma^2 - 1.302|\phi| - 0.861|\phi|\gamma - 0.037|\phi|\gamma^2)}{1 + 0.903\,\gamma + 0.224\,\gamma^2 - 0.644|\phi| - 0.473|\phi|\gamma - 0.017|\phi|\gamma^2}$$

$$r_{0.950} \approx \frac{1.645 + 1.466\,\gamma + 0.338\,\gamma^2 - 0.413|\phi| - 0.539|\phi|\gamma - 0.026|\phi|\gamma^2}{1 + 0.903\,\gamma + 0.224\,\gamma^2 - 0.407|\phi| - 0.408|\phi|\gamma - 0.029|\phi|\gamma^2}$$

$$r_{0.975} \approx \frac{1.960 + 1.845\,\gamma + 0.470\,\gamma^2 + 0.318|\phi| - 0.236|\phi|\gamma - 0.050|\phi|\gamma^2}{1 + 0.906\,\gamma + 0.239\,\gamma^2 - 0.063|\phi| - 0.259|\phi|\gamma - 0.035|\phi|\gamma^2}$$

$$r_{0.990} \approx \frac{2.3263 + 2.2846\,\gamma + 0.6161\,\gamma^2 + 0.244|\phi| - 0.039|\phi|\gamma + a_{0.990}}{1 + 0.8809\,\gamma + 0.2328\,\gamma^2 - 0.197|\phi| - 0.198|\phi|\gamma + b_{0.990}}$$

$$a_{0.990} \equiv 0.009|\phi|\gamma^2 - 0.110\phi^2 - 0.002\phi^2\gamma$$

$$b_{0.990} \equiv -0.013|\phi|\gamma^2 + 0.089\phi^2 - 0.008\phi^2\gamma$$

$$r_{0.995} \approx \frac{2.5758 + 2.6736\,\gamma + 0.7685\,\gamma^2 + 0.174|\phi| - 0.119|\phi|\gamma + a_{0.995}}{1 + 0.8864\,\gamma + 0.2362\,\gamma^2 - 0.290|\phi| - 0.254|\phi|\gamma + b_{0.995}}$$

$$a_{0.995} \equiv -0.076|\phi|\gamma^2 - 0.154\phi^2 - 0.033\phi^2\gamma$$

$$b_{0.995} \equiv -0.042|\phi|\gamma^2 + 0.185\phi^2 + 0.000\phi^2\gamma. \quad \blacksquare$$

The Pearson system contains the distributions of all shifted and scaled normal variables, gamma variables (including chi-squared), t-variables, F-variables and beta variables (including arc-sine), and others not fitting any of these descriptions. Figure 1 shows various distributions in this system.

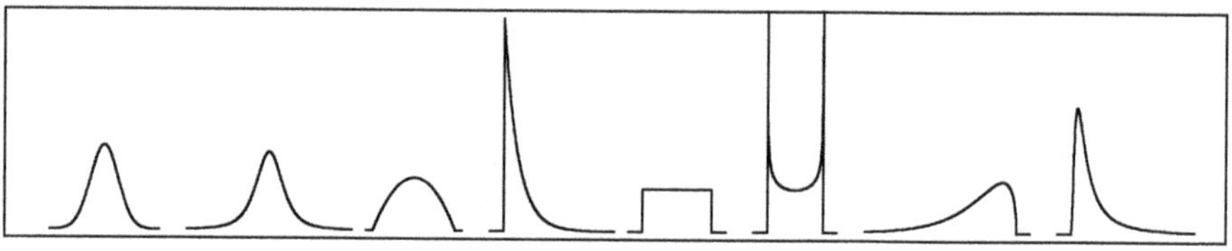

Fig. 1. Probability density functions of some variables from the Pearson system.

A standardized variable is a variable with mean 0 and standard deviation 1. The equations on the previous page give close approximations to the 0.005, 0.01, 0.025, 0.05, 0.95, 0.975, 0.99 and 0.995 quantiles of all standardized Pearson variables with skewness ϕ and excess kurtosis γ satisfying $-1 \leq \phi \leq 1$ and $-1.2+1.8\phi^2 \leq \gamma \leq 6$. The p-quantile for skewness ϕ is the opposite of the $(1-p)$-quantile for skewness $-\phi$, so some brevity is obtained using expressions that involve $|\phi|$. Thus, if $\phi \geq 0$ then the required quantile of the standardized Pearson variable is $q = r_p$ while if $\phi < 0$ then it is $q = -r_{1-p}$. (The approximations are correct to within $\pm 0.5\%$ for all combinations of $\phi^2 \in \{0, 0.01, \ldots, 1\}$ and $\gamma \in \{-1.2+1.8\phi^2, a, a+0.1, \ldots, 6.00\}$, where a is the lowest multiple of 0.1 not less than $-1.2+1.8\phi^2$.[5] The approximations with $\phi = 0$ are correct to within $\pm 0.05\%$.[4])

The approximation of tail quantiles of a general variable X from its first four moments becomes straightforward. Thus, let X be a random variable with mean μ, variance σ^2, coefficient of skewness ϕ and coefficient of excess kurtosis γ. The p-quantile of X is given approximately by $\hat{q}_p(X) = \mu + q\sigma$, where q is the p-quantile of the standardized Pearson variable with coefficient of skewness ϕ and coefficient of excess kurtosis γ found using (1) and the appropriate equation above.

4. Approximation using the Cornish–Fisher expansion

The second method involves the Cornish–Fisher expansion,[2,3] which is a series expansion for the quantile of a variable taken about the corresponding quantile of a standard normal distribution using the cumulant ratios $\{\eta_k\}$. The expansion for a general variable X can be stated as follows.

□ Let X be a random variable with mean μ, variance $\sigma^2 \equiv \kappa_2$ and kth cumulant ratio $\eta_k \equiv \kappa_k/\kappa_2^{k/2}$ for $k \geq 3$. Also, let z denote the p-quantile of a standard normal variable. Then the p-quantile of X is given by $q_p(X) = \mu + q\sigma$ where $q = z + A_3 + A_4 + A_5 + A_6 + A_7 + A_8 + \cdots$ and

$$A_3 \equiv \tfrac{1}{6}\eta_3\left(z^2-1\right)$$

$$A_4 \equiv \tfrac{1}{24}\eta_4\left(z^3-3z\right) - \tfrac{1}{36}\eta_3^2\left(2z^3-5z\right)$$

$$A_5 \equiv \tfrac{1}{120}\eta_5\left(z^4-6z^2+3\right) - \tfrac{1}{24}\eta_3\eta_4\left(z^4-5z^2+2\right) + \tfrac{1}{324}\eta_3^3\left(12z^4-53z^2+17\right)$$

$$\begin{aligned} A_6 \equiv{}& \tfrac{1}{720}\eta_6\left(z^5-10z^3+15z\right) - \tfrac{1}{180}\eta_3\eta_5\left(2z^5-17z^3+21z\right) \\ &-\tfrac{1}{384}\eta_4^2\left(3z^5-24z^3+29z\right) + \tfrac{1}{288}\eta_3^2\eta_4\left(14z^5-103z^3+107z\right) \\ &-\tfrac{1}{7776}\eta_3^4\left(252z^5-1688z^3+1511z\right) \end{aligned}$$

$$\begin{aligned} A_7 \equiv{}& \tfrac{1}{5040}\eta_7\left(z^6-15z^4+45z^2-15\right) - \tfrac{1}{432}\eta_3\eta_6\left(z^6-13z^4+33z^2-9\right) \\ &-\tfrac{1}{240}\eta_4\eta_5\left(z^6-12z^4+29z^2-8\right) + \tfrac{1}{1080}\eta_3^2\eta_5\left(16z^6-181z^4+393z^2-90\right) \\ &+\tfrac{1}{576}\eta_3\eta_4^2\left(12z^6-129z^4+271z^2-64\right) \\ &-\tfrac{1}{1296}\eta_3^3\eta_4\left(80z^6-803z^4+1513z^2-304\right) \\ &+\tfrac{1}{29160}\eta_3^5\left(960z^6-8937z^4+15062z^2-2651\right) \end{aligned}$$

$$\begin{aligned} A_8 \equiv{}& \tfrac{1}{40320}\eta_8\left(z^7-21z^5+105z^3-105z\right) \\ &-\tfrac{1}{5040}\eta_3\eta_7\left(2z^7-37z^5+160z^3-135z\right) \\ &-\tfrac{1}{1152}\eta_4\eta_6\left(z^7-17z^5+69z^3-57z\right) \\ &-\tfrac{1}{3600}\eta_5^2\left(2z^7-33z^5+132z^3-108z\right) \\ &+\tfrac{1}{5184}\eta_3^2\eta_6\left(18z^7-293z^5+1100z^3-795z\right) \\ &+\tfrac{1}{1440}\eta_3\eta_4\eta_5\left(18z^7-273z^5+974z^3-695z\right) \\ &+\tfrac{1}{3072}\eta_4^3\left(9z^7-131z^5+451z^3-321z\right) \\ &-\tfrac{1}{19440}\eta_3^3\eta_5\left(396z^7-5708z^5+18755z^3-11811z\right) \\ &-\tfrac{1}{13824}\eta_3^2\eta_4^2\left(594z^7-8193z^5+26006z^3-16367z\right) \\ &+\tfrac{1}{62208}\eta_3^4\eta_4\left(5148z^7-67004z^5+195259z^3-109553z\right) \\ &-\tfrac{1}{4199040}\eta_3^6\left(154440z^7-1887684z^5+5033714z^3-2542637z\right). \quad \blacksquare \end{aligned}$$

When the distribution is normal, $\eta_k = 0$ for $k \geq 3$ and the expansion gives the exact result $\mu + \sigma z$. Often the expansion is truncated after the term A_4. The expansion is a divergent 'asymptotic' expansion, so there is no guarantee that including very term to A_8 leads to a more accurate result. Because including more terms does not necessarily improve accuracy, we will truncate at the point after the term of even order where the magnitude of the sum of the next two terms is larger, adding half the first pair of terms omitted if there is oscillation. Thus, our estimate of $q_p(X)$ is

$$\hat{q}_p(X) = \mu + (z + \textstyle\sum_{k=3}^{r} A_k + cA_{r+1} + cA_{r+2})\sigma \tag{2}$$

with

$$\begin{array}{lll}
r=4 & c=0 & |A_3+A_4| < |A_5+A_6| \cap (A_3+A_4)(A_5+A_6) > 0 \\
r=4 & c=\frac{1}{2} & |A_3+A_4| < |A_5+A_6| \cap (A_3+A_4)(A_5+A_6) < 0 \\
r=6 & c=0 & |A_3+A_4| \geq |A_5+A_6| < |A_7+A_8| \cap (A_5+A_6)(A_7+A_8) > 0 \\
r=6 & c=\frac{1}{2} & |A_3+A_4| \geq |A_5+A_6| < |A_7+A_8| \cap (A_5+A_6)(A_7+A_8) < 0 \\
r=8 & c=0 & |A_3+A_4| \geq |A_5+A_6| \geq |A_7+A_8|.
\end{array}$$

5. Performance comparison by example

We now compare the abilities of the methods to estimate the tail quantiles of a variable X in six simple examples. In many cases, the moments or cumulants of X are described, if necessary for re-representation using the equations of Sec. 2. The examples use the facts that (i) the kth moment about the origin of the jth power of a random variable V is the $(j \times k)$th moment about the origin of V and (ii) the kth cumulant of independent variables V_1 and V_2 is the sum of the kth cumulants of V_1 and V_2. Also, where practical, an expression is given for the exact quantiles against which the approximations can be compared. The probability density functions of X in the examples are shown in Fig. 2, in which each trace depicts the outline of a histogram of a random sample of size 5×10^7.

Example 1: Let X be a continuous uniform random variable with lower limit 1 and upper limit 2, i.e., $X \sim \mathrm{U}[1,2]$. Then $m_k = (2^{k+1} - 1)/(k+1)$. The exact p-quantile of X is $q_p(X) = 1 + p$.

Example 2: Let X be the square of the variable in Example 1. The kth moment of X (about the origin) is the $2k$th moment of $\sqrt{X}$, so $m_k = (2^{2k+1} - 1)/(2k+1)$. The exact p-quantile of X is $q_p(X) = (1+p)^2$.

Example 3: Suppose $X = \exp(Y)$ where Y has the normal distribution with mean a and standard deviation b, i.e., $Y \sim \mathrm{N}(a, b^2)$. Then X has a log-normal distribution and $m_k = \exp(ka + k^2b^2/2)$. The exact p-quantile of X is $\exp(a + b\Phi^{-1}(p))$, where $\Phi(\cdot)$ is the standard normal distribution function. We set $a = 0$ and $b = 0.3$.

Example 4: Suppose that $X = Y^3$ where $Y \sim \mathrm{N}(0.9, 0.2^2)$. The kth moment of X is the $3k$th moment of Y. The cubic function is monotonic increasing, so the exact p-quantile of X is $q_p(X) = (0.9 + 0.2\Phi^{-1}(p))^3$.

Example 5: Suppose the total error E in a measurement process is modelled as the sum of independent normal, uniform and reversed exponential variables, according to the relationship $E = E_1 + E_2 + E_3$ with $E_1 \sim \mathrm{N}(0, 1)$, $E_2 \sim \mathrm{U}[-1, 1]$ and $E_3 = 1.5(1 - Y)$ with $Y \sim \mathrm{EXP}(1)$. (So the exponentially distributed component of error is negated and translated to have mean zero.) Because of the independence, the kth cumulant of E is the sum of the kth cumulants of its components, E_1, E_2 and E_3.

Example 6: Suppose the total error in a measurement process is modelled as $E = E_1 + E_2$ with $E_1 \sim \mathrm{U}[-1, 1]$ and $E_2 \sim \mathrm{U}[-0.5, 0.5]$. So E has a trapezoidal probability density function.

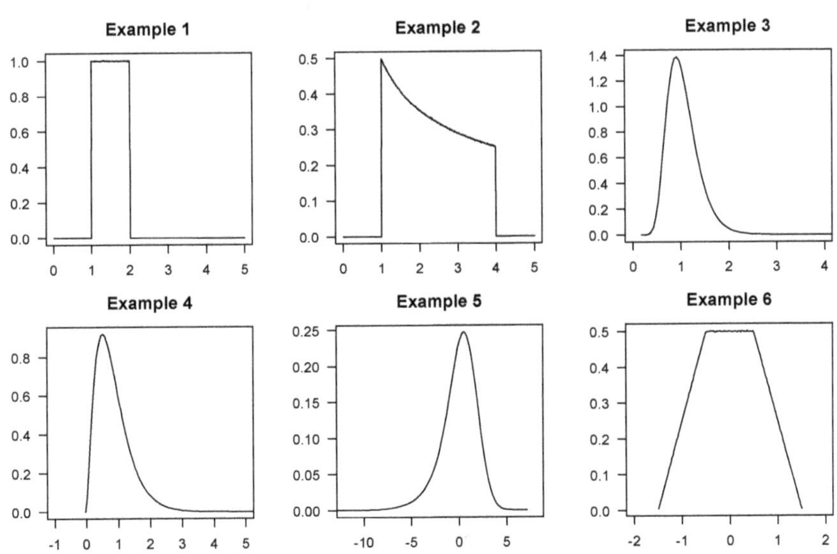

Fig. 2. Distributions of the variable X in Examples 1–6.

Table 1 shows the exact quantiles of X and the estimates obtained using the Pearson-system method, the Cornish–Fisher (CF) method and the normal approximation, in which the estimate is $\mu + z\sigma$. It also shows the values of r and c in the CF method. (All values are rounded, and the exact quantiles for Example 5 were approximated from a sample of 10^8.)

Table 1. Exact and approximate values of the p-quantile of X in Examples 1–6.

p	0.005	0.010	0.025	0.050	0.950	0.975	0.990	0.995
Example 1								
Exact	1.005	1.010	1.025	1.050	1.950	1.975	1.990	1.995
Pearson	1.005	1.010	1.025	1.050	1.950	1.975	1.990	1.995
CF	0.983	0.950	0.959	1.015	1.985	2.041	2.050	2.017
$\mu + z\sigma$	0.756	0.828	0.934	1.025	1.975	2.066	2.172	2.244
r, c	8, 0	8, 0	8, 0	6, 0	6, 0	8, 0	8, 0	8, 0
Example 2								
Exact	1.010	1.020	1.051	1.102	3.802	3.901	3.960	3.980
Pearson	1.029	1.037	1.065	1.112	3.803	3.918	3.982	4.001
CF	1.134	0.928	0.844	0.932	3.886	4.059	4.189	4.151
$\mu + z\sigma$	0.094	0.311	0.630	0.904	3.763	4.037	4.355	4.572
r, c	8, 0	8, 0	8, 0	6, $\frac{1}{2}$	8, 0	6, 0	8, 0	8, 0
Example 3								
Exact	0.462	0.498	0.555	0.611	1.638	1.800	2.010	2.166
Pearson	0.464	0.499	0.556	0.612	1.637	1.801	2.008	2.164
CF	0.468	0.500	0.550	0.612	1.640	1.803	2.023	2.173
$\mu + z\sigma$	0.219	0.299	0.417	0.518	1.574	1.675	1.793	1.873
r, c	6, 0	6, 0	8, 0	8, 0	6, $\frac{1}{2}$	6, $\frac{1}{2}$	8, 0	8, 0
Example 4								
Exact	0.057	0.082	0.131	0.186	1.856	2.157	2.545	2.834
Pearson	–	–	–	–	–	–	–	–
CF	0.054	0.081	0.132	0.188	1.858	2.161	2.550	2.840
$\mu + z\sigma$	-0.535	-0.402	-0.207	-0.039	1.713	1.881	2.076	2.209
r, c	8, 0	8, 0	8, 0	8, 0	8, 0	8, 0	8, 0	8, 0
Example 5								
Exact	-6.891	-5.852	-4.476	-3.437	2.621	3.029	3.492	3.799
Pearson	-6.763	-5.731	-4.421	-3.426	2.589	2.968	3.388	3.669
CF	-6.757	-5.698	-4.377	-3.468	2.681	3.002	3.466	3.668
$\mu + z\sigma$	-4.876	-4.404	-3.710	-3.114	3.114	3.710	4.404	4.876
r, c	8, 0	8, 0	8, 0	6, 0	8, 0	8, 0	6, $\frac{1}{2}$	6, $\frac{1}{2}$
Example 6								
Exact	-1.359	-1.300	-1.184	-1.053	1.053	1.184	1.300	1.359
Pearson	-1.347	-1.291	-1.180	-1.055	1.055	1.180	1.291	1.347
CF	-1.356	-1.320	-1.213	-1.073	1.073	1.213	1.320	1.356
$\mu + z\sigma$	-1.663	-1.502	-1.265	-1.062	1.062	1.265	1.502	1.663
r, c	8, 0	8, 0	8, 0	6, 0	6, 0	8, 0	8, 0	8, 0

In Example 1, the distribution of X is a Pearson distribution, so the Pearson-system method is only in error to the extent that the equations in Sec. 3 are approximate. In Examples 1 and 6, the variable X is symmetric. In Example 4, the coefficient of skewness in 1.29, which lies outside the stated range of validity of the equations of Sec. 3. So, although those equations would still give useful results, they have not been applied.

In general, the two proposed methods perform much better than the normal approximation. In these examples, the Pearson distribution has tended to give a closer approximation than the Cornish–Fisher expansion. This might be expected for variables with discontinuous density functions, as in Examples 1 and 2, because the Pearson system contains some such distributions. Neither method is uniformly superior to the other.

6. Discussion and conclusion

In much practice, the total error in a measured value can be modelled as the sum of many small independent errors. If none of these errors is dominant then the random variable for the total error will be approximately normal, by the Central Limit Theorem for non-identical variables. In that case, the simple normal approximation is likely to be adequate when evaluating a tail quantile of the total error. However, there are other situations, e.g., where one of the components has substantially more variance than the others, in which much greater accuracy can be obtained using higher moments of the error variables. The equations of Sec. 2 and the additive property of cumulants used in Examples 5 and 6 enable us to find the higher moments of the total error. Then either of the methods of Secs. 3 and 4 will give more accurate tail quantiles and, hence, more accurate intervals of uncertainty.

References

1. M. Kendall and A. Stuart. *The Advanced Theory of Statistics, Vol. 1, 4th edn.* (Griffin, 1977).
2. E. A. Cornish and R. A. Fisher. Moments and cumulants in the specification of distributions. *Rev. de l'Inst int. de stat.* **5**, 307–322 (1937).
3. R. A. Fisher and E. A. Cornish. The percentile points of distributions having known cumulants. *Technometrics* **2**, 209–225 (1960).
4. R. Willink. A procedure for the evaluation of measurement uncertainty based on moments, *Metrologia* **42**, 329–343 (2005).
5. R. Willink. Uncertainty analysis by moments for asymmetric variables, *Metrologia* **43**, 522–530 (2006).

F Pavese, A Bošnjaković, S Eichstädt, A B Forbes, J Alves e Sousa (eds.):
Advanced Mathematical and Computational Tools in Metrology and Testing XIII
Series on Advances in Mathematics for Applied Sciences, Vol. 94

https://doi.org/10.1142/9789819800674_0030

Linearization region in the straight-line calibration

Gejza Wimmer[1,*], Viktor Witkovský[2] and Eva Fišerová[3]

[1]*Mathematical Institute*
Slovak Academy of Sciences
Štefánikova 49, 814 73 Bratislava, Slovak Republic
**E-mail: wimmer@mat.savba.sk*

[2]*Institute of Measurement Science*
Slovak Academy of Sciences
Dúbravská cesta 9, 841 04 Bratislava, Slovak Republic
E-mail: witkovsky@savba.sk

[3]*Department of Mathematical Analysis and Applications of Mathematics*
Faculty of Science, Palacký University
17. listopadu 12, 771 46 Olomouc, Czech Republic
E-mail: eva.fiserova@upol.cz

We consider the statistical straight-line calibration model, which, as it turns out, is a nonlinear regression errors-in-variables model. Since the related measurement model to determine the straight-line parameters is generally nonlinear, the derived covariance matrix of the parameter estimators only provides approximations of the uncertainties. To address this, we introduce a statistical test that helps us determine whether the linearized statistical straight-line calibration model is suitable for the measured data. If the model demonstrates linearity, we can utilize the optimal estimators, the best linear unbiased estimators for the model parameters, along with their covariance matrix. Moreover, by including additional normality assumptions, we can calculate confidence intervals for any feasible linear combination of parameters. In particular, here we identify the linearization region of the straight-line calibration model and provide a simple sufficient condition for the model to be considered linearizable.

Keywords: linear comparative calibration; weakly nonlinear model; linearization region; best linear unbiased estimator.

1. Introduction

Calibration plays a central role in the field of measurement science and metrology and has applications across various industrial sectors and services, see, e.g., [1, 2]. According to [3], calibration is an operation that, under specific conditions, serves a dual purpose. It establishes the relationship between quantity values and the corresponding indications, including their associated measurement uncertainties. Subsequently, this information is utilized to derive the measurement result from the indication.

Comparative calibration, see, e.g., [4], is a specific type of calibration comparing two measuring instruments or measurement processes. Typically, one of these instruments is considered more precise, often referred to as the *gold standard*, while the other is less precise, often serving as a more economical alternative. A fundamental objective of a calibration experiment is to establish a mathematical relationship that expresses the true quantity values measured by a more precise measuring instrument (the response) as a function of the true quantity values measured by a less precise instrument (the stimulus).

The most commonly used comparative calibration method is straight-line calibration based on the errors-in-variables (EIV) regression model. Despite its apparent simplicity, this model must be considered as a nonlinear from a statistical perspective. The estimation of the model parameters and their associated uncertainties necessitate careful treatment, in accordance with [5–7]. At present, the prevailing approach, introduced in [1], is undergoing an update to align with ISO standards (expected to be published as ISO 28037:20XX). Moreover, it is being revised in response to the constructive feedback and suggestions outlined in [8], with the aim of ensuring that the uncertainties associated with the parameters comply with the Guide to the Expression of Uncertainty in Measurement (GUM) standards.

The primary focus of our paper is twofold. First, we aim to identify and define the conditions and boundaries for linearization within which this model can be accurately approximated by a traditional linear regression model. This simplification not only streamlines the situation but also ensures the statistical correctness of the results, in accordance with [1]. This approximation allows to estimate the model parameters via the generalized least squares (GLS) methodology, enabling to compute the optimal estimators (the best linear unbiased estimators, BLUEs), and their covariance matrix. Furthermore, with additional normality assumptions, we can establish confidence intervals for any valid linear combination of parameters.

2. EIV model for straight-line comparative calibration

The straight-line calibration function relates the true values of measured quantities using instruments $\mathcal{X}$ (a less precise calibrated instrument) and $\mathcal{Y}$ (the gold standard measuring instrument). We consider a calibration experiment involving the measurement of a set of m carefully chosen objects.

Their true values $\mu_1, \ldots, \mu_m$ in the units of $\mathcal{X}$, span the calibration range, while their true values in the units of $\mathcal{Y}$ are $\nu_1, \ldots, \nu_m$.

The EIV regression model representing the direct measurements in the calibration experiment can be expressed in the vector-matrix notation as

$$\begin{pmatrix} \boldsymbol{X} \\ \boldsymbol{Y} \end{pmatrix} = \begin{pmatrix} \boldsymbol{\mu} \\ \boldsymbol{\nu} \end{pmatrix} + \begin{pmatrix} \boldsymbol{\epsilon}_x \\ \boldsymbol{\epsilon}_y \end{pmatrix}, \tag{1}$$

where $\boldsymbol{X} = (X_1, \ldots, X_m)'$, $\boldsymbol{Y} = (Y_1, \ldots, Y_m)'$ represent the measurement results of m measurands and the vector parameters $\boldsymbol{\mu}$, $\boldsymbol{\nu}$ are related by

$$\boldsymbol{\nu} = a\mathbf{1}_m + b\boldsymbol{\mu}. \tag{2}$$

Here, the symbol $\mathbf{1}_m$ denotes a vector of m ones; a and b are the unknown parameters of primary interest representing the intercept and the slope of the straight-line calibration function. Note that the relation (2) is nonlinear in the model parameters $\boldsymbol{\mu}$, $\boldsymbol{\nu}$, a and b.

To obtain the exact null distribution of the test statistic used to assess the linearity of the model (1)–(2), an additional assumption is made, which is that the measurement errors $\boldsymbol{\epsilon}_x$, $\boldsymbol{\epsilon}_y$ follow a joint normal distribution with a zero mean and a positive definite covariance matrix $\boldsymbol{\Sigma}$.

3. Linearized calibration model

The model (1)–(2) can be expressed using a Taylor series expansion around any selected point $(\boldsymbol{\mu}_0', a_0, b_0) \in \mathbb{R}^{m+2}$. More precisely, by denoting

$$\boldsymbol{\mu} = \boldsymbol{\mu}_0 + \boldsymbol{\mu}_\Delta, \quad a = a_0 + a_\Delta, \quad b = b_0 + b_\Delta,$$

we can rewrite the model (1)–(2) in the matrix form as

$$\begin{pmatrix} \boldsymbol{X} \\ \boldsymbol{Y} \end{pmatrix} - \begin{pmatrix} \boldsymbol{\mu}_0 \\ a_0\mathbf{1}_m + b_0\boldsymbol{\mu}_0 \end{pmatrix} = \boldsymbol{F}_0 \begin{pmatrix} \boldsymbol{\mu}_\Delta \\ a_\Delta \\ b_\Delta \end{pmatrix} + \begin{pmatrix} \mathbf{0}_m \\ b_\Delta\boldsymbol{\mu}_\Delta \end{pmatrix} + \begin{pmatrix} \boldsymbol{\epsilon}_x \\ \boldsymbol{\epsilon}_y \end{pmatrix}, \tag{3}$$

where $\boldsymbol{F}_0$ is a block matrix of order $2m \times (m+2)$ defined as

$$\boldsymbol{F}_0 = \begin{pmatrix} \boldsymbol{I}_{m,m} & \mathbf{0}_m & \mathbf{0}_m \\ b_0\boldsymbol{I}_{m,m} & \mathbf{1}_m & \boldsymbol{\mu}_0 \end{pmatrix}.$$

Here, $\boldsymbol{I}_{m,m}$ is the $m \times m$ identity matrix and $\mathbf{0}_m$ is a vector of m zeros. Since the nonlinear model (1)–(2) can be expressed without terms of order higher than two in the Taylor series expansion, we refer to this model as a *weakly nonlinear model* (as seen in, e.g., [9]).

The source of nonlinearity in the straight-line calibration model lies in the term $b_\Delta \boldsymbol{\mu}_\Delta$. When this term is negligible, the calibration model becomes linear and can be represented as follows:

$$\begin{pmatrix} \boldsymbol{X} \\ \boldsymbol{Y} \end{pmatrix} - \begin{pmatrix} \boldsymbol{\mu}_0 \\ a_0 \mathbf{1}_m + b_0 \boldsymbol{\mu}_0 \end{pmatrix} = \boldsymbol{F}_0 \begin{pmatrix} \boldsymbol{\mu}_\Delta \\ a_\Delta \\ b_\Delta \end{pmatrix} + \begin{pmatrix} \boldsymbol{\epsilon}_x \\ \boldsymbol{\epsilon}_y \end{pmatrix}. \tag{4}$$

The conditions under which the nonlinear calibration model (3) can be linearized around $(\boldsymbol{\mu}_0', a_0, b_0)$ to the linear model (4) are stated in Sec. 4.

The GLS estimators for the unknown parameters $\boldsymbol{\mu}_\Delta$, a_Δ, and b_Δ in a linear model (4) are provided in accordance with [10] as follows:

$$\begin{pmatrix} \widehat{\boldsymbol{\mu}}_\Delta \\ \widehat{a}_\Delta \\ \widehat{b}_\Delta \end{pmatrix} = \left(\boldsymbol{F}_0' \boldsymbol{\Sigma}^{-1} \boldsymbol{F}_0\right)^{-1} \boldsymbol{F}_0' \boldsymbol{\Sigma}^{-1} \left[\begin{pmatrix} \boldsymbol{X} \\ \boldsymbol{Y} \end{pmatrix} - \begin{pmatrix} \boldsymbol{\mu}_0 \\ a_0 \mathbf{1}_m + b_0 \boldsymbol{\mu}_0 \end{pmatrix} \right], \tag{5}$$

and the GLS estimators of $\boldsymbol{\mu}$, a, and b are straightforwardly determined as

$$\widehat{\boldsymbol{\mu}} = \widehat{\boldsymbol{\mu}}_\Delta + \boldsymbol{\mu}_0, \quad \widehat{a} = \widehat{a}_\Delta + a_0, \quad \widehat{b} = \widehat{b}_\Delta + b_0. \tag{6}$$

Under the specified assumptions and in accordance with *the Gauss–Markov Theorem*, the GLS estimators are BLUEs, with the covariance matrix

$$\operatorname{cov}\left[(\widehat{\boldsymbol{\mu}}', \widehat{a}, \widehat{b})'\right] = \left(\boldsymbol{F}_0' \boldsymbol{\Sigma}^{-1} \boldsymbol{F}_0\right)^{-1}. \tag{7}$$

Formally, the parameter estimate (5) depends on the specific selection of the vector $(\boldsymbol{\mu}_0', a_0, b_0)$. As a natural and particularly significant choice for model linearization, we recommend to consider the following selection:

$$(\boldsymbol{\mu}_0', a_0, b_0) = \underset{\boldsymbol{\mu}, a, b}{\operatorname{argmin}} \begin{pmatrix} \boldsymbol{x} - \boldsymbol{\mu} \\ \boldsymbol{y} - (a\mathbf{1}_m + b\boldsymbol{\mu}) \end{pmatrix}' \boldsymbol{\Sigma}^{-1} \begin{pmatrix} \boldsymbol{x} - \boldsymbol{\mu} \\ \boldsymbol{y} - (a\mathbf{1}_m + b\boldsymbol{\mu}) \end{pmatrix},$$

the weighted total least squares solution in the nonlinear EIV model (1)–(2), as discussed in, for instance, [8]. Here, $\boldsymbol{x}$ and $\boldsymbol{y}$ correspond to the observed values of measurands measured by instruments $\mathcal{X}$ and $\mathcal{Y}$, respectively.

4. Criterion for linearization of calibration model

In this section, we focus on the issue, whether the linearized calibration model (4) is adequate to the measured data, or it is necessary to use the nonlinear model (3). As we have shown above, the nonlinearity arises due to the term $b_\Delta \boldsymbol{\mu}_\Delta$. To verify that this term is so small that it can be neglected, we can in the model (3) test the hypothesis

$$H_0 : b_\Delta \boldsymbol{\mu}_\Delta = \mathbf{0}_m \quad \text{against} \quad H_1 : b_\Delta \boldsymbol{\mu}_\Delta \neq \mathbf{0}_m. \tag{8}$$

The test procedure will be based on the statistic R_0^2, defined as

$$R_0^2 = \boldsymbol{V}'\left(\boldsymbol{\Sigma}^{-1} - \boldsymbol{\Sigma}^{-1}\boldsymbol{F}_0(\boldsymbol{F}_0'\boldsymbol{\Sigma}^{-1}\boldsymbol{F}_0)^{-1}\boldsymbol{F}_0'\boldsymbol{\Sigma}^{-1}\right)\boldsymbol{V}, \tag{9}$$

$$\boldsymbol{V} = \begin{pmatrix}\boldsymbol{X}\\ \boldsymbol{Y}\end{pmatrix} - \begin{pmatrix}\boldsymbol{\mu}_0\\ a_0\mathbf{1}_m + b_0\boldsymbol{\mu}_0\end{pmatrix},$$

which, in model (3), has a noncentral χ^2 distribution with $m-2$ degrees of freedom and with the noncentrality parameter δ, denoted as $\chi^2_{m-2}(\delta)$,

$$\delta = (\mathbf{0}_m', b_\Delta\boldsymbol{\mu}_\Delta')\left[\boldsymbol{\Sigma}^{-1} - \boldsymbol{\Sigma}^{-1}\boldsymbol{F}_0(\boldsymbol{F}_0'\boldsymbol{\Sigma}^{-1}\boldsymbol{F}_0)^{-1}\boldsymbol{F}_0'\boldsymbol{\Sigma}^{-1}\right](\mathbf{0}_m', b_\Delta\boldsymbol{\mu}_\Delta')',$$

following from properties of quadratic forms of normally distributed vectors.

Under the null hypothesis in (8), the statistic R_0^2 follows a central χ^2 distribution with $m-2$ degrees of freedom. Therefore, we reject the null hypothesis at the significance level α if the observed value of R_0^2 is greater than $\chi^2_{m-2,1-\alpha}$ (the quantile of the central χ^2 distribution). In this situation, the term $b_\Delta\boldsymbol{\mu}_\Delta$ in (3) cannot be ignored. Consequently, we strongly recommend exploring alternative methods for estimating the calibration model parameters and calculating their covariance matrix, rather than relying on (5)–(7), as is the Monte Carlo method outlined in [6, 7].

If we fail to reject the null hypothesis, $b_\Delta\boldsymbol{\mu}_\Delta$ in (3) can be disregarded. Its impact on the agreement between the measured data and the linearized model (4) is negligible. In such cases, we can opt (at the expense of a predefined acceptable level of risk) the linear calibration model (4) and use the optimal BLUE estimations as detailed in (5)–(7) instead of relying on the true nonlinear calibration model (1)–(2) or (3) and the related parameter estimation and the (approximate) uncertainty evaluation methods.

We will now establish a *linearization region* in the parameter space, such that the R_0^2 test will not reject the null hypothesis when the true model is in fact nonlinear but the parameters fall within this region. To maintain the probability of rejecting the null hypothesis if the true model is nonlinear with the true parameters from the specified linearization region within acceptable limits, we apply an approach based on the evaluation of the Bates and Watts intrinsic curvature, as elaborated in [11], and define the specific linearization regions, as in [9]. The Bates and Watts intrinsic curvature of the nonlinear calibration model (3) at the point $(\boldsymbol{\mu}_0', a_0, b_0)$ is

$$K^{(int)}_{(\boldsymbol{\mu}_0', a_0, b_0)} = \sup\left\{k_{int} : (\boldsymbol{\mu}_\Delta', a_\Delta, b_\Delta) \in \mathbb{R}^{m+2}\right\}, \tag{10}$$

$$k_{int} = \frac{\sqrt{4(\mathbf{0}_m', b_\Delta\boldsymbol{\mu}_\Delta')\left(\boldsymbol{\Sigma}^{-1} - \boldsymbol{\Sigma}^{-1}\boldsymbol{F}_0(\boldsymbol{F}_0'\boldsymbol{\Sigma}^{-1}\boldsymbol{F}_0)^{-1}\boldsymbol{F}_0'\boldsymbol{\Sigma}^{-1}\right)(\mathbf{0}_m', b_\Delta\boldsymbol{\mu}_\Delta')'}}{(\boldsymbol{\mu}_\Delta', a_\Delta, b_\Delta)\boldsymbol{F}_0'\boldsymbol{\Sigma}^{-1}\boldsymbol{F}_0(\boldsymbol{\mu}_\Delta', a_\Delta, b_\Delta)'}.$$

According to the approach from [9], the following theorem can be stated.

Theorem 4.1. *Let $\varepsilon > 0$ be a chosen small real number. If in the model (3) is true that*

$$\begin{pmatrix} \boldsymbol{\mu}_\Delta \\ a_\Delta \\ b_\Delta \end{pmatrix} \in \mathcal{L}_0 = \left\{ \mathbf{z} \in \mathbb{R}^{m+2} : \mathbf{z}'\boldsymbol{F}_0'\boldsymbol{\Sigma}^{-1}\boldsymbol{F}_0\mathbf{z} \leq \frac{2\sqrt{\delta_{max}}}{K^{(int)}_{(\boldsymbol{\mu}_0, a_0, b_0)}} \right\}, \tag{11}$$

where δ_{max} satisfies the equation

$$P\{\chi^2_{m-2}(\delta_{max}) > \chi^2_{m-2,1-\alpha}\} = \alpha + \varepsilon, \tag{12}$$

then the test R_0^2 rejects the null hypothesis in (8) with probability

$$P\left\{R_0^2 > \chi^2_{m-2,1-\alpha} \,|\, (\boldsymbol{\mu}_\Delta', a_\Delta, b_\Delta) \in \mathcal{L}_0\right\} \leq \alpha + \varepsilon. \tag{13}$$

The region $\mathcal{L}_0$ is referred to as the *linearization region*, as in [9]. Within $\mathcal{L}_0$, the test R_0^2 does not reject the null hypothesis in (8) with probability greater or equal to $1 - (\alpha + \varepsilon)$. Consequently, the quadratic calibration model (3) can be suitably approximated by the linear model (4). In this scenario, the probability of rejecting the null hypothesis if the true model is nonlinear with the true parameters from the linearization region does not exceed $\alpha + \varepsilon$, where ε is a predetermined small value.

The challenge lies in determining whether the true parameters fall within the linearization region. As true parameters are unknown, a feasible way is to construct a confidence region by inverting the test specified by the R_0^2 statistic within the linear model (4) and compare these two regions. If the confidence ellipsoid is a subset of a linearization region, the linear model (4) can be used instead of the nonlinear model (3). More precisely, the confidence region represents a range of parameter values that are statistically plausible based on the observed data and the chosen nominal confidence level $1 - \alpha$. In particular, the confidence ellipsoid is a set

$$\mathcal{E} = \left\{ \mathbf{z} \in \mathbb{R}^{m+2} : \left[\mathbf{z}' - (\widehat{\boldsymbol{\mu}}', \widehat{a}, \widehat{b})\right] \boldsymbol{F}_0'\boldsymbol{\Sigma}^{-1}\boldsymbol{F}_0 \left[\mathbf{z} - (\widehat{\boldsymbol{\mu}}', \widehat{a}, \widehat{b})'\right] \leq \chi^2_{m+2,1-\alpha} \right\}$$

such that, assuming the linear model (4) holds, it covers the true values of the parameters $(\boldsymbol{\mu}', a, b)$ with the probability $1 - \alpha$. The center of $\mathcal{E}$ is at the estimated point $(\widehat{\boldsymbol{\mu}}', \widehat{a}, \widehat{b})$. The region $\mathcal{L}_0$ is an ellipsoid with the center at the origin. The directions of semi-axes of both ellipsoids $\mathcal{L}_0$ and $\mathcal{E}$ are identical and are determined by the eigenvectors of the matrix $\boldsymbol{F}_0'\boldsymbol{\Sigma}^{-1}\boldsymbol{F}_0$.

The resulting linearization region in the parameter space is $\mathcal{L} = \{(\boldsymbol{\mu}_0', a_0, b_0)' + \mathcal{L}_0\}$. If the point $(\boldsymbol{\mu}_0', a_0, b_0)$ coincides with (or is close to) the estimate $(\widehat{\boldsymbol{\mu}}', \widehat{a}, \widehat{b})$, we can establish a straightforward linearization criterion,

$$\chi^2_{m-2,1-\alpha} \leq \frac{2\sqrt{\delta_{max}}}{K^{(int)}_{(\boldsymbol{\mu}'_0, a_0, b_0)}}. \tag{14}$$

If criterion (14) is met, then $\mathcal{E} \subset \mathcal{L}$, indicating that the quadratic calibration model (3) can be linearized.

In this case, the random interval with the boundaries

$$\boldsymbol{p}'(\widehat{\boldsymbol{\mu}}', \widehat{a}, \widehat{b})' \pm u_{1-\alpha/2}\sqrt{\boldsymbol{p}'(\boldsymbol{F}'_0\boldsymbol{\Sigma}^{-1}\boldsymbol{F}_0)^{-1}\boldsymbol{p}}, \quad \boldsymbol{p} \in \mathbb{R}^{m+2},$$

where $u_{1-\alpha/2}$ is the quantile of the standard normal distribution, covers the true value of a linear combination of parameters $\boldsymbol{p}'(\boldsymbol{\mu}', a, b)'$ with the probability at least $1 - \alpha - \varepsilon$.

5. Conclusion and discussion

This study delves into the straight-line calibration model, acknowledging its inherent nonlinearity. The central challenge revolves around obtaining a suitable approximation for the covariance matrix of parameter estimators, serving as a proxy for determining the required measurement uncertainties of the straight-line parameters, which are the principal parameters of interest. To address this issue, we propose a method for defining a linearization region, albeit at the cost of a predefined acceptable level of risk. When the model demonstrates linearity, we can utilize the BLUEs for parameter estimation, along with their associated covariance matrix, and compute confidence intervals, contingent upon specific normality assumptions.

The normality assumption might be viewed as a constraint within this methodology. Nevertheless, the proposed approach, which relies on the utilization of the considered test statistic and the intrinsic curvature measure, has proven to be a valuable tool for detecting departures from linearity. In our forthcoming research, we intend to relax the normality assumption and explore the distributional characteristics of the test statistic under a variety of scenarios commonly encountered in metrological applications.

Acknowledgements

The research was supported by the Slovak Research and Development Agency, projects APVV-21-0216, SK-CZ-RD-21-0109/LUASK22008, and APVV-21-0195, by the Scientific Grant Agency of the Ministry of Education of the Slovak Republic and the Slovak Academy of Sciences, projects

VEGA 2/0120/24 and VEGA 2/0023/22, by the Operational Programme Integrated Infrastructure, project 313011BWH2, co-financed by the European Regional Development Fund, and by Palacký University Olomouc, project IGA_PrF_2023_009.

References

[1] ISO/TS 28037:2010, *Determination and use of straight-line calibration functions.* (International Standards Organization, Geneva, 2010).

[2] A. M. van der Veen and M. G. Cox, *Good practice in evaluating measurement uncertainty. Compendium of Examples.* (EMUE - Examples of Measurement Uncertainty Evaluation, 2021).

[3] JCGM200:2012 (VIM), *The international vocabulary of metrology – Basic and general concepts and associated terms*, 3rd edn. (ISO, BIPM, IEC, IFCC, ILAC, IUPAC, IUPAP and OIML, 2012).

[4] C. Osborne, Statistical calibration: A review. *International Statistical Review/Revue Internationale de Statistique*, 309–336, (1991).

[5] JCGM 100:2008 (GUM), *Evaluation of measurement data – Guide to the expression of uncertainty in measurement* (GUM 1995 with minor corrections), (ISO, BIPM, IEC, IFCC, ILAC, IUPAC, IUPAP and OIML, 2008).

[6] JCGM 101:2008 (GUM S1), *Evaluation of measurement data – Supplement 1 to the Guide to the expression of uncertainty in measurement – Propagation of distributions using a Monte Carlo method.* (ISO, BIPM, IEC, IFCC, ILAC, IUPAC, IUPAP and OIML, 2008).

[7] JCGM 102:2011 (GUM S2), *Evaluation of measurement data – Supplement 2 to the Guide to the expression of uncertainty in measurement – Extension to any number of output quantities.* (ISO, BIPM, IEC, IFCC, ILAC, IUPAC, IUPAP and OIML, 2011).

[8] K. Klauenberg, S. Martens, A. Bošnjaković, M. G. Cox, A. M. van der Veen and C. Elster, The GUM perspective on straight-line errors-in-variables regression. *Measurement*, **187**, 110340 (2022).

[9] L. Kubáček and E. Tesaříková, *Weakly nonlinear regression models.* (Palacký University, Olomouc, 2008).

[10] L. Kubáček, *Foundations of estimation theory.* (Elsevier, Amsterdam, Oxford, New York, Tokyo, 1988).

[11] D. M. Bates and D. G. Watts, Relative curvature measures of nonlinearity. *Journal of the Royal Statistical Society. Series B (Methodological)*, **42**(1), 1–25 (1980).

F Pavese, A Bošnjaković, S Eichstädt, A B Forbes, J Alves e Sousa (eds.):
Advanced Mathematical and Computational Tools in Metrology and Testing XIII
Series on Advances in Mathematics for Applied Sciences, Vol. 94
 (pp. 338–348)
https://doi.org/10.1142/9789819800674_0031

Composition of t-distribution laws

I. P. Zakharov
Department of Information and Measurement Technique,
Kharkiv National University of Radio Electronics,
Kharkiv, 61166, Ukraine
E-mail: igor.zakharov@nure.ua
https://nure.ua/

O. A. Botsiura
Department of Higher Mathematics,
Kharkiv National University of Radio Electronics,
Kharkiv, 61166, Ukraine
E-mail: olesia.botsiura@nure.ua
https://nure.ua/

D. I. Diakov
Department of Precision Engineering and Measurement Instruments,
Technical University of Sofia,
Sofia, 1000, Bulgaria
E-mail: diakov@tu-sofia.bg
https://tu-sofia.bg

Using the numerical methods, coverage factors were calculated for the composition of two Student distribution laws with different numbers of degrees of freedom and the ratio of standard deviations. The number of degrees of freedom corresponding to this composition was calculated. An approximating expression for the coverage coefficient of this composition is given. The results obtained by numerical methods are compared with the results obtained using the GUM method.

Keywords: student distributions; numerical methods; composition of distribution laws; coverage factor; number of degrees of freedom; approximating expression.

1. Introduction

Repeated observations of the measurands are carried out to improve the accuracy of their evaluation. As a rule, the arithmetic mean of these repeated observations is taken as estimates of these measurands [1]. It should be noted that the arithmetic mean is a consistent, unbiased and effective estimate of the measurand only when its probability density function (PDF) obeys the normal law [2].

If the measurand Y is simply a single normally distributed quantity X that is estimated by the arithmetic mean $\overline{x}$ of n independent repeated observations x_q with the experimental standard deviation of the mean $s(\overline{x})$, then the PDF of the estimate of the measurand y can be described as a t-distribution (Student's PDF) with a number of degrees of freedom (NDF) $\nu = n - 1$ [3].

If the measurand Y is a function of N normally distributed input quantities X_1, X_2, …, X_N, $Y = f(X_1, X_2, \ldots, X_N)$, and if each input quantity X_i, $i = 1, \ldots, N$, is estimated by the arithmetic mean $\overline{x}_i$ of n_i independent repeated observations with the experimental standard deviation (SD) of the mean $s(\overline{x}_i)$, then in general the PDF of the estimate of the measurand y can be approximately described as a t-distribution with an effective NDF ν_{eff} [1].

The problem of finding an effective NDF does not have an exact solution [2]. Approximate formulas for effective NDF estimates were obtained in the works of Welch B.L. and Satterthwaite F.E. [4,5] in the forties of the last century, when numerical modelling was not widely used due to the lack of computers.

The goal of this work is to find an expression that provides the best approximation of the coverage factor and effective number of degrees of freedom.

To achieve this goal, the following tasks were solved:

1) numerical modeling of the composition (convolution) of two Student's PDF with different number of degrees of freedom ν_1 and ν_2 with SD ratio $\gamma = S_2/S_1$;

2) finding reliable estimates of the coverage factors k for the convolution of Student's PDF;

3) evaluation of the number of degrees of freedom ν of the resulting composition.

2. Determination of the coverage factors of Students PDF's composition

In the presented work, modeling of *t*-distributions and their composition was performed by the Monte Carlo method [3] using Web-based software application to evaluate the measurement uncertainty "NIST Uncertainty Machine" [6] or Matlab algorithm TDIST [7]. The latter is a reasonably fast and precise numerical algorithm for statistical applications which require computing the distribution and quantiles of a linear combination of independent Student's *t* random variables.

The use of this algorithm made it possible to find the coverage factor for the probability $p = 0.95$ for the composition of two Student's PDF with NDF ν_1, ν_2 varying from 2...20 for SD ratios γ equal to 0.2; 0.6 and 1.0.

The dependence graphs $k(\nu_1,\nu_1,\gamma)$ are shown in Figs. 1–3. All of them have the form of decreasing functions of ν_1, ν_2.

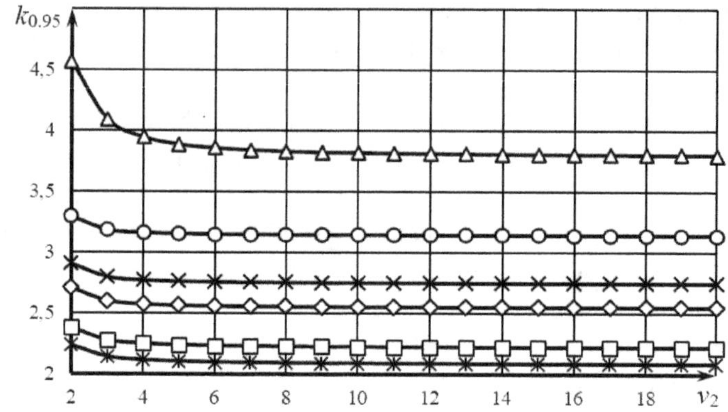

Fig. 1. Dependences $k(\nu_2)$ for $\gamma = 0.2$ and $\nu_1 = 2$ (Δ); 3 (O); 4(X); 5 (◊); 10 (□); 20 (Ж).

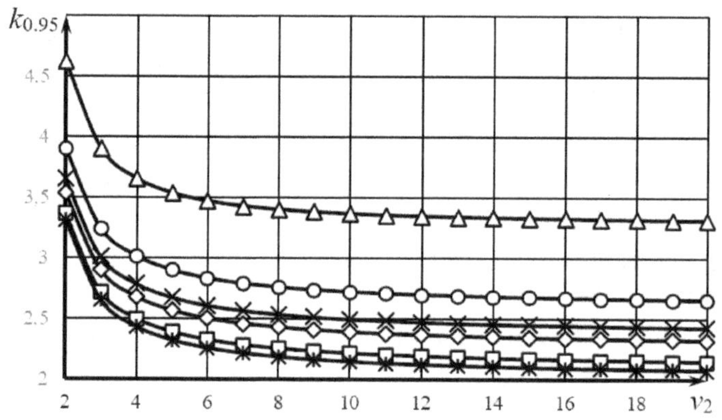

Fig. 2. Dependences $k(\nu_2)$ for $\gamma = 0.6$ and $\nu_1 = 2$ (Δ); 3 (O); 4(X); 5(◊); 10 (□); 20 (Ж).

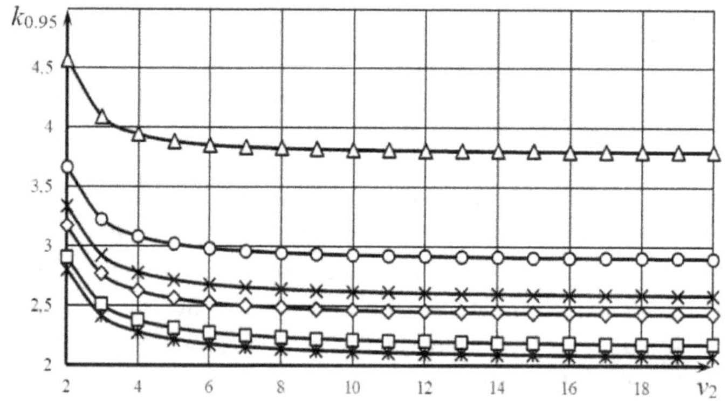

Fig. 3. Dependences $k(\nu_2)$ for $\gamma = 1.0$ and $\nu_1 = 2$ (Δ); 3 (O); 4(X); 5(◊); 10 (□); 20 (Ж).

Reliable estimation of coverage factor for the convolution of Student distribution laws k_{app} we will look used the formula proposed in [8]:

$$k_{app} = \frac{\sqrt{t^2_{0.95;\nu_1} S_1^2 + t^2_{0.95;\nu_2}(\nu_2) S_2^2}}{\sqrt{S_1^2 + S_2^2}} = \sqrt{\frac{t^2_{0.95;\nu_1} + t^2_{0.95;\nu_2}\gamma^2}{1+\gamma^2}}. \quad (1)$$

The dependences of the relative approximation errors δ_k on the parameters ν_1 and ν_2 determined by formula

$$\delta_k = 100 \cdot \frac{k_{app} - k_{0.95}}{k_{0.95}}, \%. \quad (2)$$

In Fig. 4, the dependences of the maximum relative approximation errors $\delta_k(\nu_2, \gamma)$ are presented. For $\gamma = 0.2$ maximum relative approximation errors achieved at $\nu_1 = 3$; for $\gamma = 0.6$ and $\gamma = 1.0$ – at $\nu_1 = 2$.

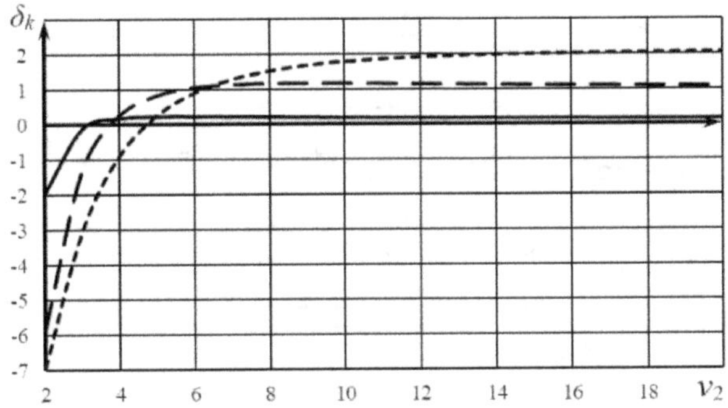

Fig. 4. Dependences of the maximum relative approximation errors $\delta_k(\nu_2, \gamma)$ for $\gamma = 0.2$ (——); $\gamma = 0.6$ (----) and $\gamma = 1.0$ (– – –).

As can be seen from Fig. 4, the relative approximation error δ_k increases with increasing γ and has maximum for $\gamma = 1.0$. Additionally, the value δ_k increases as decreases ν_1, ν_2.

3. Determination of the NDF of the convolution of the t-distribution laws

To estimate the NDF of the resulting convolution, it is necessary to inversely transform the t-coefficients into the number of degrees of freedom. To do this, we first found an approximation of the dependence of the *t*-distribution coefficients on the NDF $k(\nu)$ for the confidence level $p = 0.95$:

$$t_{0.95;\nu} = t_{0.95;\infty}\left[1+\frac{1}{0.822\nu-0.87}\right], \tag{3}$$

where $t_{0.95;\infty} = 1.96$.

The approximation error for $\nu = 2$ is 4%, for $\nu \geq 3$ no more than ± 0.22%.

After that, the desired empirical dependence was obtained from expression (3):

$$\nu = \frac{1}{0.822}\left[\frac{1.96}{t_{0.95;\nu} - 1.96} + 0.87\right]. \tag{4}$$

The dependence of the NDF of the convolution of the t-distribution laws ν on ν_1 and ν_2 for different ratios γ, found by the proposed method, is shown in Figs. 5–7.

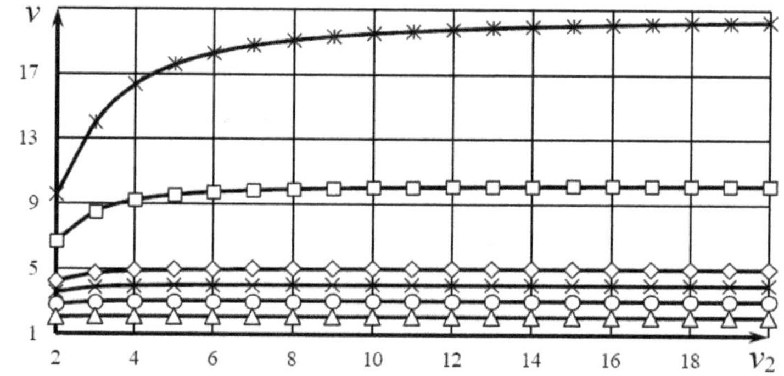

Fig. 5. Dependences $\nu(\nu_2)$ for $\gamma = 0.2$ and $\nu_1 = 2$ (Δ); 3 (O); 4(X); 5 (◊); 10 (□); 20 (Ж).

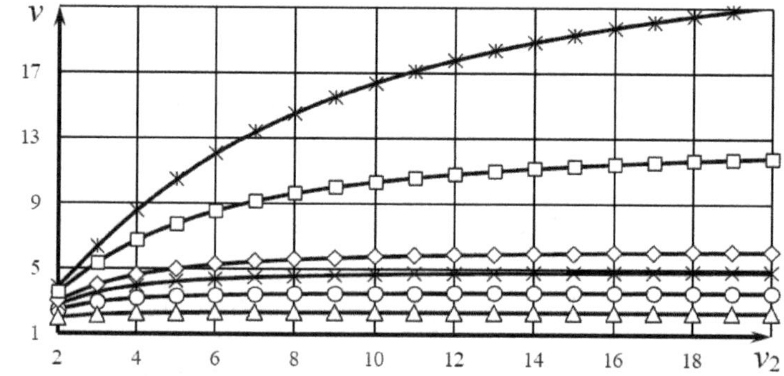

Fig. 6. Dependences $\nu(\nu_2)$ for $\gamma = 0.6$ and $\nu_1 = 2$ (Δ); 3 (O); 4(X); 5(◊); 10 (□); 20 (Ж).

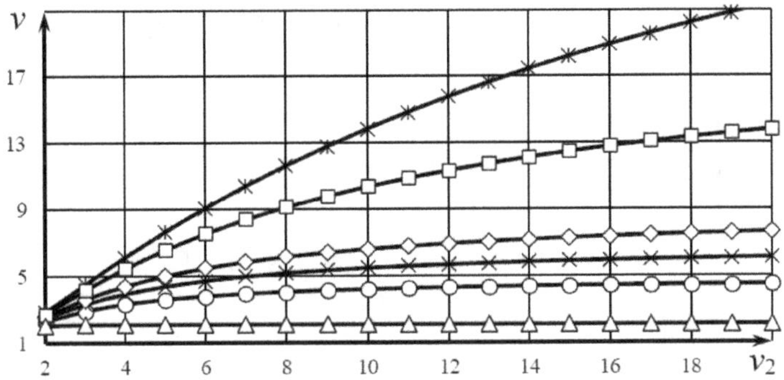

Fig. 7. Dependences $\nu(\nu_2)$ for $\gamma = 1.0$ and $\nu_1 = 2$ (Δ); 3 (O); 4(X); 5(◊); 10 (□); 20 (Ж).

The same dependences obtained by Welch-Satterthwaite formula [1]:

$$\nu_{eff} = \left(S_1^2 + S_2^2\right)^2 \Big/ \left(S_1^4/\nu_1 + S_2^4/\nu_2\right) = \left(1+\gamma^2\right)^2 \Big/ \left(1/\nu_1 + \gamma^4/\nu_2\right) \tag{5}$$

is shown in Figs. 8–10.

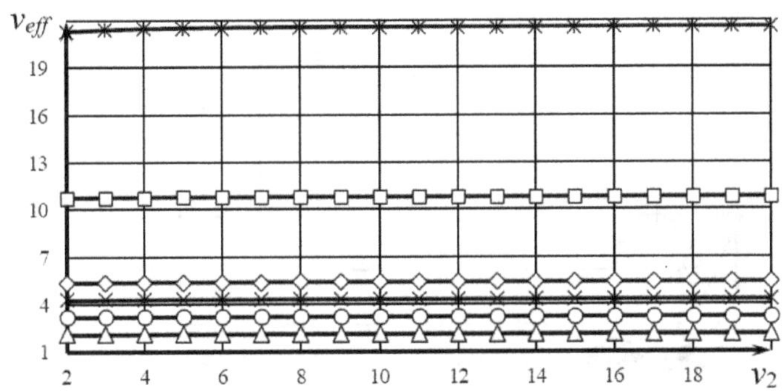

Fig. 8. Dependences $\nu_{eff}(\nu_2)$ for $\gamma = 0.2$ and $\nu_1 = 2$ (Δ); 3 (O); 4(X); 5 (◊); 10 (□); 20 (Ж).

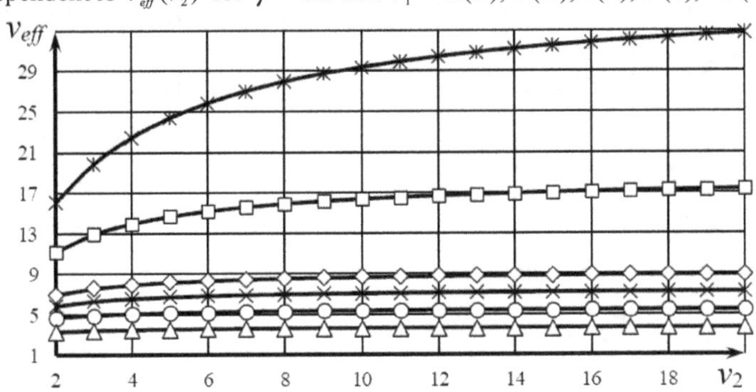

Fig. 9. Dependences $\nu_{eff}(\nu_2)$ for $\gamma = 0.6$ and $\nu_1 = 2$ (Δ); 3 (O); 4(X); 5(◊); 10 (□); 20 (Ж).

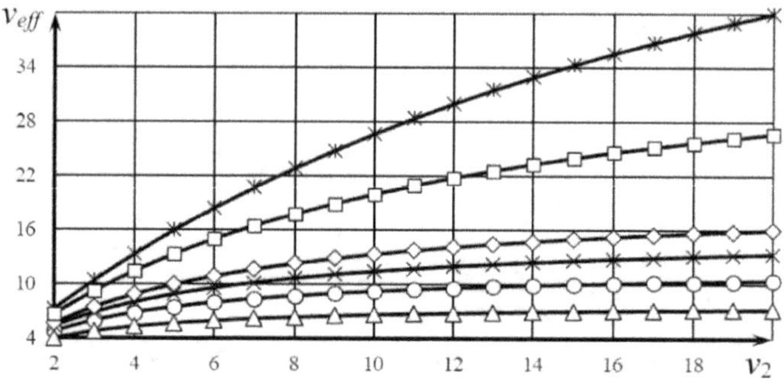

Fig. 10. Dependences $v_{eff}(v_2)$ for $\gamma = 1.0$ and $v_1 = 2$ (Δ); 3 (O); 4(X); 5(◊); 10 (□); 20 (Ж).

The relative errors in calculating NDF using the Welch–Satterthwaite formula are calculated using the formula:

$$\delta_v = 100 \cdot \frac{v_{eff} - v}{v}, \%. \tag{6}$$

The dependences of error δ_v on v_1 and v_2 for different ratios γ are shown in Figs. 11–13.

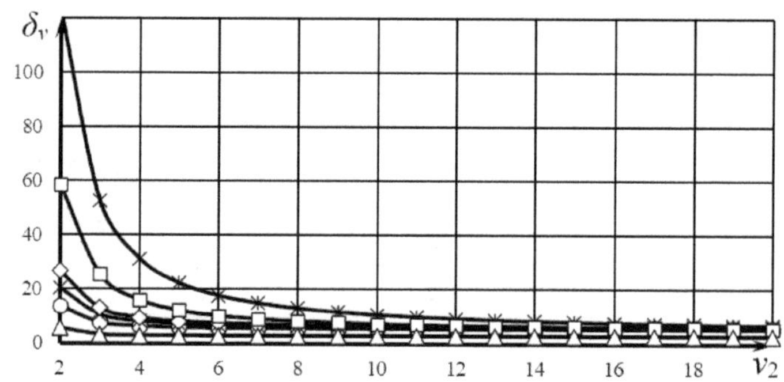

Fig. 11. Dependences $\delta_v(v_2)$ for $\gamma = 0.2$ and $v_1 = 2$ (Δ); 3 (O); 4(X); 5 (◊); 10 (□); 20 (Ж).

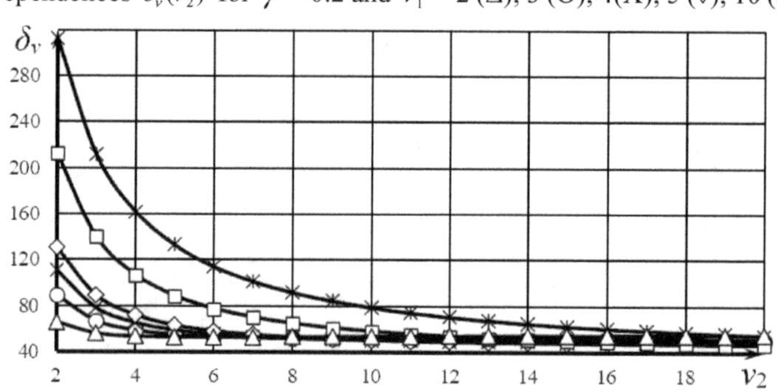

Fig. 12. Dependences $\delta_v(v_2)$ for $\gamma = 0.6$ and $v_1 = 2$ (Δ); 3 (O); 4(X); 5(◊); 10 (□); 20 (Ж).

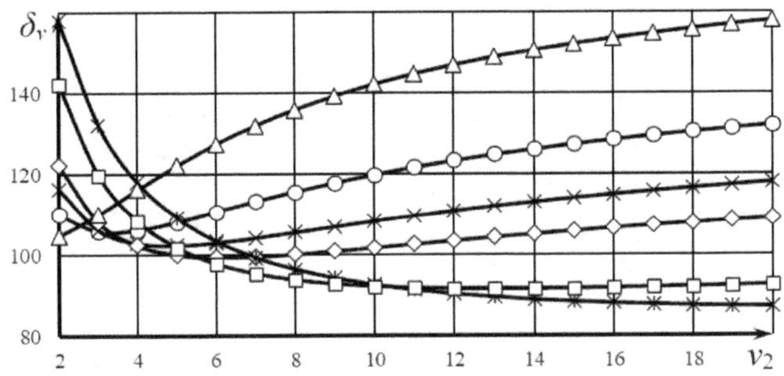

Fig. 13. Dependences $\delta_\nu(\nu_2)$ for $\gamma = 1.0$ and $\nu_1 = 2$ (Δ); 3 (O); 4(X); 5(◊); 10 (□); 20 (Ж).

Since coverage factors are calculated in GUM [1] based on the Welch–Satterthwaite formula, we will determine the error in calculation k_{GUM} in this case. The relative approximation errors δ_{GUM} was calculated by the formula:

$$\delta_{\text{GUM}} = 100 \cdot \frac{k_{\text{GUM}} - k_{0.95}}{k_{0.95}}, \% \tag{7}$$

and is shown in Figs. 14–16.

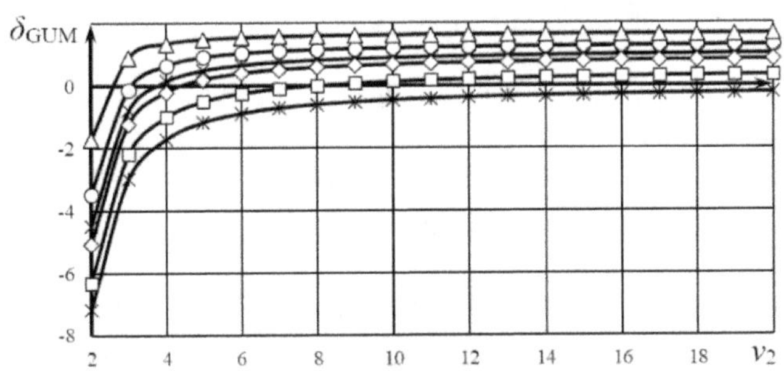

Fig. 14. Dependences $\delta_{\text{GUM}}(\nu_2)$ for $\gamma = 0.2$ and $\nu_1 = 2$ (Δ); 3 (O); 4(X); 5 (◊); 10 (□); 20 (Ж).

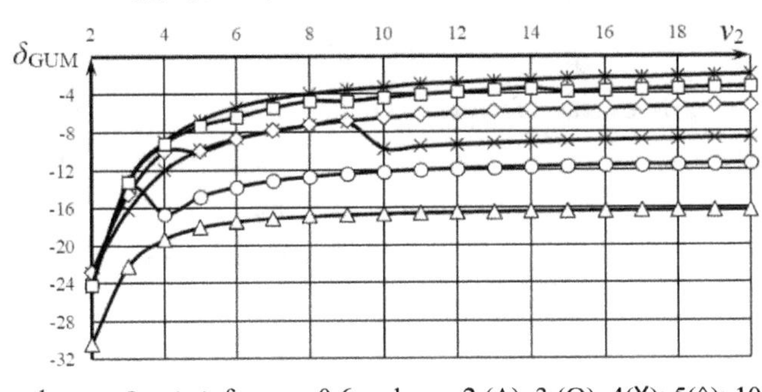

Fig. 15. Dependences $\delta_{\text{GUM}}(\nu_2)$ for $\gamma = 0.6$ and $\nu_1 = 2$ (Δ); 3 (O); 4(X); 5(◊); 10 (□); 20 (Ж).

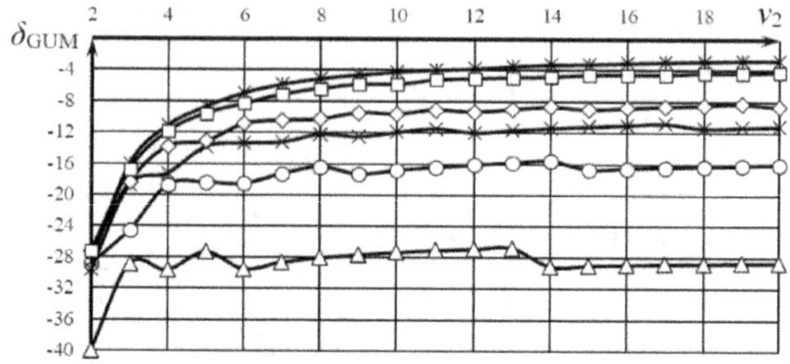

Fig. 16. Dependences $\delta_{GUM}(\nu_2)$ for $\gamma = 1.0$ and $\nu_1 = 2$ (Δ); 3 (O); 4(X); 5(◊); 10 (□); 20 (Ж).

4. Law of Propagation of Expanded Uncertainty

The results obtained can be used to estimate the expanded uncertainty for the number of repeated observations of input quantities $n \geq 2$ based on the Law of Propagation of Expanded Uncertainty (LPEU) [9].

The expression for calculating the expanded uncertainty for confidence levels $p = 0.95$ or $p = 0.9545$ in this case has the form:

$$U(y) = \sqrt{U_A^2(y) + U_B^2(y)}, \tag{8}$$

where $U_A(y)$, $U_B(y)$ are expanded uncertainties of type A and B of the measurand, respectively. Here:

$$U_A = \sqrt{\sum_{i=1}^{N} t_{(p;\nu_i)}^2 c_i^2 \cdot s^2(\overline{x}_i)}, \tag{9}$$

where $s(x_i)$ and c_i are SD of arithmetic mean and sensitive coefficient of ith input quantity, respectively; $t_{(p;\nu_i)}$ is Student's coefficient for $\nu = (n-1)$.

The expanded uncertainty of type B is calculated as:

$$U_B = k_B \cdot u_B(y), \tag{10}$$

where k_B is type B coverage factor, calculated by the kurtosis method [9] using the formulas:

$$k_p = \begin{cases} 0.1085\eta^3 + 0.1\eta + 1.96, & \text{for } p = 0.95; \\ 0.12\eta^3 + 0.1\eta + 2, & \text{for } p = 0.9545; \end{cases} \tag{11}$$

$u_B(y)$ is type B standard uncertainty of the measurand, calculated as:

$$u_B(y) = \sqrt{\sum_{i=1}^{N} c_i^2 u_B^2(x_i)}. \tag{12}$$

The study of this method shows that the difference between the estimates of the expanded uncertainty obtained by the LPEU differs from the estimates obtained by the MCM by no more than ±4.5%.

5. Conclusions

1. Composite of two Student's PDF with different NDF ν_1 and ν_2 and SD ratio γ is obtained by numerical simulation.
2. The approximating formula (1) for the coverage factor k_{app} of the Student's PDF convolution is obtained and the approximation errors are determined.
3. The NDF of the convolution of two Student's PDF is estimated and the error in effective NDF calculation using the Welch–Satterthwaite formula is evaluated.
4. The relative error of using the Welch–Satterthwaite formula when calculating the coverage factors for the composition of two Student's PDF can reach 40%.
5. The practical application of the proposed expanded uncertainty type *A* evaluation method for expanded uncertainty evaluation of the measurand based on the LPEU is shown.

References

1. JCGM 100:2008. *Evaluation of measurement data – Guide to the expression of uncertainty in measurement.*
2. Rabinovich S.G. *Measurement Errors and Uncertainties. Theory and Practice.* 3rd ed., Springer, 313 p.
3. JCGM 101:2008. *Evaluation of measurement data – Supplement 1 to the "Guide to the expression of uncertainty in measurement" – Propagation of distributions using a Monte Carlo method.* JCGM, 2008, 90 p.
4. Welch B.L. The generalization of Students problem when several different population variances are involved. *Biometrica*, 1947, v. 34, pp. 28–35.
5. Satterthwaite F.E. An distribution approximate of estimates of variance components. *Biometrics Bulletin 2*: 1946. pp. 110–114.
6. https://uncertainty.nist.gov/.
7. V. Witkovský, Matlab algorithm TDIST: The distribution of a linear combination of Student's t random variables. *In COMPSTAT 2004 Symposium*. 2004: Physica-Verlag/Springer 2004, Heidelberg, Germany.

8. Zakharov, I.P. Estimating measurement uncertainty on the basis of observed and logical correlation. *Measurement Techniques*, 2007, Vol. 50, No. 8, pp. 808–816.
9. COOMET R/GM/35:2022 *Expression of the expanded measurement uncertainty (method of kurtosis).*

F Pavese, A Bošnjaković, S Eichstädt, A B Forbes, J Alves e Sousa (eds.):
Advanced Mathematical and Computational Tools in Metrology and Testing XIII
Series on Advances in Mathematics for Applied Sciences, Vol. 94

https://doi.org/10.1142/9789819800674_bmatter

Keyword Index

F Pavese, A Bošnjaković, S Eichstädt, A B Forbes, J Alves e Sousa (eds.):
Advanced Mathematical and Computational Tools in Metrology and Testing XIII
Series on Advances in Mathematics for Applied Sciences, Vol. 94

https://doi.org/10.1142/9789819800674_bmatter

Author Index

Series on Advances in Mathematics for Applied Sciences

Aims and Scope

This series reports on new developments in mathematical research related to new analytical and numerical methods, mathematical modeling in the applied physical and natural sciences as well as in economics, and quantitative and qualitative analysis approaches for the mathematical models. Topics covered include modelling, constitutive theories, fluid and solid mechanics, kinetic and transport theories. The series ranges from monographs to lecture notes, quality conference proceedings and collections of papers. The high quality of research, the novelty of mathematical tools, and the potential for frontier problems will be the guidelines for the selection of the content for this series.

Instructions for Authors

Submission of proposals should be addressed to the editors-in-charge or to any member of the editorial board. In the latter, the authors should also notify the proposal to one of the editors-in-charge. Acceptance of books and lecture notes will generally be based on the description of the general content and scope of the book or lecture notes as well as on sample of the parts judged to be more significantly by the authors.

Acceptance of proceedings will be based on relevance of the topics and of the lecturers contributing to the volume.

Acceptance of monograph collections will be based on relevance of the subject and of the authors contributing to the volume.

Authors are urged, in order to avoid re-typing, not to begin the final preparation of the text until they received the publisher's guidelines. They will receive from World Scientific the instructions for preparing camera-ready manuscript.

Series on Advances in Mathematics for Applied Sciences

ISSN: 1793-0901

Published:*

Vol. 79 From Genetics to Mathematics
eds. M. Lachowicz and J. Miękisz

Vol. 80 Inelasticity of Materials: An Engineering Approach and a Practical Guide
by A. R. Srinivasa and S. M. Srinivasan

Vol. 81 Stability Criteria for Fluid Flows
by A. Georgescu and L. Palese

Vol. 82 Applied and Industrial Mathematics in Italy III
eds. E. De Bernardis, R. Spigler and V. Valente

Vol. 83 Linear Inverse Problems: The Maximum Entropy Connection
by H. Gzyl and Y. Velásquez

Vol. 84 Advanced Mathematical and Computational Tools in Metrology and Texting IX
eds. F. Pavese et al.

Vol. 85 Continuum Thermodynamics
Part II: Applications and Examples
by B. Albers and K. Wilmanski

Vol. 86 Advanced Mathematical and Computational Tools in Metrology and Testing X
eds. F. Pavese et al.

Vol. 87 Mathematical Methods for the Natural and Engineering Sciences (2nd Edition)
by R. E. Mickens

Vol. 88 Road Vehicle Dynamics: Fundamentals of Modeling and Simulation
by G. Genta and A. Genta

Vol. 89 Advanced Mathematical and Computational Tools in Metrology and Testing XI
eds. A. B. Forbes et al.

Vol. 90 Advanced Mathematical and Computational Tools in Metrology and Testing XII
eds. F. Pavese et al.

Vol. 91 Fractal Analysis: Basic Concepts and Applications
by C. Cattani, A. B. Mabrouk and S. Arfaoui

Vol. 92 Chaos, Complexity, and Nonlinear Economic Theory
by Wei-Bin Zhang

Vol. 93 Mathematical Modelling: Selected Objects and Processes
by A. A. Martynyuk & B. Radziszewski

Vol. 94 Advanced Mathematical and Computational Tools in Metrology and Testing XIII
eds. F. Pavese et al.

*To view the complete list of the published volumes in the series, please visit:
https://www.worldscientific.com/series/samas

www.ingramcontent.com/pod-product-compliance
Lightning Source LLC
LaVergne TN
LVHW011703200926
848484LV00004B/1056